INDUSTRIAL ENGINEERING AND OPERATIONS RESEARCH

INDUSTRIAL ENGINEERING AND OPERATIONS RESEARCH

David M. Miller
University of Alabama

J.W. Schmidt
Virginia Polytechnic Institute and State University

With contributions by **R.J. Craig,** *Virginia Polytechnic Institute and State University*; **T.J. Greene,** *Virginia Polytechnic Institute and State University*; and **D.L. Kimbler,** *University of South Florida*

Library of Congress Cataloging in Publication Data:

Miller, David M., 1945-
Industrial engineering and operations research.

Includes index.
1. Industrial engineering. 2. Operations research.
I. Schmidt, J. William (Joseph William) II. Title.

T56.M49 1984 658.4'034 84-7474
ISBN 0-471-05027-X

Printed in the United States of America

10 9 8 7 6 5 4 3 2 1

Preface

The origins of industrial engineering (IE) may be traced to the experience of English industrialists during the Industrial Revolution. In the United States, the pioneering work of Frederick W. Taylor in the early part of this century led to a more scientific approach to the analysis of problems faced by the industrial engineer. Following the scientific approach exemplified by Taylor, Frank and Lillian Gilbreth and Henry Gantt led the development of industrial engineering practice, establishing a methodology for industrial problem solving which prevailed into the 1940s.

The term "Operations Research" (OR) was first applied to efforts in England and the United States to solve tactical and strategic military problems during World War II. During this period the OR approach to the problem solving usually called for the formation of a highly capable team of scientists and military personnel. The scientific contribution resulted in the application of the principles of psychology as well as quantitative techniques of analysis such as mathematics, probability, and statistics. The success of the OR approach during World War II led to its application in both private and public sectors of the U.S. economy after the war.

Today the OR approach to problem solving is a fundamental part of industrial engineering. Much of traditional IE practice, established prior to World War II, has been modified by incorporation of OR techniques. In addition, the activities of the industrial engineer are no longer restricted to problems relating to product manufacture. The principles of industrial engineering and operations research are now applied in an integrated fashion throughout the manufacturing and service industry.

Whether industrial engineering and operations research are or should be a single discipline is a matter of diverse opinion, and it is not our intent to argue the point one way or the other. We do believe, however, that problems involving the productive use of people and materials can be best addressed through an integrated application of the principles of both areas. Our purpose in writing this textbook is to present a discussion of problems commonly encountered by the industrial engineer and approaches to their solution which integrate the techniques of traditional industrial engineering and operations research. We have organized the text to reflect this integration.

Specifically, we begin by presenting a history of the two areas and one philosophy for viewing their relationships. Our introduction continues with a chapter on models, tools essential to the successful application of both IE and OR concepts and techniques.

The next section of the book covers the primary functional activities traditionally engaged in by industrial engineers. A separate chapter is devoted to describing the meaning and mechanics of each of the following activities: Product and Process Planning, Work Methods and Work Measurement, Engineering Economy, and Job Evaluation and Wage Incentives. A section of the text is then devoted to examining the primary techniques used in OR studies. Included are chapters on Classical Optimization Methods, Mathematical Programming, Queueing Theory, and Simulation.

The fourth section of the book addresses problem areas in which both the OR tools described in the third section and the traditional IE methods discussed in the second section are used to make key management decisions. This section includes a discussion of problems in Facility Design, Production Planning and Scheduling, Inventory Control, and Quality Control. The final section of the text attempts to put the use of IE and OR tools into a simple perspective—as productivity analysis and improvement tools.

The treatment of IE/OR in this text is intended for students in the sophomore or junior year of college. Our approach is a quantitative one, and assumes that the student has had at least one year of differential and integral calculus. If so, the material included in the text can be covered in one quarter/semester. It should be of interest to students in all fields of engineering, business, economics, mathematics, computer science, and the physical sciences.

The topics treated herein and the manner of their presentation are based upon the combined industrial, teaching, research, and consulting experience of the authors. Although the influence of former students, teachers, and writers has been strong, we cannot express our indebtedness to each individually. There are a few, however, whose influence has been direct. We would first like to express our gratitude to Marvin H. Agee, whose encouragement, critique, and philosophy of teaching have influenced both the topical selection and the style of our presentation. We are grateful to Harry L. Snyder and Robert D. Dryden for their encouragement and for providing a working atmosphere conducive to the successful completion of a task such as this. This text could not have been completed without the contributions of Robert J. Craig, Timothy J. Greene, and Delbert L. Kimbler. The chapters that they have contributed to this text are essential to any overview of IE/OR. Finally, we would like to thank Laura Moran-Lopez, Paula Davis, and Tammy Holtkamp, who patiently typed and retyped the manuscript for this text.

Table of Contents

SECTION ONE INTRODUCTION

SECTION TWO INDUSTRIAL ENGINEERING FUNCTIONAL AREAS 43

SECTION FOUR THE USE OF IE AND OR TOOLS IN TRADITIONAL APPLICATION AREAS 311

Chapter 11 Facility Design 314

Chapter 12 Production Planning and Control 354

Chapter 13 Inventory Control 385

Chapter 14 Quality Control 407

SECTION FIVE TYING IT ALL TOGETHER THROUGH PRODUCTIVITY 435

Chapter 15 Productivity 436

INDUSTRIAL ENGINEERING AND OPERATIONS RESEARCH

SECTION ONE

INTRODUCTION

Chapter 1

Introduction

1.1 INDUSTRIAL ENGINEERING AND OPERATIONS RESEARCH

There are many factors that affect the long-term success of a company. Marketing strategies, product research, employee relations policies, and many other aspects of how an organization runs its business all make some impact on the "bottom-line" health of that firm. But perhaps the biggest impact is made by the effectiveness and efficiency with which the organization converts input resources into finished products and services—that is, how productive or cost-effective the company is in using labor, equipment, materials, capital, and other resources. Firms that find ways to produce products at lower costs than their competitors are usually the firms that prosper and grow.

Finding ways to utilize input resources in a more cost-effective manner is what industrial engineering and operations research are all about. These two scientific disciplines originated out of the need of businesses and military organizations to invent better methods for accomplishing a job or mission—methods to reduce the labor cost for assembling components, ways to control the quality of products being manufactured, procedures to reduce the time required to fabricate a part, and so forth.

Industrial engineering and operations research use engineering concepts, mathematics, economics, and principles of human behavior to design and implement more efficient, more productive systems. It is this blended use of math, economics, and behavioral considerations that distinguishes industrial engineering and operations research from other scientific and engineering disciplines. And it is the emphasis on finding more productive uses of resources that gives industrial engineering and operations research their high place of importance in the various job functions in an organization.

The purpose of this book is to present an overview of the fundamental concepts of industrial engineering and operations research. To this end we shall discuss those functions of primary concern to the industrial engineer and the operations research analyst and the methods used to successfully achieve their execution. Owing to space limitations, the extent of the discussion of each topic covered must be succinct; however, we shall highlight the salient features.

In the remainder of this chapter we describe or characterize the activities of industrial engineering and operations research and place each discipline in historical perspective by briefly discussing the development of each. The historical development of the two disciplines will lead us to a discussion of their roles in today's society as well as their relationship to one another.

1.2 HISTORY OF INDUSTRIAL ENGINEERING

Initial work in industrial engineering (IE) is usually credited to applied economists and industrialists in England around 1800. The Scottish economist Adam Smith made pioneering observations as to the economies inherent in the division of labor in manufacturing jobs. He argued that by dividing a production job, such as building a pushcart, into small specialized tasks and assigning a single worker to each task, the overall job could be completed less expensively than if one man were assigned the entire job. Smith claimed that such economics were due to (1) the development of new skills when a single task is performed repetitively, (2) the saving of time lost in changing from one task to another, and (3) the invention of new, special-purpose tools and equipment. Smith's theories and observations are summarized in the classic manuscript *The Wealth of the Nations* published in 1776 (see Reference [8]).

Charles Babbage extended the work of Smith. Babbage recognized that by specializing labor task assignments it would not be necessary to pay for skill levels only used during a fraction of the total job. Those portions of the job which do not require high skills or extensive experience can be organized as individual, separate jobs carrying lower wage rates. Thus, the overall cost of production can be reduced. Babbage presented his conclusions in *On the Economy of Machinery and Manufactures* in 1835 [5].

Around 1795, Matthew Boulton and James Watt developed a modern, closely integrated factory to produce steam engines in Soho, England [8]. They implemented sophisticated management systems to aid in operating the plant, including standards for jobs in the factory. They maintained some 22 books of standards for use in detecting waste and inefficiency. They also utilized methods for forecasting, plant location, plant layout, and wage incentives. Their factory system is generally recognized to have been 100 to 150 years ahead of its time.

Developments in industrial engineering in America began with Frederick W. Taylor in the early 1900s. In fact, Taylor is often called the "father of industrial engineering." Whereas Adam Smith and Charles Babbage were primarily observers and writers, Taylor was a "doer," developing principles and concepts through experimentation. He was concerned with developing scientific methods of both performing work and managing a production facility. His theories are summarized in *The Principles of Scientific Management* [12].

From the beginning of the Industrial Revolution until Taylor's era, work was carried out on a rule-of-thumb basis. There were, generally, no standards as to how a particular job was to be performed or what results should be achieved. Managerial planning and analysis procedures were almost nonexistent. Taylor made a significant contribution toward changing this condition to one in which management is a scientific activity rather than a "happenstance." His efforts were centered around four guidelines for managing an organization:

1. Develop a science for each element of a man's work, thereby replacing existing rule-of-thumb methods.
2. Select the best worker for each particular task, rather than allowing the worker to select his own task.
3. Divide work into almost equal shares between management and labor, each taking over the work for which he is best suited (rather than the former condition in which most of the work and responsibility were thrust on labor).
4. Develop a spirit of cooperation between management and labor in order that work be carried out according to the procedures referred to in 1 and 2.

Early applications of Taylor's theories were in such diverse areas as metal-cutting processes, organizational structure, wage incentive systems, work methods, work measurement systems, and production planning and control. However, his major contribution was in the integration of his philosophy as a practical approach to management, leading to what was referred to as scientific management.

Taylor was not alone in his recognition of the need for a scientific approach to management. Frank and Lillian Gilbreth, working as a husband and wife team, were responsible for much of the theory and techniques used today in work methods and in work measurement. They performed many experiments in time and motion studies of various work operations in order to develop principles to aid in designing more efficient job procedures. Henry Gantt was also concerned with the theory of work methods, but made more significant contributions in the area of production planning and control. He developed a systematic method of scheduling production work orders for processing customer jobs on required equipment to account for plant operations as a whole, including interdependencies among work-order processing requirements. The basic tool used in Gantt's approach was a graphical representation of the simultaneous status of all work orders in the plant, called a "Gantt chart." This chart is still used today by many firms as their primary scheduling aid.

There was rapid growth of industrial engineering during the two decades prior to 1930. Significant developments took place in the areas of work methods, work measurement, facility design, wage incentives, job evaluation, organization theory, human factors, and production planning and control. However, with these advances came misuse of the techniques and principles developed by Taylor and others, creating a good deal of ill will toward industrial engineering as a discipline. This was especially true in the application of work measurement. Unrealistic time standards were often set for jobs, and standards were frequently set without informing the workers involved. Such practices led workers to resent the "efficiency expert" involved in developing the standard, and to mistrust the use of standards in general in evaluating worker performance. Even today, there is a tendency for a person with a clipboard and a stopwatch on a plant floor to be looked upon with some misgivings.

Until the late 1940s, developments in industrial engineering followed the tradition established by Taylor, Gantt, and the Gilbreths and constituted refinements of their work. However, during the late 1940s and early 1950s the techniques of operations research were introduced to industry in the United States and Britain. Since that time, much of the growth of industrial engineering has been in the development and application of operations research techniques for the solution of problems in both the public and private sectors.

1.3 HISTORY OF OPERATIONS RESEARCH

The impetus for the origination of most of the activities of operations research was World War II. In fact, the term *operations research* (OR) was coined as a result of "research on (military) operations" during the war. Groups of mathematicians, economists, and other scientists were organized to analyze various military operational problems. These groups were formed in England and the United States, with the U.S. Navy employing more than 70 analysts. A variety of problems were successfully addressed, such as where to locate radar installations, how to search for enemy submarines, how to deploy aerial mines in the seas around Japan, determination of the optimal size of merchant convoy fleets, and the development of maneuver strategies for ships under enemy attack.

Although the term operations research was not coined until World War II, the scientific approach to operational problems can be traced to the work of Taylor and Gantt already cited. McCloskey and Trefethen [9] point to the work of F. W. Lanchester and Thomas Edison in World War I as forerunners of operations research. However, the OR effort during the war was largely independent of these earlier activities and was based on new applied mathematical concepts and techniques.

Operations research applications in World War II were characterized by a team approach to operational problems, with initial efforts carried out in Britain. For example, Professor P. M. S. Blackett was assigned the problem of analyzing the coordination of radar equipment at gun sites. The group he assembled (called Blackett's "circus") included physiologists, physicists, mathematicians, an army officer, and a surveyor. The mixed-team approach was also adopted in the United States following the example set by the British effort. On both sides of the Atlantic, OR teams included mathematicians, statisticians, physical scientists, engineers, biologists, and psychologists.

The success of OR efforts during the war encouraged post-war industrial firms in England and the United States to attempt to apply a similar approach to their operational and managerial problems. Many of the same issues that people such as Taylor and Gantt had addressed were subjected to analyses using more quantitative, systems-oriented procedures than had traditionally been applied. A key figure in the post-war development of OR was George Dantzig. He is generally credited with the development of linear programming, one of the most powerful and commonly used OR tools. Dantzig is often referred to as "the father of linear programming." Besides linear programming, early developments in OR were in areas such as statistical quality control, dynamic programming, queueing analysis, and inventory control. A schematic diagram illustrating the development of some of these initial activities as given by Dantzig [4] is shown in Figure 1.1.

Recent developments in OR have included refinements in these initial areas, as well as the creation of new analytical techniques such as geometric programming,

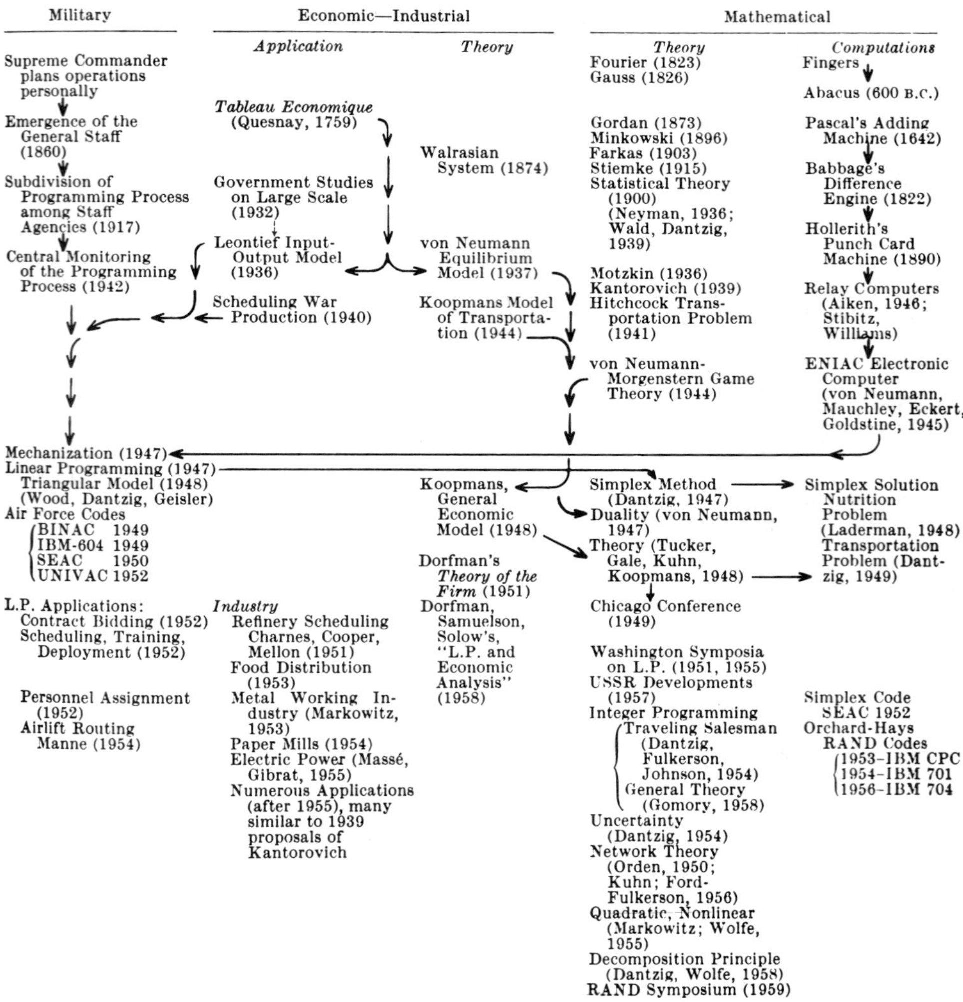

Figure 1.1 Development of Pioneering OR Activities. (By Permission of the Princeton University Press.)

fractional programming, goal programming, and simulation. However, developments in OR methodology have, to a large extent, been dependent on advances in computers and computer science. Most problems attacked by OR techniques are large in scale and require significant computing resources for their analyses. In fact, many OR techniques would be essentially useless in analyzing practical problems if the techniques were not implemented and executed on an efficient computer system. It is this dependence on computing power that has led the development of OR activities to parallel the development of computer systems.

1.4 INDUSTRIAL ENGINEERING VERSUS OPERATIONS RESEARCH

There is a diversity of opinion as to whether or not industrial engineering and operations research (IE/OR) is really a single discipline or two separate disciplines. Although they have had somewhat separate histories as we have seen, IE and OR are nevertheless considered by many practitioners to be part of the same overall discipline. This is primarily due to the mission or "reason for being" that they share in common, that of providing effective, efficient answers to questions relating to design, analysis, and evaluation. This common mission is reflected in many of the formal definitions offered for the two areas. For example, consider the definition of industrial engineering as published by the Institute of Industrial Engineers [3]: "Industrial Engineering is concerned with the design improvement, and installation of integrated systems of men, materials, and equipment. It draws upon specialized knowledge and skill in the mathematical, physical, and social sciences together with the principles and methods of engineering analysis and design, to specify, predict, and evaluate the results to be obtained from such systems." The essence of this definition is that of a scientific discipline concerned with the productive use of men and equipment.

The concerns and methods of operations research are similar to those of industrial engineering. Wagner [13] defines operations research as "a scientific approach to problem-solving for executive management." He goes on to state that:

> An application of operations research involves:
>
> - Constructing mathematical, economic, and statistical descriptions or models of decision and control problems to treat situations of complexity and uncertainty.
> - Analyzing the relationships that determine the probable future consequences of decision choices, and devising appropriate measures of effectiveness in order to evaluate the relative merit of alternative actions.

Churchman [3] characterizes OR as "the application of scientific methods, techniques, and tools to problems involving the operations of systems so as to provide those in control of the operations with optimum solutions to the problems." From these definitions, we can conclude that both IE and OR are concerned with the application of scientific techniques to operational problems. The underlying systems are often the same, as are the study objectives. The primary differences are usually in the scope of the analysis and the type of models and methodology used in the study.

For instance, the early developments in industrial engineering were associated with manufacturing plants, and were heavily based on systematic common-sense approaches to problems rather than on a predominate use of mathematics. This included activities such as process planning, methods improvement, setting job time standards, and job evaluation, which were collectively recognized as industrial engineering. Although these activities are just as important and commonly applied today as in the past, they have been labeled "traditional" or "classical" industrial engineering activities.

However, in the last 30 or so years, the majority of the developments in industrial engineering have arisen through the derivation and application of analytical techniques based on applied mathematical concepts. These are operations research techniques and have been developed by industrial engineering specialists as well as by economists, mathematicians, physicists, chemical engineers, and civil engineers, to mention but a few.

Because of the contrast in the mathematical orientation between traditional industrial engineering and operations research, it may be useful to think of these two areas as a spectrum or continuum of activities with traditional IE at one end and OR at the other. Traditional IE tends to be more applicable to problems in a manufacturing environment. On the other end of the spectrum, OR has a broader scope, being oriented toward more macro level problems in a wide variety of application areas of which manufacturing is only one. Operations research also places heavier reliance on mathematical concepts, especially mathematical models, than does traditional IE.

Under this continuum view of industrial engineering and operations research, activities such as work methods and job evaluation are at one end of the spectrum and linear programming and queueing analysis are at the other. However, there is a great deal of overlap in the middle of the spectrum of activities where problems that have been an integral part of traditional industrial engineering have recently been analyzed using the mathematical methods of operations research.

An example of this type of activity is facility design. Industrial engineers have traditionally been concerned with facility design issues such as determining the best layout for a plant or the best location of a distribution warehouse. Traditional analytical tools have been used, such as flow diagrams, templates, and guidelines checklists. However, more recently OR techniques such as queueing analysis and mathematical programming have been successfully applied in resolving these same facility design problems. In addition, the scope of facility design has been broadened beyond the industrial context to include layout and location problems for post offices, airports, sports organizations, and other service industries. Other activities and problem areas in the middle of this IE/OR spectrum include inventory control, forecasting, production scheduling, and quality control.

An additional area in the IE/OR spectrum is human factors. This area also is a traditional industrial engineering activity, but one in which mathematical and statistical methodologies recently have assumed a more important role. However, these methodologies are not operations research techniques such as mathematical programming. Rather, they include experimental design and analysis techniques, and digital and analog computer process control and analysis.

The distinction between IE and OR is treated in a somewhat different fashion by

Barish [1]. He suggests that OR is the *applied science* for managerial systems, whereas IE is the *engineering* of managerial systems. This distinction is analogous to the relation between medicine and medical science, or chemical engineering and the science of chemistry. Barish believes that "while the industrial engineer should be primarily concerned with the application and modification of existing techniques to the problems as they occur, the operations researcher should be endeavoring to discover fundamental relations and to develop new approaches, models, and techniques. Obviously, the two fields overlap in many areas." Barish is essentially defining a continuum or spectrum similar to the one discussed earlier. However, the spectrum is defined by Barish to have at one end the pure application of existing IE/OR concepts and tools, and at the other end research activity directed at the development of new concepts and techniques. IE activities are at the application end of the spectrum, while OR activities are at the research end.

Each student of industrial engineering and operations research will undoubtedly develop his own philosophy of the relationship between these two areas. The important factor in this philosophy is not so much the exact distinction between the two areas as it is the recognition of their complementary nature. Both areas share in common productivity-oriented goals and provide the analyst with powerful tools useful in their appropriate setting.

1.5 INDUSTRIAL ENGINEERING AND OPERATIONS RESEARCH ACTIVITIES

In the previous sections of this chapter, we have attempted to provide a general reference point as the development and interpretation of IE/OR as a discipline. In the remainder of this chapter, we present an overview of what this discipline is "all about." Specifically, we shall give a few illustrations of typical activities that industrial engineering and operations research analysts have been engaged in within organizations. Hopefully, these examples will serve to give the reader a feel for the variety of IE/OR task assignments. In addition, we shall indicate the more common, basic tasks or activities in which IE/OR analysts are engaged regardless of their organizational environment. These basic activities are routinely used by most organizations in industry and government alike.

1.5.1 Basic Activities

There is a wide range of basic IE/OR activities, from designing wage payment incentive systems to developing solution procedures for complex management decision problems. A list of several common activities is given in Table 1.1. Notice that these activities are applicable to a variety of organizations, including manufacturing firms, banks, hospitals, fire departments, and military installations. This broad scope is a major reason for the popularity of IE/OR as an engineering discipline.

The descriptions given for the activities in Table 1.1 are general in nature. They

Table 1.1 Typical IE/OR Activities

Description[a]	Activity	Text Chapter Reference
1. Determining the most appropriate manufacturing operations and tooling to use to produce a particular product	Manufacturing processes	3
2. Determining a standard for a given work task	Work measurement	4
3. Designing efficient and effective methods for work tasks	Work methods	4
4. Evaluating the economic costs and benefits of one or more investment alternatives	Economic analysis (or engineering economy)	6
5. Designing the best layout of a facility	Facility layout	11
6. Determining the most appropriate location for one or more facilities	Facility location	11
7. Determining the best system for moving goods within a set of facilities	Material handling	11
8. Designing a wage incentive system for a group of employees	Wage incentives	5
9. Designing a system to evaluate the relative worth of each of a set of jobs	Job evaluation	5
10. Developing mathematical models of decision problems in order to derive improved decisions for problems involving allocation of scarce resources	Mathematical programming	8
11. Analysis of problems involving waiting lines	Queueing analysis	9
12. Determining optimal reorder quantities and reorder points for inventoried items	Inventory control	13
13. Predicting the future state of a system or process, such as sales of a product	Forecasting	12
14. Determining when to process each of a set of work tasks on each of a set of processing facilities	Scheduling	12
15. Designing testing procedures to ascertain the quality level of a production process	Quality control	14

[a]These descriptions are not intended to be complete definitions; see the associated text chapter for more comprehensive explanations.

refer to the *classes* of problem situations, such as "designing the best layout of a facility." In order to get a better understanding of exactly what these activities refer to, it is helpful to consider some specific illustrations of the *application* involved in these activities. Several such illustrations are cited below:

1. Determining the most efficient procedures for a worker to follow in assembling a bicycle.

2. Designing the layout of a social security office so that walking distance is minimized.
3. Determining the method of cutting the maximum number of shirt patterns from a large piece of cloth so as to minimize the wasted trim.
4. Determining the locations of the fewest number of fire stations required to provide a response time of no greater than 5 minutes to a fire anywhere in a city.
5. Setting standards for the time to complete various production jobs, such as casting an engine block or welding two end plates together.

These projects center around basic IE/OR activities or analytical approaches that can be applied in any context. For instance, project 1 above involves the use of work methods, a fundamental IE/OR activity that is concerned with finding the most efficient and effective way of accomplishing a given work task. The task could be a product assembly job, a machining operation, or a service activity such as a policeman writing a traffic ticket. Regardless of the specific task or its context, the fundamental activity being performed is "work methods." Similarly, projects such as 2 and 4 involve a basic IE/OR activity, that of facility layout or facility design. Project 3 involves mathematical programming, while project 5 illustrates the use of work measurement.

The primary purpose of this book is to describe the basic or fundamental activities in IE/OR, such as those mentioned above. While each activity is a complete subject by itself, there is a significant interrelationship between these topics as will be pointed out in future chapters.

1.5.2 Classification of Activities

There are many ways to classify or group IE/OR activities. One approach is to identify those activities which involve *descriptive* analysis and those which involve *normative* analysis. By descriptive analysis, we mean an activity that is directly or immediately concerned with obtaining information about a process or system to aid in describing or understanding the characteristics of that process or system. For example, queueing analysis is a mathematical approach for obtaining information of interest for waiting-line situations, such as the expected number of units in the waiting line. Such information would aid in understanding the operations of the waiting-line system. Therefore, queueing analysis can be thought of as a descriptive type of activity. Examples of other IE/OR activities that can also be classified as involving descriptive analysis are work methods, work measurement, process charting, job evaluation, simulation, forecasting, and quality control.

In contrast, other IE/OR activities are immediately concerned with designing processes, systems, or decision strategies in which criteria of "goodness" are used in selecting one alternative design over another. Whereas descriptive analysis is concerned with developing information to help *describe* an alternative design, normative analysis is concerned with developing the *best* design or solution. For example, an IE/OR activity that can be classified as a normative analysis is facility layout. This activity involves designing the interior layout of a structure, such as a plant or warehouse. There are a large number of alternative layouts that could be used; the one that is the best in some sense is chosen. Facility layout is not concerned with merely describing the characteristics of a particular layout, but rather with designing the layout itself. Other IE/OR activities that involve normative analysis are mathematical programming, facility location, inventory control, and wage incentives. It should be noted that many of these normative analysis activities utilize descriptive analysis results as input data or as part of the design process.

1.5.3 IE/OR Tools

It is important to recognize the distinction between the basic IE/OR *activity* being conducted, such as work methods or queueing analysis, and the underlying mathematical and statistical *tools* and *concepts* that are being employed. For example, the activity of work methods relies on various graphs, charts, and diagrams to aid in improving a specific work task, as well as relying on statistical theory to determining adequate observation sample sizes. In this case, the statistical concepts and the graphs, charts, and diagrams are the "tools" used by the IE/OR analysis to conduct the work methods study.

Other statistical tools include regression analysis, analysis of variance, and Markov chain analysis. For example, regression analysis is used in forecasting, wage incentives, and work measurement. Applied mathematical techniques are another set of tools, which include classical calculus, matrix algebra, and calculus of variations.

Perhaps the most commonly used IE/OR tool is a model. As discussed in Chapter 2, models are used to represent the physical process or system that the IE/OR analyst is investigating. Models can be of a purely descriptive nature or can include normative criteria for direct evaluation of alternative designs. Models are used in virtually all IE/OR activities. Both traditional IE and OR rely heavily on the use of models. For example, work methods makes extensive use of charts and diagrams (i.e., "descriptive" models) to represent the layout of a workplace and the manner in which a job is performed. Mathematical programming utilizes "normative" mathematical models to represent and solve decision problems faced by a person or by an organization.

1.6 OVERVIEW

As we have already mentioned, the purpose of this textbook is to introduce the reader to the fundamental concepts of industrial engineering and operations research. This first chapter provides the reader with an orientation as to what IE/OR is all about, including a brief history of the separate development of IE and OR, and several philosophies as to their uniqueness as a single discipline today.

The remainder of this book is divided into three sections, the first containing the problem areas of traditional concern to the industrial engineer and the time-tested techniques that have been applied to their resolution. This material is presented in

Chapter 3 through Chapter 6. The topics in this first section are a collection of the major activities at the traditional end of the IE/OR spectrum mentioned earlier.

In the next section of the text, Chapter 7 through Chapter 10, we describe a sample of the major activities at the other end of the IE/OR spectrum, specifically the OR analytical approaches and techniques that are most commonly applied by the industrial engineering and the operations research analyst. In each of these chapters the focus of our attention is on the development and analysis of models for certain generalized problem structures. That is, we are more concerned there with mathematical modeling and solution methodology than with the specific problems to which the methodology may be applied. However, special applications are treated to illustrate computational procedures and interpretation of results in a variety of problem settings.

The final portion of the book contains material on activities in the middle or overlap portion of the IE/OR spectrum, such as facility design, and production planning and control. The topics discussed in these chapters, Chapter 11 through Chapter 14, are *problem areas* that have been analyzed by traditional industrial engineers since Taylor's era. However, as mentioned earlier, they are also areas that have been successfully analyzed with more quantitative OR techniques such as mathematical programming and queueing theory. Our discussion of these topics emphasizes the overlap and integration of traditional approaches with the appropriate operations research techniques.

Before presenting these three major portions of the text, a preliminary chapter discusses an important topic that links most of the content in the remainder of the book together. Specifically, Chapter 2 is devoted to the formulation, validation, and use of *models*. As mentioned earlier, models are perhaps the most common and most important IE/OR tools. The discussion in Chapter 2 serves a secondary purpose of providing a common basis for the understanding of the relationship between the analysis inherent in traditional industrial engineering and that inherent in operations research. A concluding chapter discusses the topic of *productivity*—how the productivity performance of an organization can be measured, and how it can be improved. After reading this book, it should become clear that analyzing and improving the productive use of resources is what IE/OR is all about.

REFERENCES AND SUGGESTED READING

1. N. N., Barish, "Operations Research and Industrial Engineering: The Applied Science and its Engineering."

2. R. N., Blair, and C. W., Whitston, *Elements of Industrial Systems Engineering*, Prentice-Hall, Englewood Cliffs, NJ 1971.

3. C. W., Churchman, R. L. Achoff, and L. E., Arnoff, *Introduction to Operations Research*, Wiley, New York, 1957.

4. G. B., Dantzig, *Linear Programming and Extensions*, Princeton University Press, Princeton, NJ, 1963.

5. W. C., Giffin, *Introduction to Operations Engineering*, Irwin, Homewood, IL, 1971, Chapter 1.

6. R. W., Hovey, and H. M., Wagner, ''A Sample Survey of Industrial Operations Research Activities, *Op. Res.*, **6**, 880, (1958).

7. E. A., Johnson, ''Long-Range Future of Operations Research'', *Op. Res.*, **8**, No. 2 (1960).

8. H. B., Maynard, *Industrial Engineering Handbook*, McGraw-Hill, New York, 1963, Section 1.

9. J. F., McCloskey and F. N., Trefethen, *Operations Research for Management*, Johns Hopkins Press, Baltimore, 1954.

10. P. E., Overcast, ''This Profession—Industrial Engineering,'' *J. Ind. Eng.*, **8**, 301, (1957).

11. G. M., Parks and R. B., Collins, ''Two Hundred Years of Industrial Engineering,'' *Ind. Eng.*, July, 14–25 (1976).

12. F. W., Taylor, *Scientific Management*, Harper & Row, New York, 1947.

13. H. M., Wagner, *Principles of Operations Research*, Prentice-Hall, Englewood Cliffs, NJ, 1969.

Chapter 2

Modeling

One of the most common and important aids used in industrial engineering and operations research activities is a *model*. A model is simply a representation of reality. That is, a model is an abstraction of a physical system or a property of that system, or of a concept (such as a mathematical operation). Graphs, charts, formulas, and drawings are all examples of models.

The utility of a model stems from its use as a convenient vehicle for analyzing the system being represented (Here, "system" is defined very broadly to mean any collection of components—such as a manufacturing process, a sales generation process, a work task, or a decision-making process.) We can work with the model rather than having to deal directly with the actual system. Often it is very expensive or impractical to experiment with the actual system, so that using an accurate representation or model as a test vehicle allows analyses to be performed that might otherwise be impossible. For example, consider the case of a plant manager who is under pressure to eliminate bottlenecks as parts are moved throughout the plant and to reduce material handling time and costs. One option he has is to shut the plant down for a few days and rearrange the equipment and facilities in the plant in hopes that the new layout will be an improvement. Obviously, this is an expensive way to check out the plant manager's guesstimate as to a better arrangement of facilities. A more practical way would be to build a *model* of the plant and to play "what if" games with this model to evaluate several alternative arrangements. Use of, say, a scale model of equipment, machines, storage bins, and so forth, like that shown in Figure 2.1, could eliminate the expenses of moving actual equipment around as well as avoid the need of unproductive downtime in the plant itself.

Models have performed a critical role in traditional industrial engineering work over the years. For example, the Gantt chart developed by H. L. Gantt in the early 1900s is a descriptive model of the dynamic operation of facilities (e.g., a group of machines in a job shop), and has been a basic tool for job scheduling in many firms since its creation. The templates and drawings used by layout analysts for many years

Figure 2.1 Example of a Three-Dimensional Template Iconic Model.

are also examples of models that have been a vital asset to traditional industrial engineering activities.

Models also play an important role in operations research. Formulas and equations are commonly used to represent underlying processes and operations in order to adequately describe them for analysis purposes. These ''mathematical'' models are used to perform such functions as predicting the future state of a production process, describing how product cost changes as production volume changes, or representing how a random variable takes on different numerical values.

Models other than formulas and equations are also frequently used in operations research. The same types of graphs, scale models, and charts that are used in traditional industrial engineering are also used in operations research activities. In fact, the formulation and use of models is perhaps the predominant link between industrial engineering and operations research. Both areas of activity rely heavily on the use of models. Models are important components in the ''tool kit'' of both the traditional industrial engineering analyst and the operations research analyst.

In the remainder of this chapter, we discuss the various types of models that are used in industrial engineering and operations research activities, as well as the construction and use of these models. The development and use of models play a fundamental part in all of the remaining chapters.

2.1 TYPES OF MODELS

There are several different ways to classify models. One common scheme is to distinguish between (1) iconic models, (2) analogue models, and (3) symbolic models. Each of these types of models is described below.

2.1.1 Iconic Models

A model that uses or consists of the actual properties and characteristics of the system being modeled is called an iconic model. Such a model has the same shape and appearance as the actual system, but is usually a scaled version of the system. Examples of iconic models are pilot plants for manufacturing processes, three-dimensional templates representing machines and workers, and scale models of a pipeline network. An illustration of a template model is given in Figure 2.1.

2.1.2 Analogue Models

A model that consists of real properties, but ones that are different from those of the system being modeled is called an analogue model. Unlike iconic models, analogue models do not necessarily look like the system being represented; however, both types of models are generally used to represent only real systems, as opposed to concepts or ideas. Examples of analogue models are the contour lines used to represent graphically the elevations of a hill, a ''histogram'' in which bars of different lengths are

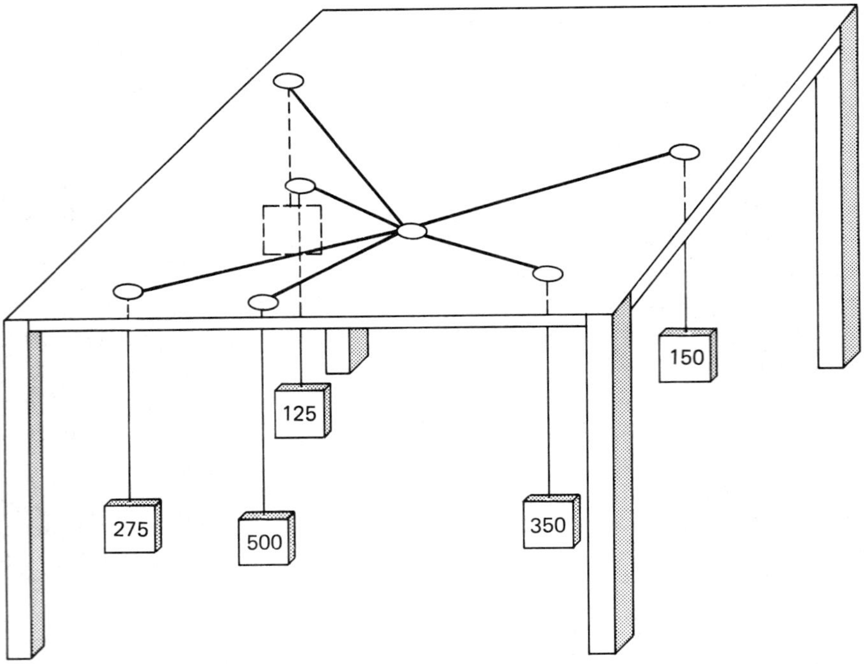

Figure 2.2 Example of an Analogue Model: A Location Table.

used to represent the relative frequency of occurrence of various events, or an analog computer representation of the level of inventory of an item. Another example of an analogue model useful in locating a new facility, such as a plant or fire station, with respect to several existing demand points is the "location table" illustrated in Figure 2.2. Note that strings are being used to represent (i.e., as an *analogue* for) the distance from the new facility to a demand point, while weights attached to the end of the strings are used to model the importance of each customer, that is, the customer's annual demand. The system being modeled consists of the demand points, the new facility, and the relationships among the weighted distances traveled between the new facility and each of the demand points.

2.1.3 Symbolic Models

Symbolic models use letters, numbers, and other symbols to represent the properties or characteristics of the system being modeled. Examples of symbolic models are equations, written sentences, and charts and diagrams such as the "Left-Hand, Right-Hand" chart shown in Figure 2.3, which is used to represent the hand movements involved in a particular work activity.

Symbolic models are the most abstract type of model, and the most commonly used in industrial engineering and operations research activities. In fact, operations

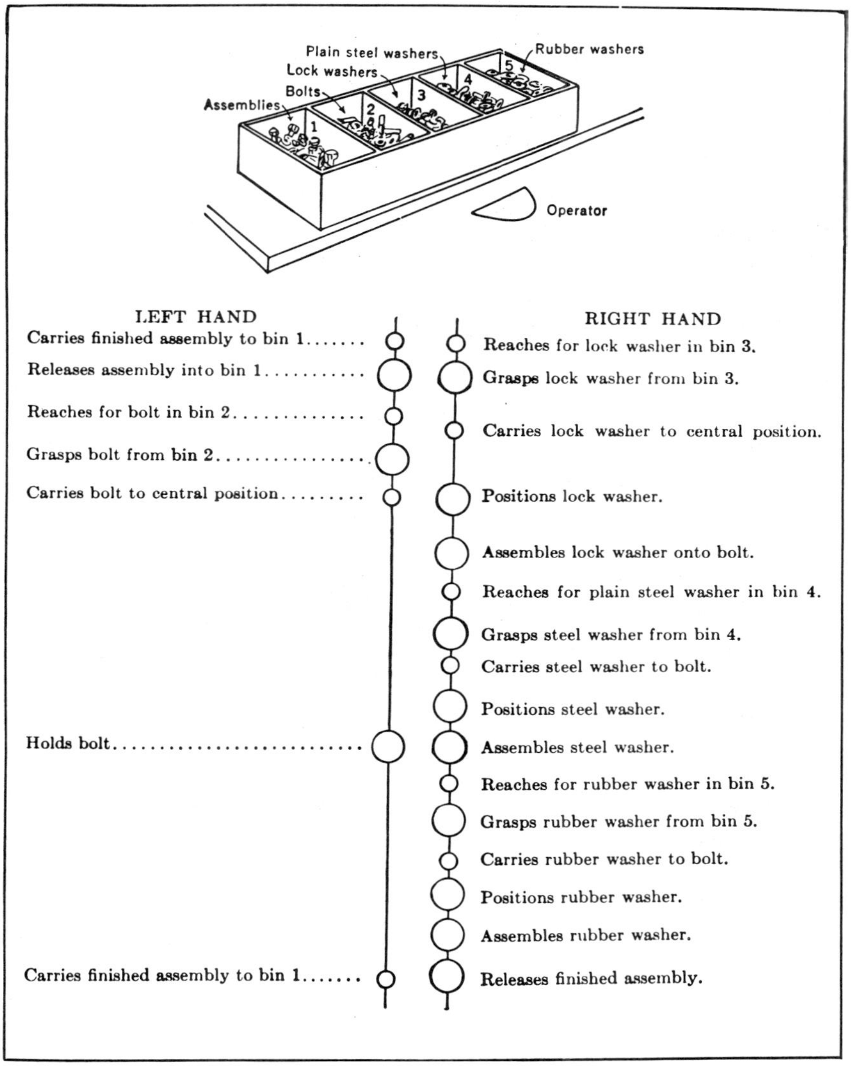

Figure 2.3 Illustration of a Symbolic Model: A "Left-Hand, Right-Hand" Chart.

research is often considered to be synonymous with the formulation and use of a special form of symbolic models called *mathematical* models. Mathematical models involve the use of only symbols that are letters, numbers, and mathematical operators; for example, "$+$," "$\div$," and "$\leq$." These symbols are put together to represent a mathematical expression or relationship. For example, $F = M \times A$ is a mathematical model that represents the relationship between the force exerted by a body in motion (represented by F), the mass of the body (represented by M), and its acceleration (represented by A). This equation depicts not only the factors or variables involved, that is, F, M, and A, but also the specific form of the relationship between them. That is, the model or equation implies that force is equal to the *product* of mass and acceleration (as opposed to, say, their sum).

Descriptive Mathematical Models

There are two types of mathematical models that are frequently used in industrial engineering and operations research activities. *Descriptive* mathematical models simply describe some aspect of the system being modeled, such as its future state or an operating characteristic. Equations are the most common illustrations of descriptive mathematical models. For example, $d_t = a + bt$ is a model that describes how demand for a product changes over time. Specifically, the model implies that demand in period t, as represented by d_t, is a constant value of a plus a change that is directly proportional to the number of time periods of interest. Note that the same information concerning how this demand changes over time can be obtained by using a graphical representation or model rather than the descriptive mathematical model. Such a representation is given in Figure 2.4.

Decision Problem Models

The other type of mathematical model is a *normative* model. As pointed out in Chapter 1, a major function of operations research activities and, to some extent, of industrial engineering activities is the analysis and resolution of decision problems. To aid in this analysis, a mathematical model of the decision problem itself is often constructed. Such a model is a normative model, or a *decision problem* model. As is evident in the discussions in Chapters 8 through 16, decision problem models are fundamental to most operations research analysis and methodology, whether the model is used directly as in the case of "math programming" (see Chapter 8), or used indirectly as in the case of "queueing theory" or "simulation" (see Chapters 11 and 12).

Decision problems are simply issues that are to be resolved by one or more "decision makers." For example, in designing the layout of a bank, the management

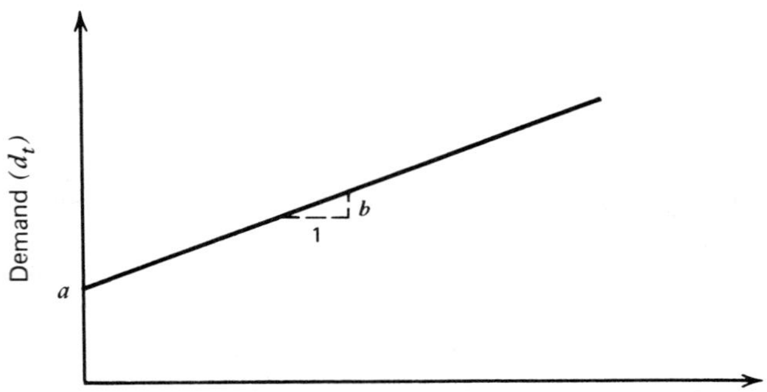

Figure 2.4 Graphical Representation of the Change in Demand Over Time.

of the bank must decide how many drive-in teller facilities to install. In making the decision, management has some objective or objectives in mind, such as minimizing the average waiting time of customers who use the drive-in facilities. However, management may be constrained by limited resources, such as an expense budget, so that the "ideal" number of teller facilities cannot be constructed.

Organizations are confronted with decision problems every day. Some are long-range planning-type decisions, as in the case of the bank example previously mentioned, whereas others are day-to-day issues, such as whether or not to accept a lot or batch of raw material with a certain quality when the vendor promised a higher quality level. In fact, we all face decision problems daily, such as whether to go to a particular class or what to eat for lunch.

There are certain basic *components* present in every decision problem model. These are (1) decision variables, (2) objective function, (3) constraints, (4) functional relationships, and (5) parameters.[1] *Decision variables* are the issues to be determined. That is, they are the variables or factors in the problem for which answers (usually in the form of numerical values) are sought. In the bank teller example cited earlier, the decision variable was "the number of drive-in teller facilities" to have.

The *objective function* is a representation of (a) the criterion (or criteria if more than one is involved) that expresses the decision maker's manner of evaluating the desirability of alternative values of the decision variable(s), and (b) how that criterion is to be optimized (usually, this optimization operator is "minimize" or "maximize"). For example, the objective function in the teller problem is "minimize the average waiting time of customers." The criterion being used here is the average customer waiting time, with *minimize* as the optimization operator. This implies that the bank decision makers feel that a "good" layout design is one that produces as small an amount of average customer waiting time as possible. Note that the decision makers could well have considered many other criteria as relevant and used one of those instead of average waiting time, such as "maximum customer waiting time," "average number of customers in the system at any time," or "maximum number of customers in the system at any time."

Constraints are restrictions on the values that the decision variables can be assigned. These restrictions can arise for a variety of reasons. For example, in the teller example budget limitations may restrict the maximum number of teller facilities that can be built and used to, say, two facilities, owing to the cost of constructing each facility. Also, limits on the space available may restrict the number of facilities that can be built to, say, three. While some decision problem models have no constraints included, there are many instances of models with hundreds and even thousands of constraints.

In formulating a decision problem model, the relevant objective function and constraints are usually converted into descriptive mathematical models so that the end result is a complete mathematical statement of the decision problem. These descriptive models are the *functional relationships* that relate the decision variables to the criteria

---

[1]Some decision problems involve no constraints.

measures, and to the measures of the restricting conditions in the constraints. For example, in the teller problem the criterion of average waiting time might be modeled using the following functional relationship.

$$W(x) = \frac{P_0(\lambda/\mu)^x \rho}{\lambda(1 - \rho)^2 x!} \tag{2.1}$$

where

x = the number of teller facilities used (i.e., the decision variable),

$W(x)$ = the average waiting time of a customer, given that x teller facilities are used,

λ = the expected arrival rate of customers,

μ = the expected rate of serving customers at a teller facility,

ρ = $\lambda/\mu x$, and

P_0 = the probability of having no customers waiting in line or being served.

This descriptive model, Eq. (2.1), depicts the manner in which we assume that average customer waiting time will change as we change the number of teller facilities to be used, x. Notice that other models could have been used instead of Eq. (2.1) to represent this relationship. Equation (2.1) is based on several underlying, *implied* assumptions, such as that by "waiting time" we mean "time spent in a line in front of one of the teller facilities." If, instead, we define waiting time to be "time spent in a line in front of one of the teller facilities *plus* time spent being served," then the model could be formulated as:

$$W(x) = \frac{P_0\,(\lambda/\mu)^x\rho}{\lambda(1 - \rho)^2 x!} + \frac{1}{\mu} \tag{2.2}$$

The fifth component of a decision problem model is the set of *parameters* involved. Parameters are the "constants" in the functional relationships. For example, in the model of the average waiting time $W(x)$, the parameters are P_0, ρ, λ, and μ. (P_0 is actually a function of x as well as of λ and μ, as is ρ, so that the right-hand side of Eq. (2.1) has only x's, λ's, and μ's in it.)

Parameters can be deterministic or probabilistic in nature. A deterministic parameter is one whose value is assumed to occur with certainty. For example, in the $W(x)$ model the parameters P_0, λ, and μ may be treated as fixed, known numerical values such as 0.1, 2, and 4. In this case, the parameters are considered to be deterministic quantities and model (2.1) can be expressed in *numerical* form as follows.

$$W(x) = \frac{(0.1)(2/4)^x(2/4x)}{2(1 - 2/4x)^2 x!} \tag{2.3}$$

However, if λ and μ are not treated as fixed constants but are considered directly or explicitly as random variables (i.e., the *average* rate is random), then they are probabilistic parameters. For instance, we might assume that λ can realize any of three possible values, say 10, 20, or 40, with equal likelihood (i.e., λ may realize a value of 10 with a probability of $\frac{1}{3}$, or a value of 20 also with a probability of $\frac{1}{3}$, or a value of 40 with a probability of $\frac{1}{3}$).

A decision problem model expresses the objective function, decision variables, and constraints simultaneously in a single, normative mathematical model (along with the imbedded functional relationships and their parameters). For example, a decision problem model of the teller problem might be formulated as follows (before substituting numerical values for parameters).

$$\underset{x}{\text{Minimize}}\ W(x) = \frac{P_0(\lambda/\mu)^x\rho}{\lambda(1-\rho)^2x!} \tag{2.4}$$

$$\text{subject to: } c_1x \leq B \tag{2.5}$$

$$c_2x \leq F \tag{2.6}$$

$$x = 0, 1, 2, \ldots \tag{2.7}$$

In this model, P_0, λ, μ, ρ, x, and $W(x)$ remain as previously defined, c_1 is the cost to construct a teller facility, B is the amount of money available for building teller facilities, c_2 is the space that a facility requires, and F is the total space available for construction of teller facilities.

There are several observations that should be made concerning this model. First, note that the decision variable x is shown under the optimization operator. This is merely a notational convenience. Second, the verbal or narrative version of the model would be "find the value of x that minimizes $W(x)$, subject to the constraints that. . . ." Third, other formulations of the imbedded descriptive models could have been used. For example, the first constraint could have been formulated (perhaps more realistically) as $c_0 + c_1x \leq B$, where c_0 is a fixed site preparation cost that is independent of the number of facilities built. Fourth, the last constraint implies that the number of teller facilities built and used must be an integer number. An alternative way of expressing this constraint would be "x integer." Finally, the model is shown in general symbolic form. If values are obtained for the parameters involved, the model could be expressed in numerical form. For example, if $\lambda = 2$, $\mu = 4$, $c_1 = \$4000$, $c_2 = 500$ ft^2, $B = \$10{,}000$, and $F = 1500$ ft^2, then the model would appear as:

$$\underset{x}{\text{Minimize}}\ W(x) = \frac{(0.1)(2/4)^x(2/4x)}{2(1-2/4x)^2x!}$$

$$\text{subject to: } 4{,}000x \leq 10{,}000$$

$$500x \leq 1500$$

$$x \text{ integer}$$

Models of different decision problems have different forms, such as different objectives, different numbers of constraints, or different functional relationships for the same criteria. However, the basic structure of every decision problem model is the same. In fact, a general version of a decision problem model can be written as:

$$\underset{\mathbf{x}}{\text{Maximize}} \text{ (or } \underset{\mathbf{x}}{\text{minimize}}\text{) } f(\mathbf{x}) \tag{M1}$$

$$\begin{aligned} \text{subject to: } g_1(\mathbf{x}) &\leq b_1 \\ g_2(\mathbf{x}) &\leq b_2 \\ &\vdots \\ g_m(\mathbf{x}) &\leq b_m \end{aligned}$$

where $\mathbf{x}$ is a vector of decision variables $(x_1, x_2, \ldots, x_n)$; $f(\mathbf{x})$ is the criterion to be optimized; and $g_i(\mathbf{x}) \leq b_i$ is the ith constraint (note that b_i can be interpreted as the fixed amount of the ith scarce resource to be used). As relates to the teller problem model, $\mathbf{x} = x$, $f(\mathbf{x}) = W(x)$, $g_1(\mathbf{x}) = 4000x$, $b_1 = 10{,}000$, $g_2(\mathbf{x}) = 500x$, $b_2 = 1500$, and $g_3(\mathbf{x}) \leq b_3$ is interpreted to be "x integer."

The general decision problem model (M1) can be classified in several different ways. For example, if there are one or more constraints, the model is said to be a *constrained model* and is referred to as a "mathematical programming problem" (see Chapter 8 for a discussion of such problems); otherwise, the model is said to be an *unconstrained model* or, equivalently, an "unconstrained optimization problem" (see Chapter 7 for a discussion of this class of models). If all of the parameters and functional relationships are treated as deterministic in nature, then the model is classified as a *deterministic model*; otherwise, it is a *probabilistic* or stochastic *model*. Also, a decision problem model is referred to as a *linear model* if all of functional relationships $f(\mathbf{x})$, $g_1(\mathbf{x})$, $g_2(\mathbf{x})$, . . . , $g_m(\mathbf{x})$ are of a linear form; otherwise, if one or more of the relationships are nonlinear, the model is said to be a *nonlinear model*. These classifications of decision problem models are further discussed in Chapter 8.

In addition to mathematical models, *simulation* models are frequently used in industrial engineering and operations research activities. A simulation model is a descriptive model representing the way in which a system operates, and is intended to be used to simulate or imitate these operations. A simulation model can be an iconic model, an analogue model, or a collection of mathematical models. For example, a pilot process (i.e., an iconic model) is often used to simulate the operations of the associated full-scale process in order to determine the feasibility of full-scale operations. Mathematical equations that represent how workers wait for service at a plant's tool crib can form a simulation model.

The most common type of simulation model used in industrial engineering and operations research activities is a *computerized* model. This is simply a computer program that is a collection of coded statements or equations that describe how the system being modeled operates. Usually, the program also contains control statements to carry out the actual simulation itself. Manual simulations using mechanical devices are also useful, especially when "sellable" methodology is of importance. Mechanical simulators, such as a magnetic "Gantt chart" board used to simulate alternative shop scheduling strategies, are excellent for providing a decision maker with a tangible, relatively easily understood visualization of *how* the simulation was conducted.

It is important to note that a distinction is being made here between a simulation model and the process or act of simulation. The activity of simulation is the actual conduct of the imitation of the system's operation. That is, "simulation" involves first constructing a simulation model, then using that model to experiment with or imitate the operations of the system. This distinction should become more evident after reading Chapter 10.

2.2 USE OF MODELS

In general, models are used as an aid in the analysis of a problem. However, a model can serve a number of specific functions or purposes. For example, according to Elmaghraby [2], a model can be helpful in any of five basic functions of an industrial engineering or operations research analysis. First, a model can serve as an *aid in thought*. That is, descriptive mathematical models, iconic models, and analogue models can help the analyst to visualize a system so that he can better understand its structure or operation. The model serves to stimulate the analytical thought process.

Second, a model can serve as an *aid to communication*. Specifically, a model represents how a system or problem is conceptualized by someone. Therefore, the model can serve to transmit ideas and visualizations among people in the organization, such as between analysts, or between an analyst and a decision maker. For example, the Left-Hand, Right-Hand Chart shown in Figure 2.3 might model how a particular analyst feels the job involved should be performed. Therefore, the chart can serve as an aid for the analyst to communicate his idea of a "good" work method to others in the organization for their scrutiny.

Third, a model can be used for *training* and *instruction*. In this capacity, the model can serve as a communication device, or as a training tool. For example, Figure 2.3 can function as a document for instructing a worker as to how he is to perform the job in question. On the other hand, a simulation model that provides expected outcomes for various inventory planning strategies can serve to train a production manager in how to properly set inventory policies.

Fourth, models can be used as *tools for prediction*. That is, certain types of models can be used to forecast or predict the future state of a process or system. Descriptive mathematical models and simulation models are especially useful for predictive purposes. For example, the descriptive mathematical model given in Eq. (2.3) can be used to predict the average waiting time of customers as the number of teller facilities changes. Alternatively, Eq. (2.3) can be used to predict the consequences of increasing the rate of serving of customers; that is, the effect on average waiting time of increasing μ. The same simulation model used to train the production manager in inventory management can be used to predict the consequences of implementing alternative inventory policies. In fact, the primary use of a simulation model is in predicting the outcome of a prescribed set of strategies so as to provide needed input–output information to solve a decision problem.

Finally, the fifth primary function that a model can serve is as a direct tool for *control* or *decision-making purposes*. Decision problems regarding controllable aspects of a system influence the design or operation of the system by changing or controlling the value of some decision variables. Therefore, decision problem models can serve directly in the control process by aiding in the solution of the decision problem. Decision problem models are "solved" by finding the values of the decision variables that (1) meet or satisfy all constraints simultaneously (such a set of values is called a *feasible solution* to the model), and, at the same time, (2) achieve the stated objective; that is, result in the maximum or minimum value of the objective function when compared to all other feasible solutions. Such a solution is called an *optimal solution*. Other types of models such as descriptive mathematical models and simulation models, can serve *indirectly* by providing information required in the decision problem model.

In summary, models are used in industrial engineering and operations research activities in the analysis, evaluation, and documentation of a system or decision problem. The model, such as a work shop layout or a Left-Hand, Right-Hand Chart, may serve as an end product for use as documentation or communication. Or, the same model may be used as an intermediate aid in resolving a decision problem, as in the case of a Left-Hand, Right-Hand Chart with its associated "standard time" serving to evaluate a particular alternative work method (the decision problem here would be to find the most "efficient" work method, perhaps subject to some constraints on expenditures for fixtures and tools).

2.3 FORMULATION OF MODELS

Formulating or constructing a model involves two basic activities: developing the *form* and type of model to use, and determining the values of the various *parameters* of the model, that is, the numerical values for the "constants" in a descriptive mathematical model. For any given modeling situation, the first requirement is to determine the type of model to use. For example, if it is desired to forecast future company sales demand, a number of different types of models could be used as an aid, such as a graph like Figure 2.4 or a descriptive mathematical model like $d_t = a + bt$. If a descriptive mathematical model is used, we still must determine what form the equation should have. A linear, quadratic, sinusoidal, or similar form could be used.

After deciding what type and form of model to use, we must obtain all necessary data required to transform the model into a useable, numerical form. For example, in the Left-Hand, Right-Hand Chart of Figure 2.3, specific numerical values must be determined for each item, such as "Releases assembly into bin 1."

In the mathematical model shown in Eq. (2.3), numerical values must be assigned to parameters μ and λ. Often, determining these values—that is, *parameterizing* the model—is the most critical and difficult part of formulating the model.

Constructing models is an art rather than an exact science. That is, common sense and experience are the primary tools used in formulating a model, rather than a specific step-by-step procedure. However, there are several guidelines or key points that should be considered when formulating a model. Five such guidelines are discussed in the remainder of this section, along with several illustrative examples.

The first key point is that *assumptions* are built into the model, and must be recognized and accepted by the user. Assumptions are typically made for such factors as the form of various components of the model, whether parameters are deterministic or probabilistic, and, in the case of a model of a decision problem, what criteria and constraints are relevant to the problem. Some assumptions are explicitly stated by the analyst prior to constructing the model, whereas others are only implied by the model itself. For example, a sales manager may say, ''Since our future sales volume is related to the price of our products, a good predictor of our sales for widgets is $s = 10 + 2p$, where s = sales volume and p = price per widget.'' Note that the manager explicitly *assumed* that sales were directly related to price, but that the specific linear form of this relationship was *implied* by the model itself.

Second, the model should be *validated* to measure its accuracy. That is, in order for a model to be useful, the degree to which it actually represents the system or problem being modeled must be established. (Notice that this definition excludes the inclusion of the ''value'' of the model in the validation process.) A descriptive model (whether it is a chart, diagram, mathematical model, or a simulation model) is valid if it properly depicts the system being represented. A decision problem model is valid or accurate if (1) it contains all of the objectives, constraints, and decision variables relevant to the problem; (2) those objectives, constraints, and decision variables included in the model *are all relevant* to, or actually part of, the problem; and (3) the functional relationships are valid, that is they are accurate, descriptive mathematical models—including representative forms and accurate parameter values. For example, if a model is being formulated of the decision problem of where to locate machine centers in a plant, and the layout criterion of safety is not included *either* in the objective function or the constraints, then if management deems safety to be an important consideration, the model is not valid.

The third key point pertinent to model building or formulation is that the developed model represents a *trade-off between the accuracy of the model and its utility*. As applied to descriptive models, utility means the usefulness of the model in aiding the user to represent or describe the system being modeled. For example, a complex equation used to forecast company sales that involves sinusoidal and higher-order polynomials may have no utility, even though it may be an accurate predictor, if the sales manager who must make decisions based on the results of using the model cannot understand it and hence has no confidence in its use. In the case of normative or decision problem models, utility means primarily the ease with which the model can be solved. A very large decision problem model with many nonlinear constraints has no utility to the decision maker if the model cannot be solved within the manpower and computer resources available. (Some large, complex decision problem models would require *days* of continuous computer time to solve using the most efficient solution procedure available.)

Therefore, even though a model is an excellent and correct representation of a system, it can be useless as an analytical or decision-making tool. On the other hand, a model that is easily understood by the user, and is readily solvable in the case of a normative model, may also have no value if it is not an accurate model. Hence, the "best" model to use would be one that achieves a balance between accuracy and utility. Obviously, the version of a model that strikes this balance in one situation will not necessarily produce the desired balance in another situation. For example, the sales manager of one company using a forecasting equation may have no mathematical background or training, so that a complex but accurate equation may have no use for him. On the other hand, the sales manager of a competing company may have a very quantitative background so that the same model does in fact have utility for him.

A fourth guideline that should be adhered to in constructing a model in "garbage in, garbage out (*GIGO*)." In other words, no matter how accurate or potentially useful the model is *before* its parameters or components are assigned numerical values, if those numbers are worthless then so will be the final model itself. For example, suppose a police department is trying to develop a model to predict how crime rate changes as the average income level of the city changes. If Figure 2.5*a* depicts historical data on this relationship, then a straight line such as that shown in Figure 2.5*b* would be a good predictor in the range of interest. The associated mathematical model is $c(r) = a - br$, where a and b are parameters to be assigned values, r is the average income level in the city, and $c(r)$ represents the crime rate. However, if a value of 22 is assigned to a and a value of $1/75$ is assigned to b, then the model is rendered useless since it gives very poor predictions of $c(r)$ as shown in Figure 2.5*c*—even though the *form* of the $c(r)$ model was accurate. Even though it is sometimes tedious and time-consuming to do so, the data and information used to formulate the model should be carefully checked for accuracy.

A fifth key point is that the development of a model should be an *iterative process*, based on feedback as to the model's accuracy and utility. That is, the first version of a model should not automatically be accepted as the final version. After validating

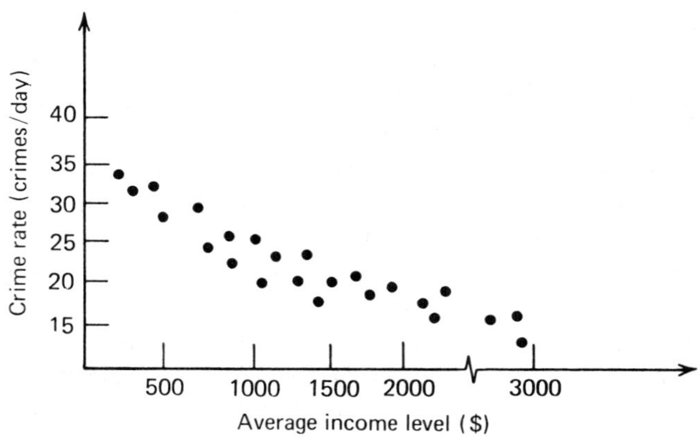

(*a*)

Figure 2.5(*a*) Historical Data on Crime Rate versus Income Level.

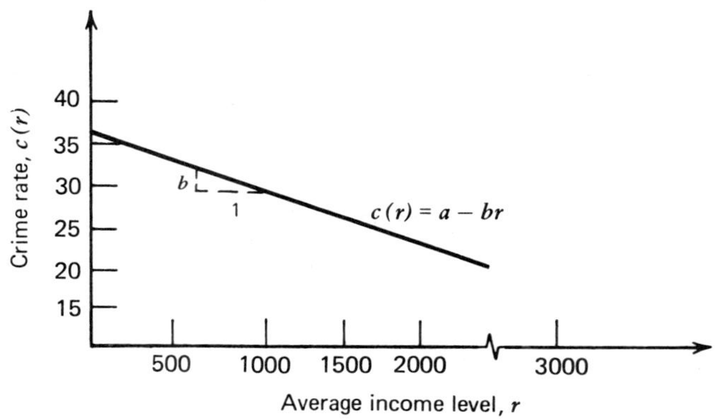

(b)

(b) Form of Model for Predicting Crime Rate.

the initial model and establishing its utility, the model should be revised to reflect the user's desires as to the accuracy–utility trade-off. The process of obtaining this feedback on a preliminary version of a model and revising the model is depicted in Figure 2.6.

The ability to formulate "good" models is something that is, to a large extent, developed with experience and practice. The five guidelines mentioned above can serve as a useful aid in model building, and should be explicitly considered whenever a model is constructed. Several examples of modeling situations will now be given to illustrate use of these guidelines as well as the model formulation process itself.

Example 2.1 Formulation of a Descriptive Model

The JLK Corporation manufacturers bolt and washer assemblies. They have their production facilities in Opp, Alabama, and their corporate office in Dallas, Texas. Management has decided to promote cost reductions throughout the company, and has

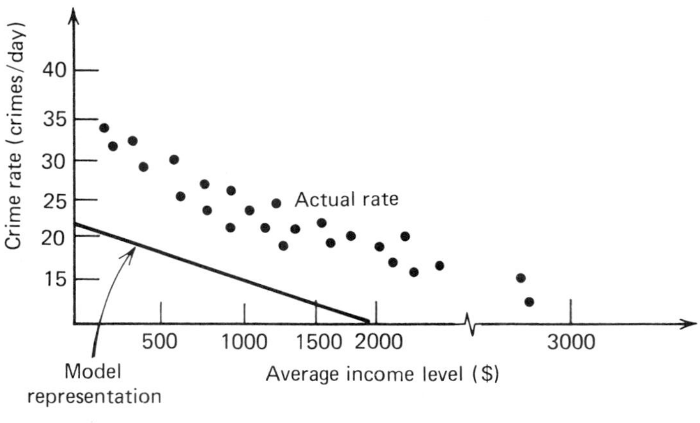

(c)

(c) Use of Inaccurate Parameter Values in Accurate Model Form.

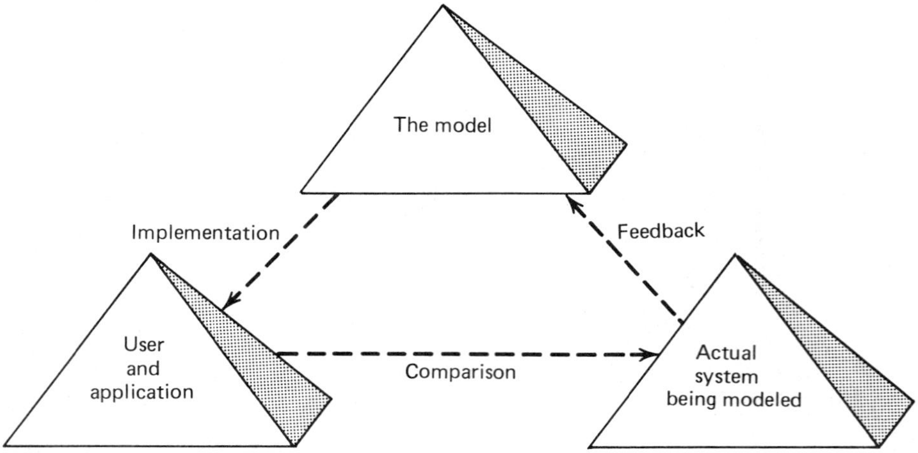

Figure 2.6 Iterative Process of Model Building.

instructed the Industrial Engineering Department to improve the efficiency of all work operations. The head of the Industrial Engineering Department has assigned one of the industrial engineering analysts the job of reviewing the operation of final assembly and recommending improvements. The Industrial Engineering Department is located in the Dallas office, while the assembly operation is performed in the plant at Opp. Unfortunately, a recommendation is needed within a week and there is no more money available in the department's travel account, so that the analyst must develop his recommendations without first-hand observation of the operation.

To carry out his assignment, the analyst realizes that the first and most important step is to visualize exactly how the assembly operation is currently being conducted. That is, his first task is to construct a model to represent the current assembly operation. (The "system" being modeled here is the assembly "method," including all tasks that take place and their sequence and time duration, along with the tools used in the operation and the layout of the work area.) After considering alternative types of descriptive models including photographs, motion pictures, and a variety of charts and diagrams, the analyst decides to develop a Left-Hand, Right-Hand Chart as the basic model to use in his analysis. To construct the chart, he needs to obtain the following information: (1) the individual movements of the assembly operator's left and right hands, (2) when each movement takes place, (3) the distance involved in each movement, (4) the duration of each movement, (5) auxiliary tools used, and (6) the arrangement or layout of the work area. This information constitutes the data required to develop a numerical version of the model.

Because of the time and budget limitations imposed on him, the analyst decides to obtain the necessary information by phoning personnel at the Opp plant. After talking to the shift foreman, the analyst constructed the model shown in Figure 2.3. Using this description of the assembly operation, the analyst made several "obvious" improvements and calculated that the new method decreased the time to assemble one clamp by 75 percent, with a cost of $50 for work place improvements. A model of the new method is given in Figure 2.7.

Let us review the guidelines for model construction that were or were not applied in this case study. First, note that the model, Figure 2.7, has several implicit assumptions associated with it. For example, it is implied that the movements shown are in fact the only movements performed in the assembly operation, and that they are conducted in the particular sequence shown *each time* a bolt set is assembled. Also, it is implied that the underlying distances and times are also invariant from one assembly cycle to another.

Second, notice that the analyst did not validate the model. This could be done (given enough time and expense budget) by comparing the model with the assembly operation as it is being conducted at the plant. Discussions with plant personnel besides the shift foreman would have also aided in establishing the validity or accuracy of the model developed. Third, Figure 2.7 is a simple but useful model of the assembly operation, and represents the analyst's decision as to a good balance between accuracy and utility. Other models might be more accurate in certain regards; for example, a motion picture would give the *dynamic* aspects of the operation, or use of more microlevel breakdowns of movements—such as reach, grasp, release—would provide a more accurate representation of the movements. However, they would not have the same simple, direct usefulness as Figure 2.7 for the analyst in his "hurried" effort to find improvements.

Fourth, the data used in constructing Figure 2.7 were obtained over the phone from an indirect worker, so that their validity is questionable. That is, it is very likely that the formulation of this model is an example of GIGO. Regardless of the accuracy of the breakdown of the movements shown in the model, if the time and distance numerical data obtained from the foreman are in significant error, the model's usefulness is greatly diminished. For example, how much faith can management place in the 75 percent savings cited for the new method? Finally, note that Figure 2.7 is a "single-pass" model; that is, it is not the result of an iterative modeling procedure in which feedback is used to improve either its accuracy or its utility. Additional versions of the model could readily be obtained by discussing the merits of the preliminary version with other analysts, plant personnel, and management. ■

Example 2.2 Formulation of a Decision Problem Model

A new bulk mail distribution center is to be opened to serve six regional post offices in the Southeast. The bulk mail center receives packages from the regional post offices and redistributes them to other regional post offices in the area for delivery. Postal service analysts have developed the "location table" shown in Figure 2.2 as an analogue model of the decision problem of determining where to locate the new bulk mail center. What is the equivalent mathematical model of this decision problem?

1. ***Decision Variable.*** The decision variables are x and y, the coordinates of the location of the new center.

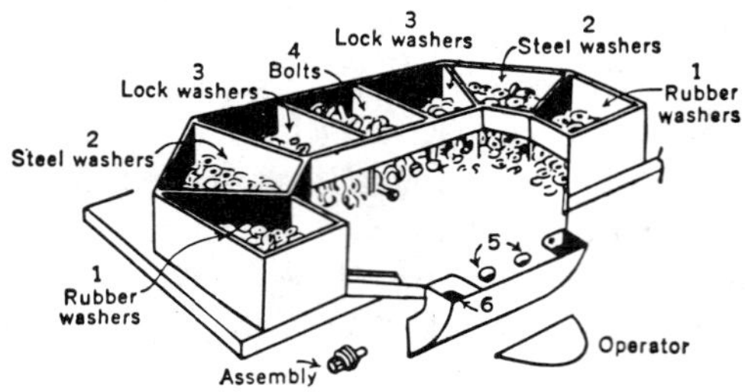

LEFT HAND	RIGHT HAND
Reaches for rubber washer in bin 1......	Reaches for rubber washer in bin 1.
Grasps rubber washer from bin 1.......	Grasps rubber washer from bin 1.
Slides rubber washer to countersunk hole	Slides rubber washer to countersunk hole.
Positions rubber washer in countersunk hole 5............................	Positions rubber washer in countersunk hole 5.
Reaches for plain steel washer in bin 2..	Reaches for plain steel washer in bin 2.
Grasps steel washer from bin 2.........	Grasps steel washer from bin 2.
Slides steel washer to countersunk hole..	Slides steel washer to countersunk hole.
Positions steel washer in countersunk hole 5............................	Positions steel washer in countersunk hole 5.
Reaches for lock washer in bin 3........	Reaches for lock washer in bin 3.
Grasps lock washer from bin 3.........	Grasps lock washer from bin 3.
Slides lock washer to countersunk hole..	Slides lock washer to countersunk hole.
Positions lock washer in countersunk hole 5............................	Positions lock washer in countersunk hole 5.
Reaches for bolt in bin 4..............	Reaches for bolt in bin 4.
Grasps bolt from bin 4................	Grasps bolt from bin 4.
Carries bolt to washers at 5............	Carries bolt to washers at 5.
Positions bolt preparatory to inserting it into washers at 5.................	Positions bolt preparatory to inserting it into washers at 5.
Assembles bolt and washers............	Assembles bolt and washers
Lifts bolt and washers, carries to left and releases into top of chute 6.......	Lifts bolt and washers, carries to RIGHT and releases into top of chute 6.

Figure 2.7 Operations Process Chart of Improved Bolt-and-Washer Assembly.

2. ***Objective Function.*** The implied objective function is

$$\min_{(x,\,y)} f(x, y) = \text{the total distance traveled per week in delivering packages between the center and the six regional post offices}$$

or,

$$\min_{(x,\,y)} f(x, y) = \sum_{i=1}^{6} w_i[(x - a_i)^2 + (y - b_i)^2]^{1/2}$$

where

w_i = the number of deliveries required per week of packages being shipped to the ith post office and from the ith post office,
a_i = the horizontal locating coordinate of the ith post office, and
b_i = the vertical location coordinate of the ith post office.

3. ***Constraints.*** The only restriction on the values that the decision variables can take on that is implied is that (x, y) be in the Southeast region, as defined by the area on the location board. Mathematically, this constraint can be modeled by defining the feasible region as shown in Figure 2.8. Then, the constraint is:

$$0 \le x \le 20 \qquad \text{and} \qquad 0 \le y \le 20$$

4. ***Parameters.*** The parameters in the model are $w_1, w_2, \ldots, w_6$ (i.e., w_i, for $i = 1, \ldots, 6$), $a_1, a_2, \ldots, a_6, b_1, b_2, \ldots, b_6$, and the 0 and 20 values defining the ranges of feasible values of x and y. The location board model implies that all the parameters have been assigned numerical values. For example, w_1 might have been assigned a value of 275 and w_2 a value of 500.

5. ***Functional Relationships.*** There are two functional relationships involved in the model, (a) the objective function *criterion* of total distance traveled, and (b) the metric or measure used for travel distance. The criterion relationship is the descriptive model for $f(x, y)$. The distance metric relating the (x, y) decision variables and the resulting distance to a specific (a_i, b_i) location is $[(x - a_i)^2 + (y - b_i)^2]^{1/2}$.

Is this a valid model? That is, does it accurately represent the associated decision problem? One obvious source of inaccuracy is the descriptive model used to represent distance (i.e., the distance metric). The *form* of this model is applicable for straight-line movement between two points (i.e., Euclidean distance), whereas actual travel to and from the bulk mail center will follow road or rail networks and will not be Euclidean. Some other model form might provide a more accurate representation. For example, a rectilinear metric might be used, so that the distance model is $|x - a_i| +$

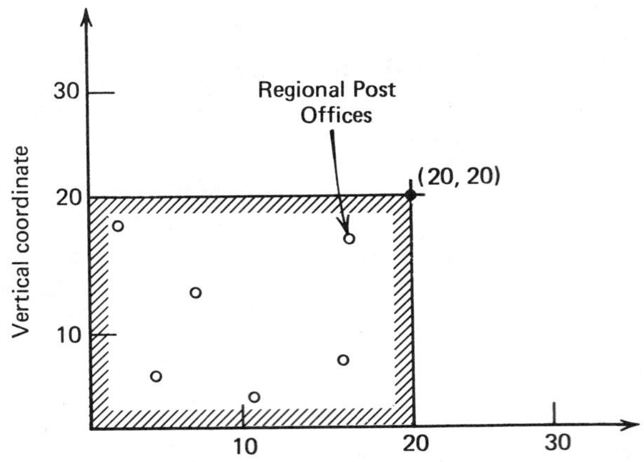

Figure 2.8 Feasible Region for the Bulk Mail Center Location Problem in Example 2.2.

$|y - b_i|$. (This model would be particularly accurate if travel were restricted to an urban area with a predominantly perpendicular road network.) However, if it is decided that the Euclidean model is accurate, we must still consider the consequential difficulty of solving the decision problem model with the descriptive model imbedded in the objective function. As it turns out, the square-root term in the distance model causes the decision problem model to be relatively hard to solve.

If we were willing to sacrifice some degree of accuracy, we could use a less representative distance model that would lead to a more readily solvable decision problem model. For example, the "squared-Euclidean" metric could be used, so that distance is modeled as $(x - a_i)^2 + (y - b_i)^2$. The decision problem model is easily solved when this alternative metric is used. Notice that the choice of a specific distance model implies that a decision has been made as to the trade-off between the model's accuracy and its utility. ■

Example 2.3 Formulation of a Decision Problem Model

The Hanley Construction Company is a large commercial building contractor that typically has four or five construction projects going on at any one time. The company has a pool of 20 utility men who can be assigned as extra labor to speed up the progress of any particular project. Historical records of the effectiveness of the use of various crew sizes of utility men added to different types of projects are given in Table 2.1. The company would like to decide how to allocate this pool of men to their current projects at the beginning of each week so that they reduce the projected finish dates of the current projects as much as possible. However, regardless of the effectiveness of a crew, the company has an obligation to assign at least three utility men to each project at the beginning of the week. In addition, union contracts call for twice as

Table 2.1 Effectiveness of Utility Crews in Example 2.3

Project	Number of Utility Men (Crew Size)						
	1	3	4	8	15	16	18
Rough construction	0.5[a]	3.0	4.0	5.0	6.0	6.2	6.8
Concrete work	1.0	3.2	4.6	7.0	10.5	12.0	12.5
Lumber work	0.2	0.6	0.8	1.6	1.9	2.0	2.1
Finish work	0.1	0.3	0.6	0.7	1.2	1.3	1.3

[a]Units of effectiveness are "days reduced from finished date."

many utility men to be assigned to rough construction as are assigned to finish work. Formulate a mathematical model of this decision problem.

1. ***Decision Variables.*** The decision issue is the number of utility men to assign to each project at the beginning of the week. Therefore, let

   x_{ij} = the number of utility men to assign to the ith project of the jth type construction; $j = 1, 2, \ldots, 4$ and $i = 1, 2, \ldots, n_j$ (n_j = the number of projects of the jth type).

   For example, if there are 2 projects of construction type 1 to be scheduled, then $n_1 = 2$, and x_{11} and x_{21} can be interpreted as the number of men to assign to the first and second projects of type 1 construction.

2. ***Objective Function.*** The criterion to be optimized in this problem is the total number of days reduced in the finished date of all projects under construction during the week in question. Letting

   $h_j(x_{ij})$ = the number of days reduced in the finish date of the ith project of type j construction when x_{ij} utility men are assigned to that project,

   then the objective function is

$$\min_{\mathbf{x}} f(\mathbf{x}) = \sum_{j=1}^{4} \sum_{i=1}^{n_j} h_j(x_{ij})$$

3. ***Constraints.*** There are three types of constraints involved in this problem. They are:

   a. At least three men must be assigned to each project, that is,

$$x_{ij} \geq 3 \qquad \forall j \text{ and } \forall i$$

b. Twice as many men must be assigned to rough-construction-type projects (i.e., $j = 1$) as assigned to finish-work-type projects (i.e., $j = 4$).

$$\sum_{i=1}^{n_1} x_{i1} = 2 \sum_{i=1}^{n_4} x_{i4}$$

or, alternatively,

$$\sum_{i=1}^{n_1} x_{i1} - 2 \sum_{i=1}^{n_4} x_{i4} = 0$$

c. Only integer numbers of men can be assigned to any project.

$$x_{ij} \text{ integer} \qquad \forall j \text{ and } \forall i$$

4. ***Parameters.*** All parameters are assumed to be deterministic in nature. The parameters of $h_j(x_{ij})$ are as yet unspecified. In addition, in the constraint that $x_{ij} \geq 3$, "3" is the numerical value of the parameter "minimum allowable number of men to be assigned to any project."

5. ***Functional Relationships.*** The only unspecified relationship is $h_j(x_{ij})$. One method of developing a descriptive model for $h_j(x_{ij})$ is to plot the historical observations of $h_j(x_{ij})$ versus values of x_{ij}, and to "eyeball" the *form* of the model. Then, a statistical curve-fitting procedure can be used to develop the best estimates of the parameters in the model (see Ref. 5 for a discussion of such curve-fitting procedures). For example, a plot of $h_1(x_{i1})$ is shown in Figure 2.9, along with the form and numerical version of a "best-fitting"

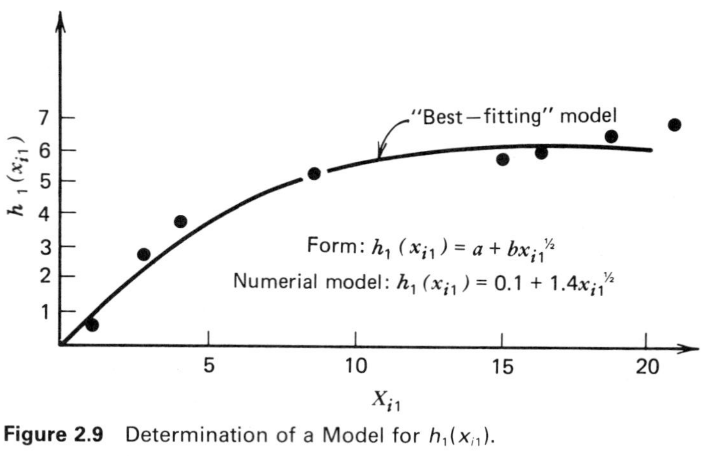

Figure 2.9 Determination of a Model for $h_1(x_{i1})$.

model. As given in the figure, the model for $h_1(x_{i1})$ for each $i = 1, 2, \ldots, n_1$ is

$$h_1(x_{i1}) = 0.1 + 1.4\, x_{i1}^{1/2}$$

Similar models can be developed from the other $h_j(x_{ij})$ function, with results as follows.

$$h_2(x_{i2}) = 0.4 + 2.5\, x_{i2}^{1/2} \qquad \text{for } i = 1, 2, \ldots, n_2$$

$$h_3(x_{i3}) = 0.05 + 0.5\, x_{i3}^{1/2} \qquad \text{for } i = 1, 2, \ldots, n_3$$

and

$$h_4(x_{i4}) = 0.04 + 0.3\, x_{i4}^{1/2} \qquad \text{for } i = 1, 2, \ldots, n_4$$

6. ***Complete Model.*** Putting all of the components together, the complete decision problem model is:

$$\min_{\mathbf{x}} f(\mathbf{x}) = \sum_{i=1}^{n_1} (0.01 + 1.4x_{i1}^{1/2}) + \sum_{i=1}^{n_2} (0.4 + 2.5x_{i2}^{1/2}) + \sum_{i=1}^{n_3} (0.05 + 0.5x_{i3}^{1/2}) + \sum_{i=1}^{n_4} (0.04 + 0.3x_{i4}^{1/2})$$

$$\text{S.T.:} \sum_{i=1}^{n_1} x_{i1} - 2\sum_{i=1}^{n_4} x_{i4} = 0$$

$$x_{ij} \geq 3 \qquad \forall j \text{ and } \forall i$$

$$x_{ij} \text{ integer} \qquad \forall j \text{ and } \forall i$$

■

2.4 SUMMARY

This chapter is concerned with the definition, use, and construction of models that are used in industrial engineering and operations research activities. Models are a vital tool to both the traditional industrial engineering analyst and the operations research analyst, and constitute perhaps the largest common link between these two areas. The types of models that are most common in industrial engineering and operations research activities are descriptive mathematical models; descriptive charts, graphs, and diagrams; normative or decision problem models; and simulation models.

Models are used in industrial engineering and operations research activities in the analysis, evaluation, and documentation of a system or decision problem. Models can serve specific functions, such as an aid in thought, an aid in communication, an aid in training, an aid in prediction, or as an aid in control or decision-making purposes.

The construction or formulation of a model is more of an art than a science.

However, there are guidelines or key points that can be useful in developing a "good" model:

1. Recognize the assumptions associated with the model.
2. Validate the accuracy of the model.
3. Explicitly make a trade-off between the accuracy of the model and its utility.
4. GIGO.
5. Model building should be an iterative process in which feedback as to the accuracy and utility of preliminary versions of the model is used to improve the model formulation.

REFERENCES AND SUGGESTED READING

1. R. L. Ackoff, *Scientific Method, Optimizing Applied Research Decisions,* Wiley, New York, 1962, Chapters 2, 3, 4, 13.

2. S. E. Elmaghraby, "The Role of Modeling in IE Design," *The Journal of Industrial Engineering,* **XIX,** No. 4, 292–305. (June 1968).

3. S. I. Gass, "Decision-Aiding Models: Validation, Assessment, and Related Issues for Policy Analysis," *Operations Research,* **31,** No. 4, 603–631. (July–August 1983).

4. R. V. Hartley, *Operations Research: A Managerial Emphasis,* Goodyear, 1976, Chapter 2.

5. W. W. Hines and D. C. Montgomery, *Probability and Statistics in Engineering and Management Science.* Ronald Press, 1972, Chapter 12.

6. P. Rivett, *Principles of Model Building,* Wiley, New York, 1972.

7. H. A. Taha, *Operations Research, An Introduction* (2nd ed.), Macmillan, New York, 1976, Chapter 1.

REVIEW QUESTIONS AND PROBLEMS

2.1 What is the utility of a model in problem solving?

2.2 Define and give an example of each of the following:

(a) Model
(b) Iconic model
(c) Analogue model
(d) Mathematical model
(e) Normative model

2.3 Compare and contrast descriptive mathematical models with decision problem models.

2.4 Briefly describe each of the five major components of a decision problem model and give an example of each.

2.5 What is meant by a descriptive mathematical model being imbedded in a decision problem model?

2.6 What is the difference between a simulation model and a decision problem model?

2.7 What is the difference between a deterministic model and a probabilistic model?

2.8 How can a model be useful to an industrial engineer or operations research analyst in (a) aiding communication? (b) training? (c) prediction?

2.9 Why is "parameterizing" a model often the most critical and difficult part of formulating a model?

2.10 Why is it important to explicitly recognize the assumptions that are built into a model?

2.11 How could the location table model shown in Figure 2.2 be validated in an actual application? How could the model in Figure 2.3 be validated?

2.12 Discuss what is meant by "the trade-off between the accuracy of the model and its utility."

2.13 Consider the decision problem model given in Example 2.3. Critique this model in terms of each of the four model-building guidelines presented in Chapter 2.

2.14 Critique the manner in which the model in Example 2.1 was validated.

2.15 Discuss how the trade-off between accuracy and utility could be affected by reformulating the model in Example 2.2

2.16 Could a mathematical model have been useful in the analysis referred to in Example 2.1?

2.17 Formulate a descriptive model, using the format of Figure 2.3, of the job of sharpening five pencils in a wall-mounted pencil sharpener. Critique your model in terms of the basic model-building guidelines.

2.18 Develop a mathematical model to act as an alternative or substitute for the analogue model given in Figure 2.2. Compare and contrast the two models.

2.19 List the descriptive models associated with the decision problem model given in Example 2.3.

2.20 Show the correspondence between the decision problem model of Example 2.3 and the general version of a decision problem model, Eq. (M1).

2.21 Give a numerical illustration of what the general decision problem model (M1) would reduce to if all the functions involved were linear.

2.22 Consider the problem of positioning the machines in a certain plant. Develop a diagrammatic model that would be useful in this layout problem. Develop a descriptive mathematical model for this problem. Compare the utility of the two models.

2.23 Consider the two diagrammatic models given in Figure 2.10. Compare the models in terms of (a) accuracy, (b) utility, and (c) data requirements to parameterize.

2.24 In "work sampling" a sample of N observations is taken of a process under investigation. The larger the sample, the more accurate will be the conclusions drawn about the nature of the process. In fact, the relationship between the accuracy of or confidence in the conclusions (as measured in terms of a "percentage confidence interval" I, and the sample size) can be modeled as follows:

$$I = 2(1.645)\sqrt{\frac{\bar{P}_i(1 - \bar{P}_i)}{N}}$$

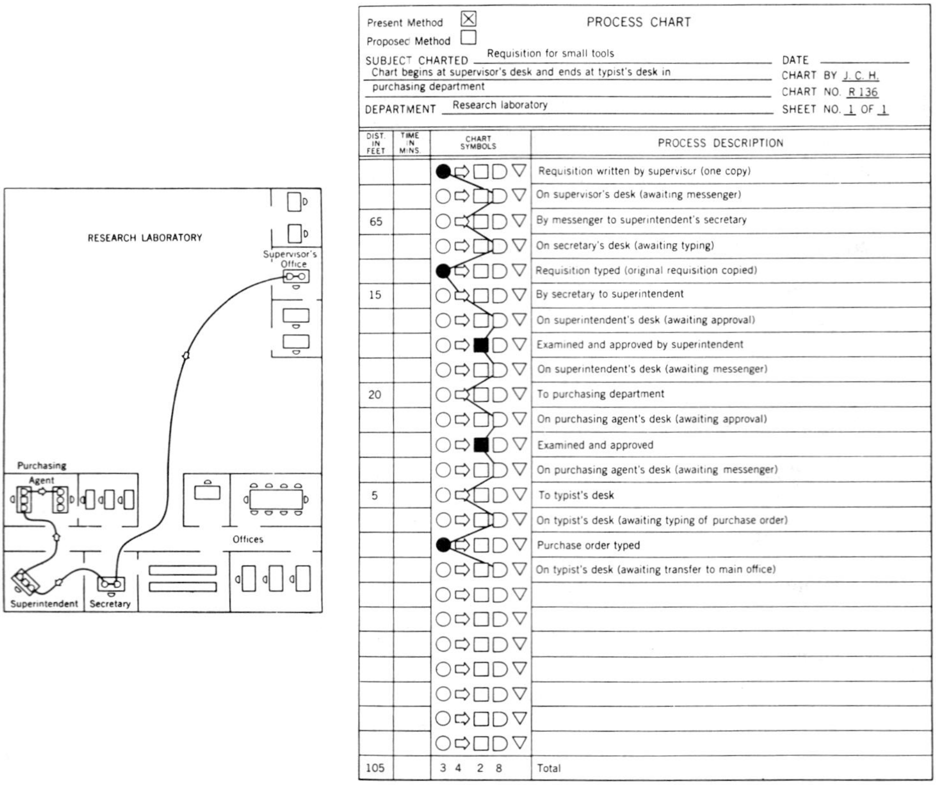

Figure 2.10 Two Alternative Models for Representing a Certain Office Procedure.

where $\bar{P}_i$ is the proportion of observations in the sample that were of activity i, and the confidence interval is set to provide a 90 percent confidence. An alternative model of this relationship is given by the "alignment chart" in Figure 2.11. Discuss the utility of these two alternative models.

2.25 Smithville, U.S.A., is in the process of adding another fire station to reduce the overall response time to fires.

(a) Describe how an analogue model such as that shown in Figure 2.2 could be developed for this application.

(b) Consider the mathematical model given below as an alternative to aid in the analysis of this fire station location problem.

$$\underset{\mathbf{x}}{\text{minimize}}\ Z = \sum_{j=1}^{n} x_j$$

$$\text{S.T.} \quad \sum_{j=1}^{n} a_{ij} x_j \geq 1, \quad \forall i$$

$$x_j = 0, 1, \forall_j$$

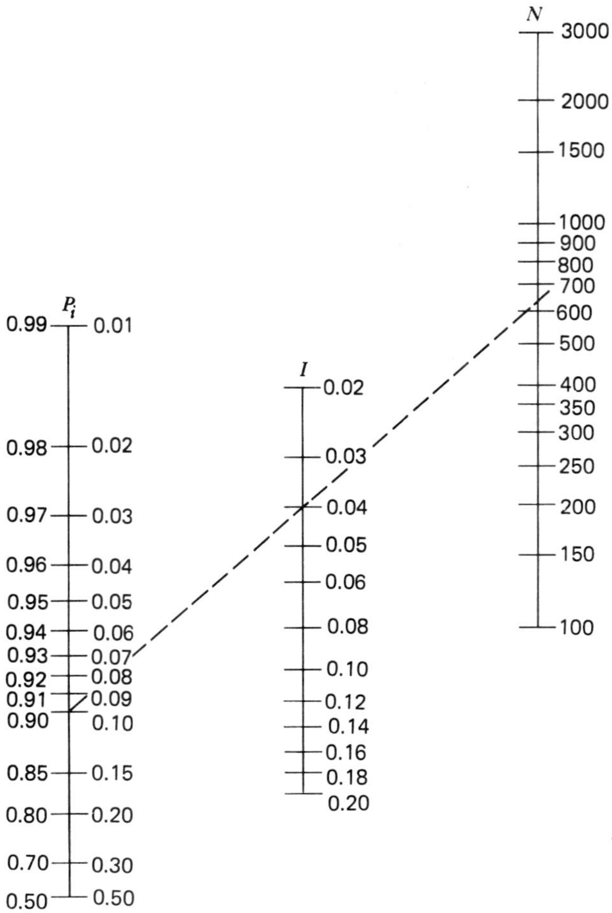

Figure 2.11 An Alignment Chart to Model the Relationship Between I and N in Work Sampling.

where $x_j = 1$ if a station is built at alternative location site j, 0 otherwise; and $a_{ij} = 1$ if a fire truck responding from station site j to a fire at "demand point" i could reach the fire within a goal of 5 minutes, 0 otherwise. Discuss the applicability and usefulness of this model in terms of (1) how it represents the actual decision problem, (2) its data requirements, (3) its utility in being understood by the Smithville fire chief, and (4) how it compares to the analogue model developed in (a) above.

2.26 Consider the classic "One-Potato, Two-Potato" problem: A frozen-food company processes three varieties of potato products—French Fries, Hash Browns, and Flakes. The same raw material is used for all three products and can be purchased from two alternative outside vendors. However, the quality and yield from the two suppliers is different. Specifically, vendor 1 supplies potatoes that provide a 20 percent yield of French Fries, a 20 percent yield of Hash Browns, and a 30 percent yield of Flakes, with the remaining

30 percent waste. Vendor 2 can provide potatoes that yield 30 percent French Fries, 10 percent Hash Browns, and 30 percent Flakes. The company has a limited market, so that the amount of each type of product it can sell is restricted as follows: 5.0 hundred pounds of French Fries per week, 3.2 hundred pounds of Hash Browns, and 4.1 hundred pounds of Flakes. Based on the sales revenues associated with the yields for the separate products, and on the purchase cost of potatoes, it is estimated that the relative profit contribution associated with purchases from vendor 1 is $10 per hundred pounds and $12 per hundred pounds for vendor 2. The firm would like to know how much to purchase from each of the two vendors.

(a) Formulate a mathematical model of this decision problem.
(b) Classify this decision problem model.
(c) How could this model be validated?
(d) Is "GIGO" likely to be true for this model?
(e) How could your model in (a) be improved?

2.27 Two descriptive models that are sometimes used in forecasting are as follows.

(a) $y_t = b_0 + b_1 t$

(b) $y_t = b_0 + b_1 \sin\left(\frac{\pi t}{12}\right)$

where y_t is the state of the process at time t, and b_0 and b_1 are constants. Describe the form of a series of data points over time for which (a) would be a more accurate model than (b).

2.28 The Fast Freeland Burger Stand (FFBS) is a fast food restaurant. They stock hamburger meat in 5-pound cases to use in making all of their hamburger products. This meat is ordered from a local supplier whenever the stock on hand falls below seven cases. It usually takes three days for the reordered supply to be delivered. FFBS is trying to determine the most economical number of cases to order whenever a resupply order is placed. They consider two costs as significant: the cost of orders placed during a year, and the cost of carrying meat in storage. Order costs are thought to vary directly as the number of orders placed during the year (which is approximately the annual demand, 1200 cases, divided by the size of the resupply order quantity), with the cost of placing a single order being around $25. Inventory carrying costs seem to vary directly with the average stock level (which is approximately half the size of the resupply order quantity). Specifically, FFBS management considers that every case in the average level of inventory costs them $60 to carry.

(a) Formulate a mathematical model of this decision problem.
(b) Classify this decision problem model.
(c) How could this model be validated?
(d) Is "GIGO" likely to be true for this model?
(e) How could your model in (a) be improved?

SECTION TWO

TRADITIONAL INDUSTRIAL ENGINEERING FUNCTIONAL AREAS

This section of the book is concerned with those activities in which industrial engineers have been involved since the beginning of the IE/OR discipline. These are the activities, such as job evaluation and work measurement, that are at the micro end of the scope spectrum mentioned in Chapter 1. That is, they tend to be oriented to the individual worker and his immediate workplace, as opposed to a broader system such as a collection of plants and warehouses. Also, these activities tend to rely more on systematic common sense than on mathematics, just the opposite of operations research techniques.

However, as pointed out in Chapter 2, all IE/OR activities—whether part of traditional industrial engineering or of mathematical operations research—rely very heavily on the development and use of *models*. Models are the most common, most basic tools used in IE or OR studies. In traditional IE activites, iconic and analogue models such as graphs and diagrams are used more than mathematical models. On the other hand, OR studies almost always utilize some form of mathematical model. As you read the material in the next four chapters, pay close attention to the use of models, even though the models may not be explicitly formulated or cited.

The topics discussed in this section are organized by functional areas or types of activity rather than by the specific techniques or tools used. This is in keeping with the format of typical IE textbooks and course curricula. On the other hand, in Section Three, the discussion of operations research will be organized according to the *techniques* used in OR studies. In making this distinction, it should be noted that IE terms like job evaluation, work measurement, or engineering economy refer to specific functional *areas* of activities or IE applications, rather than directly to specific methodologies or techniques. For example, the "payback period" as used in engineering economy refers to a specific project evaluation method, while "engineering economy" refers to a wide collection of such techniques as well as to their use in IE studies.

The first chapter in this section, Chapter 3, is concerned with the manufacturing processes that underlie the problems addressed in traditional IE studies. The IE analyst must have a good understanding of *what* work operations are involved and *how* they are being performed before he can improve their efficiency and (or) their effectiveness. The discussion in this first chapter is intended to give the reader an overview of this background material as it relates to the most common application industry—metal processing.

The next chapter is concerned with designing and improving the industrial jobs and work stations needed to carry out production operations and processes such as those mentioned in Chapter 3. Chapter 4 discusses ways in which the amount of work required by a particular job or activity can be measured. This work measurement activity is basic to improving the efficiency of a job since it provides a quantitative assessment of the job's work requirements—such as how many minutes it usually takes a normal operator to assemble a valve. We have to be able to accurately assess these requirements in order to know if a suggested change in the job is an improvement or not.

The other half of Chapter 4 provides an overview of several different ways in which work is commonly measured, including stopwatch time studies and work sampling. The result of applying IE tools such as work sampling is, in part, the creation of *data* for a variety of uses. For example, the standard time required for a job is a basic input for analyses to improve the operation of a plant, such as the layout of the machines in the plant or the scheduling of customer orders on these machines. As will be pointed out in Sections Three and Four, these data are a fundamental input in applying most operations research techniques as well as in using traditional IE tools.

One traditional IE area requiring use of the output of work measurement studies is the design of wage incentive programs. This is the topic of Chapter 5, along with the procedures used to evaluate the requirements of jobs for pay purposes. ''Job evaluation'' is similar to work measurement in that it is an IE function concerned with assessing the work content of a job. However, whereas work measure involves determing how much work is required in a job, job evaluation is used to determine how much a job is *worth* to the company compared to other jobs.

The second topic of Chapter 5, wage incentives, is another salary administration function that is usually handled by IEs. Wage incentive systems are pay schemes intended to motivate workers to be more productive. These systems provide monetary compensation to a worker in proportion to his or his group's output. Most incentive systems are based on a comparison of actual output with a standard output. This standard is usually derived through the application of the work measurement techniques discussed in Chapter 4.

The last chapter in this section is concerned with the measurement of the cost and benefits of capital investments, the purchase of new facilities, suggested work methods improvements, and the like. This cost–benefit assessment is often called engineering economy. The techniques used in engineering economy rely on descriptive mathematical models of the measurement of the economic worth of a series of cash flows. Unlike most of the other IE/OR tools discussed in this book, engineering economy techniques are used by non IEs throughout an organization just as frequently as by IEs.

There are other areas of traditional IE practice beyond those surveyed in this section. Many of these, such as plant layout, job shop scheduling, and inventory control are discussed in Section Four. The primary reason for separating these topics is that the functional areas in Section Four use OR techniques as well as traditional analytical tools. On the other hand, the activities discussed in Section Two use only traditional approaches. The use of significant mathematical techniques is not common in areas such as work measurement and job evaluation.

A final point that should be noted is that even though the emphasis in the discussions of this section is on basic, nonrigorous methodologies, there are areas in traditional IE activities in which sophisticated mathematical techniques can and have been applied. The reader should look for such extensions as he reviews the material in each of the next four chapters.

Chapter 3

Manufacturing Processes

3.1 INTRODUCTION

A manufacturing process is the operation of converting a material or product into a new and different material or product. A manufacturing process is used to change one or more characteristics of an input product, such as size, shape, or strength. There are thousands of different manufacturing processes currently in use in industrial companies. These range from simple operations such as drilling a hole in a block of steel, to complex processes such as converting atomic fuels into energy.

Other examples of manufacturing processes are: assembling individual products to make a completed automobile, casting engine blocks from molten metal, sawing wood for furniture products, and heat-treating wire to remove its brittleness. Manufacturing encompasses nearly every industry conceivable including: electronics, petrochemical, steel, textiles, woodworking, fishing, agriculture, forestry, automotive, aircraft, space, power, machinery, and many others. It should be noted that the service industries are not usually considered manufacturing. Service industries include federal, state, and local governments; transportation (airlines and railroads); and health care systems.

Industrial engineers are concerned with manufacturing processes for three basic reasons. First, in order to successfully apply the various IE/OR tools discussed in Chapter 1, a thorough understanding is necessary of the environment in which these applications take place. That is, if the industrial engineer is to improve the efficiency of a manufacturing facility's process through the use of a tool such as motion and time study or perhaps linear programming, then the IE needs to be familiar with the attributes and constraints of these manufacturing processes. The better the understanding of the processes, the better the chance for analyzing and improving the operations involved.

The second basic reason that industrial engineers are concerned with manufacturing processes is that the IE/OR analyst is often called on to assist in the design of new processes or in redesigning and improving existing ones. In such cases, the IE/OR analyst is responsible for finding ways to make the overall collection of processes as efficient as possible. This activity is often called process planning. In entails the use of analytical tools such as flow charts, as well as productivity enhancements via automation and robots, to eliminate unnecessary manufacturing operations and reduce the overall cost to implement and operate the processes in question. Typically, when working in this capacity, the industrial engineer is in a "staff" position. A staff position does not require hourly or daily involvement with the manufacturing process. Rather, the IE/OR analyst is seeking a long-term impact on the process.

The third basic reason that industrial engineers are concerned with the manufacturing process is that many IEs function as "line" manufacturing engineers. These people are responsible for the hourly and daily manufacturing operation. Typically they are in a position where they manage the first-line supervisors or foremen. In many instances these engineers are promoted to the position of plant manager, responsible for the operations of a multimillion-dollar facility. Therefore, an industrial engineer operating as a manufacturing engineer must have a sound knowledge of manufacturing processes as well as all other industrial engineering disciplines.

In view of these three motivating reasons, this chapter presents a description of the fundamental aspects of manufacturing processes. This includes a discussion of some *primary* manufacturing processes such as forming, casting, and forging, as well as two major *secondary* processes: chip removal and metal fabrication processes. The discussion of these processes centers on what transformations actually take place during the process, as well as on the equipment and machines typically used. In addition, process planning tools used to improve the efficiency and productivity of manufacturing processes systems will be described. Finally, automation and robotics are discussed.

3.2 PRIMARY PROCESSES

A primary manufacturing process includes the initial operations required to convert raw materials into a basic but more usable product. Table 3.1 lists several primary processes, the input raw material, and the resulting product. A product can progress through several primary processes. An example of this would be the manufacturing of steel where iron ore, lime, and coke are combined in a blast furnace to make an iron billet. The billet then undergoes a second primary process when it is heated and alloying agents are added to make steel. A third primary process would be forming the steel into thin sheets by applying rolling and cutting processes.

Since it would be impossible to detail all the primary processes here, three major primary processes are discussed. All three processes are directed toward the metals, mainly iron and steel. These three have been chosen because they constitute major portions of the manufacturing industry and because many secondary processes utilize, as their raw materials, the finished product of these three primary processes.

Table 3.1 Examples of Primary Processes

Industry	Typical Raw Material	Typical Product
Steel	Coke Iron ore Lime	Iron and steel in various shapes
Forestry	Timber	Plywood Lumber Pulp
Petroleum	Crude oil	Gasoline Fuel oil Petroleum by-products
Textile	Petroleum by-products Sulfuric acid Cotton Wool	Cloth
Paper	Pulp	Paper Cardboard
Agriculture	Seed Fertilizer	Grains Vegetables Vegetable oils Fruits
Mining	Property	Iron ore Coke Coal
Seafood	None	Fish Clams Shrimp

3.2.1 Forming

Forming can be defined as altering a metal's shape from an unusable slab or billet into plates, sheets, rods, or wire. Metal can be formed by rolling, drawing, or extruding. When metal is rolled, it is passed between a series of paired rolls that squeeze the material down in thickness and thereby increases the length or width. When metal is rolled at elevated temperatures (800°–2200°F) this process is called hot-rolling, and when it is rolled at ambient temperature, it is called cold-rolling. Figure 3.1 illustrates the rolling processes necessary to manufacture different finished products.

Hot-rolling requires considerable energy to elevate the metals temperature. Once the metal is hot, the rolling does not require excessive force. With hot-rolling, rather complex shapes can be obtained such as railroad track and I-beam structural products. When hot-rolled metal cools, a disfiguring scale is present on the surfaces, and it is very difficult to hold straightness and tight tolerances.

Cold-rolling requires considerable force by the rollers to alter the original shape. Because of the high forces necessary, the amount of change caused by a single pass through a pair of rollers is much less than for hot-rolling. Therefore cold-rolling requires

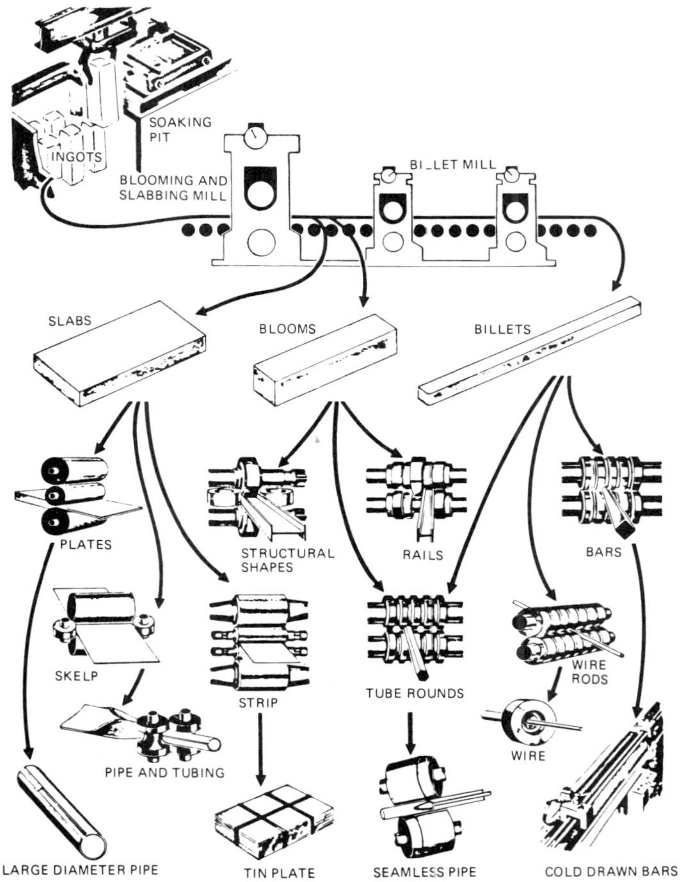

Figure 3.1 Primary Manufacturing Processes for the Forming of Steel (By Permission of McGraw-Hill)

more passes and subsequently a longer processing time. Cold-rolled metal does not have the scale found on hot-rolled, and considerably better straightness and closer tolerances can be obtained. Cold-rolling also provides increased hardness because the metal's crystal structure has been altered in a cold state and cannot return to an unstressed state.

Cold-drawing is similar to cold-rolling except that the metal is pulled, "drawn," through a set of dies rather than rollers. These dies limit the size of the material as it exits the die. Typically, cold-drawing is used to manufacture round, hexagonal, or square rod or wire. The properties and difficulties inherent with cold-rolling are also found in cold-drawing. Figure 3.2 depicts a cold-drawing operation.

Extrusion is a combination of hot-rolling and cold-drawing. Usually extruding is performed with the material at an elevated temperature as with hot-rolling. The material is forced through a die to obtain the proper shape similar to cold-drawing. Unlike

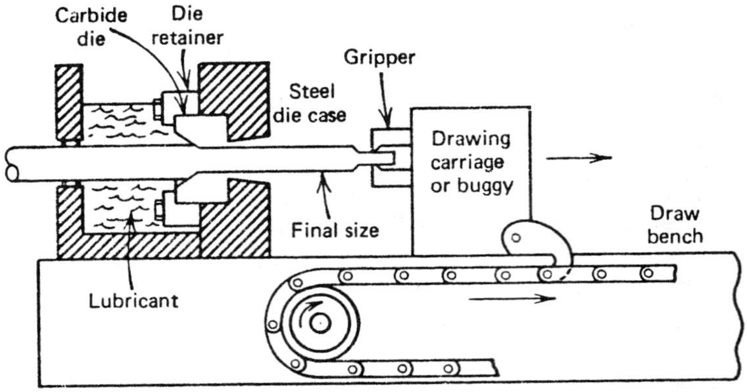

Figure 3.2 A Schematic of a Cold-Drawing Operation (By Permission of McGraw-Hill)

cold-drawing, extrusions pass through only one die and the amount of reduction is considerably greater. Extrusion allows for products with very complex cross sections such as window jams, helicopter rotors, and picture frames. Figure 3.3 depicts examples of extruded products. Nearly all extruded products require a secondary process to obtain the desired length.

3.2.2 Casting

Casting can be defined as pouring a liquid material into a previously prepared cavity or mold and permitting the material to solidify, thereby acquiring a desired shape. Molten iron is the most common material used, although steel, aluminum, brass,

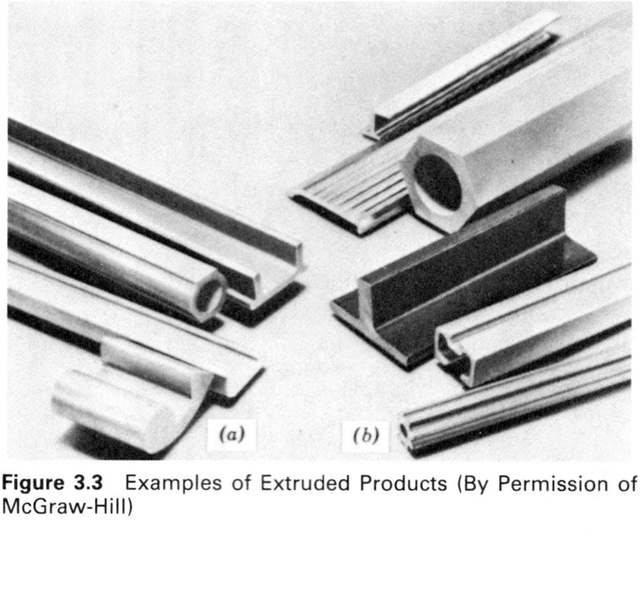

Figure 3.3 Examples of Extruded Products (By Permission of McGraw-Hill)

bronze, some polymers, and other materials can be cast. Iron is favored over other materials because of its fluidity, low shinkage, strength, and rigidity.

Cast products can range from a fraction of an inch, weighing less than an ounce (teeth in a zipper), to 30 or more feet in length, weighing many tons (propellors on an oceangoing ship). Figure 3.4 illustrates several different castings with their associated molds. Material can be cast in virtually any configuration the designer requires. A casting can have hollow sections, contain irregular surfaces, and in general be very complex. Because of this capability, many products once cast need no further secondary processes.

In all casting processes, six basic factors are involved. First, a mold cavity must be produced. The mold's internal shape must exactly exemplify the shape of the desired cast product. Second, the raw material must be in a liquid state. For iron this requires heating to more than 2200°F. Third, the liquid material must be introduced into the mold in such a manner that all air or gas in the mold escapes. Allowing air to remain in the mold will cause defects in the casting. Fourth, provisions must be made to allow for shrinkage that occurs when the material solidifies. Fifth, the mold must be designed

Figure 3.4 Examples of Castings with Their Associated Molds (By Permission of McGraw-Hill)

such that it is possible to remove the casting from the mold. In many cases, the molds are designed with centered, parting surfaces to allow for repeated mold usage. Sixth, a casting usually requires a secondary process to remove excess material that was necessary in the casting process.

Table 3.2 presents seven major casting processes. In general, the overall advantages of casting are: the liquid's ability to fill a complex-shaped mold, the economy to manufacture a large number of identical products, and the ability to employ numerous different raw materials. The two major casting disadvantages are the shrinkage and the relatively limited structural strength. These disadvantages can sometimes be minimized by employing a forging rather than a casting.

3.2.3 Forging

Forging can be defined as shaping a metal under impact or pressure to a desired shape, and through the process, improving the product's mechanical properties by controlled plastic deformation. When forging, a metal blank approximately equal to the volume of the finished product is placed between two die halves. The two die halves when joined create a cavity with the shape of the finished product. The die halves are mounted in the forging machine, which is designed to supply large amounts of force to close the die halves. After placing the blank between the die halves, the die halves are driven together by the forge, thereby squeezing the metal into the cavity's shape. Therefore, the objective of forging is to shape the input blank to a size and shape close to that of the finished product. Figure 3.5 depicts a large multi-ton forge in operation.

Forged products range in weight from 1/2 lb to more than 2 tons. Most forging is performed with the metal in a red hot, heated state but not in a molten state. With the metal in a red-hot state, the forces necessary to forge are reduced. In addition, the metal will flow more easily into the cavity. This flow causes the metal's grain to align with the product's shape, thereby giving excellent strength and ductility. Therefore, products that require above-average strength are usually forged. Typical examples of forged products include engine crankshafts, railroad track equipment, aircraft landing gear parts, and farm tractor axles.

Because most forged products are forged hot, a furnace is required to preheat the metal. A material handling device is also required to remove the red-hot product from the furnance and place it between the die halves, which are attached to the forge. There are many different forge types, but nearly all types employ a "hammer" to quickly apply a large amount of force to squeeze the die halves together. That force ranges from a ton to thousands of tons.

Forging has a specific set of advantages and disadvantages associated with it. Forging's advantages include:

1. Excellent finished-product strength and ductility because of the metals grain flow, especially when compared to casting.
2. Good uniformity of finished product size and shape from unit to unit.

Table 3.2 Major Casting Processes

Sand casting	–Sand is used as the mold material. –The small sand particles can pack into thin sections. –Sand may be used in large quantities so that products covering a wide range of sizes and details can be made. –A new mold must be made for each casting. –Gravity is used to cause the molten metal to flow into the mold.
Permanent-mold casting	–Mold is made of metal or graphite. –Molten metal is gravity fed into the mold. –Mold can be used many times.
Die casting	–Metalic mold or die. –Molten metal is forced into die cavity under pressure.
Centrifugal	–Metal or sand mold. –Centrifugal forces are used to cause the molten metal to assume a certain shape within the mold, ensuring a uniformly dense metal structure.
Investment casting	–Wax or frozen mercury pattern. –Mold is made from a plaster or other ceramic mix. –Good for small intricate parts that must be made with high-dimensional accuracy.
Shell casting	–A relatively thin shell is built up to about $\frac{1}{8}$-in. thickness by pouring a mixture of sand and plastic resin around a heated metal pattern. –This shell is then baked to cause the resin to set, forming a semirigid cavity. –Two half shells are glued or clamped together to form the mold cavity.
Plaster-mold casting	–Mold material is a modified gypsum plaster. –Plaster will produce fine detail and has almost no shrinkage, therefore has good dimensional control.

3. Forging can be automated for high-volume manufacturing.
4. Forged products can easily undergo secondary manufacturing processes.

Forging's disadvantages include:

1. Hard to obtain close tolerances.
2. Die sets are very expensive.
3. The forge and furnaces are expensive to purchase and maintain, especially when compared to casting.

3.3 SECONDARY PROCESSES

After a material or product has been altered via a primary process, it may require further manufacturing processes before it is ready for the consumer. These secondary

Figure 3.5 A Multiton Forge in Operation (By Permission of McGraw-Hill)

processes, like the primary processes, convert an input material or product into a new and different product. Also, like the primary processes, a product may undergo several secondary processes.

Table 3.3 lists by industry typical input materials and finished products for several representative secondary processes. The actual processes are not listed because to go from input material to finished product may require tens or hundreds of secondary processes, much more numerous than can be detailed here. Every industry has a large inherent, and sometimes unique, set of secondary processes. There are some processes that are common to a large number of industries such as hole drilling and polishing (automotive, household goods, furniture, and others). But some secondary processes are unique to their industry such as weaving (textiles) and typesetting (printing).

Two secondary processes have been chosen as illustrations. Chip-removal and metal-fabricating processes are two sets of major secondary processes. Each contains processes that are inherent to many industries and therefore should be understood by all industrial engineers.

Table 3.3 Examples of Products Undergoing Secondary Processes

Industry	Typical Input Material	Typical Product
Household goods	Steel Petroleum by-products Polymers	Toasters Ovens Refrigerators
Automotive	Steel Petroleum by-products	Automobiles Trucks Buses
Furniture	Wood Steel	Chairs Desks
Textile	Cloth	Pants Dresses
Printing	Paper Cardboard	Books Calendars Magazines
Human consumption	Grains Vegetables	Bread Frozen foods
Aircraft	Aluminum Titanium	Military fighters Commercial transports

3.3.1 Chip Removal

The chip-removal processes are by far the largest subset of all secondary processes. A chip-removal process is an operation that, through material removal in the form of chips, reduces the initial workpiece's dimensions to the final desired dimensions. Typically the initial workpiece's shape is created by one of the primary processes. The initial shape's alteration is performed on one or more of the machines described in the following four subsections. On any one machine a multiple number of specific, different processes (operations) can be performed. Because the machines' functions describe the processes, it is logical to dwell on them.

The following subsections describe three major families of chip-producing machinery. A fourth subsection briefly enumerates several of the other less common types of chip-removal machinery. For more detailed information concerning this machinery, refer to the Reference list at the chapter's end.

Lathes

A lathe imparts a cylindrical, conical, or spherical shape to a rotating workpiece by removing material from the workpiece in the form of scrap chips. On a lathe, processes that can be performed on a product include drilling, boring, reaming, tapping, turning, facing, threading, and cutoff. Table 3.4 summarizes some of these processes. For a cylindrical part, the lathe is the most versatile manufacturing machine.

Table 3.4 Examples of Chip-Removal Processes

Process	Description	Machine
Drilling	Creates a cylindrical hole in a product	Lathe, drill press
Boring	Enlarges and straightens a hole	Lathe, boring machine
Reaming	Provides a smooth finish in a hole	Lathe, reaming machine
Turning	Reduces the diameter of a cylinder and/or creates a cylindrical shape	Lathe
Facing	Reduces a product's length	Lathe, mill
Slotting	Creates a slot in a product of and shape	Mill
Planing	Creates a flat surface	Mill
Tapping	Creates internal threads in a hole	Lathe, drill press, tapping machine

There are many types of lathes, each designed to obtain a specific objective. Some of the more common lathe types are:

1. Engine lathes.
2. Manual turret lathes.
3. Automatic turret lathes.
4. Computer numerical control lathes.

An engine lathe is the simplest lathe type. Figure 3.6 depicts an engine lathe and details its important parts. It does require considerable operator attention and skill. It is a low-volume machine intended to produce from one to ten products of a specific shape. The manual turret lathe is designed to produce from approximately 10 to 1000 products economically. A turret lathe differs from the engine lathe by the addition of a rotating turret on the tailstock that holds the tooling necessary to complete a part without manually changing the tooling. Figure 3.7 illustrates a Warner and Swasey manual turret lathe and details some of its parts. In addition, the turret lathe has preset stops that allow the operator to make a part without having to measure it. The machine still requires an operator to select and actuate the appropriate tooling and controls during the entire operating sequence.

The automatic turret lathe is similar to the manual turret lathe except that the selection of tooling and feeding of raw material into the machine is performed automatically, actuated by a set of cams, gears, and levers. This allows the machine to operate independently of the operator, and the machine is designed to make approximately 1000 to 1,000,000 parts per each time it is set up. A setup is the time necessary to arrange the tooling, cams, and levers. Setup time varies from four hours to a full day. Therefore automatics are impractical for small numbers of parts. A Brown and Sharpe automatic turret lathe is shown in Figure 3.8.

Numerically controlled (NC) lathes are automated machines in which the automation is supplied by a microprocessor or minicomputer controller. Figure 3.9 illustrates a computer numerical control (CNC) lathe. An NC or CNC lathe can be pro-

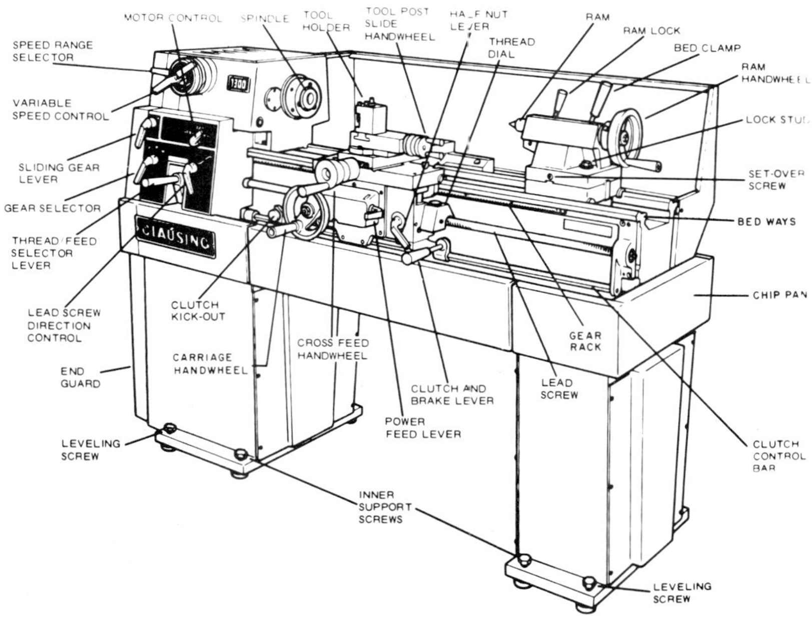

Figure 3.6 A Basic Engine Lathe with Its Parts Detailed

Figure 3.7 A Warner and Swasey Manual Turret Lathe

Figure 3.8 A Brown and Sharpe Automatic Turret Lathe (By Permission of McGraw-Hill)

grammed and set up in much less time than an automatic lathe. In addition, the NC machine can easily be reprogrammed for many different products. Therefore, an NC lathe allows for economical automation when making from one to thousands of products. It should be noted that NC lathes are much more expensive than engine lathes or manual turret lathes and still require an operator nearly full time. But NC lathes usually provide better quality and manufacture products faster than conventional machines.

Figure 3.9 A Computer Numerical Control Lathe (By Permission of McGraw-Hill)

Milling Machines

In general a milling machine is designed to create flat or slotted surfaces with a cylindrical, rotary cutter. While the cutter revolves, the work material moves past the cutter. Milling machines can be classified into four major categories.

1. Knee-and-column milling machines.
2. Production milling machines.
3. Planer milling machines.
4. Numerically controlled machining centers.

Knee-and-column milling machines are designed for light cuts and precision machining. The term "knee-and-column" is derived from the physical appearance of the table "knee" resting on the supporting "column." Figures 3.10 and 3.11 illustrate vertical and horizontal (the direction refers to the rotational axis of the milling cutter) knee-and-column milling machines. They are not intended for heavy production, because they do not have the necessary rigidity, but both variants have the capability of moving the part that is attached to the milling machine's table in all three Cartesian directions.

The production (bed) milling machine is designed as the name implies, for manufacturing large numbers of products, and is shown in Figure 3.12. The machine is designed to be very rigid so that heavy cuts can be made repeatedly. Production milling machines can be either vertical or horizontal, which again reflects the cutter's orientation. Planer milling machines are designed to perform heavy cuts on very large parts as shown in Figure 3.13. These machines are relatively specialized and are very

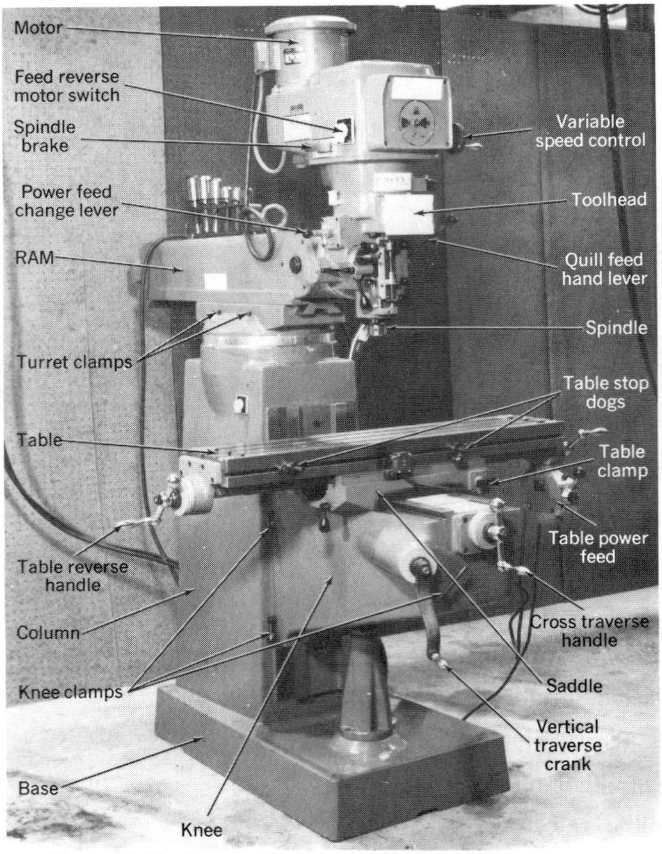

Figure 3.10 Vertical Knee-and-column Milling Machine

expensive, and thus not very common. Typically, planer milling machines are found in the aircraft and heavy earth-moving industries.

Numerically controlled machining centers are extremely flexible and can mill, drill, tap, ream, and bore. Figure 3.14 illustrates a large NC machining center. Although they are generally more expensive than knee-and-column or production milling machines, their flexibility and automatic controls result in better machine utilization.

Drill Presses

Drilling machines or drill presses are probably the most common chip-removal machines. They can be found in nearly all high school and home shops. Their main purpose is to create a cylindrical hole, and their secondary purposes are tapping and reaming. The term "press" in drill press comes from the fact that a drill is pressed

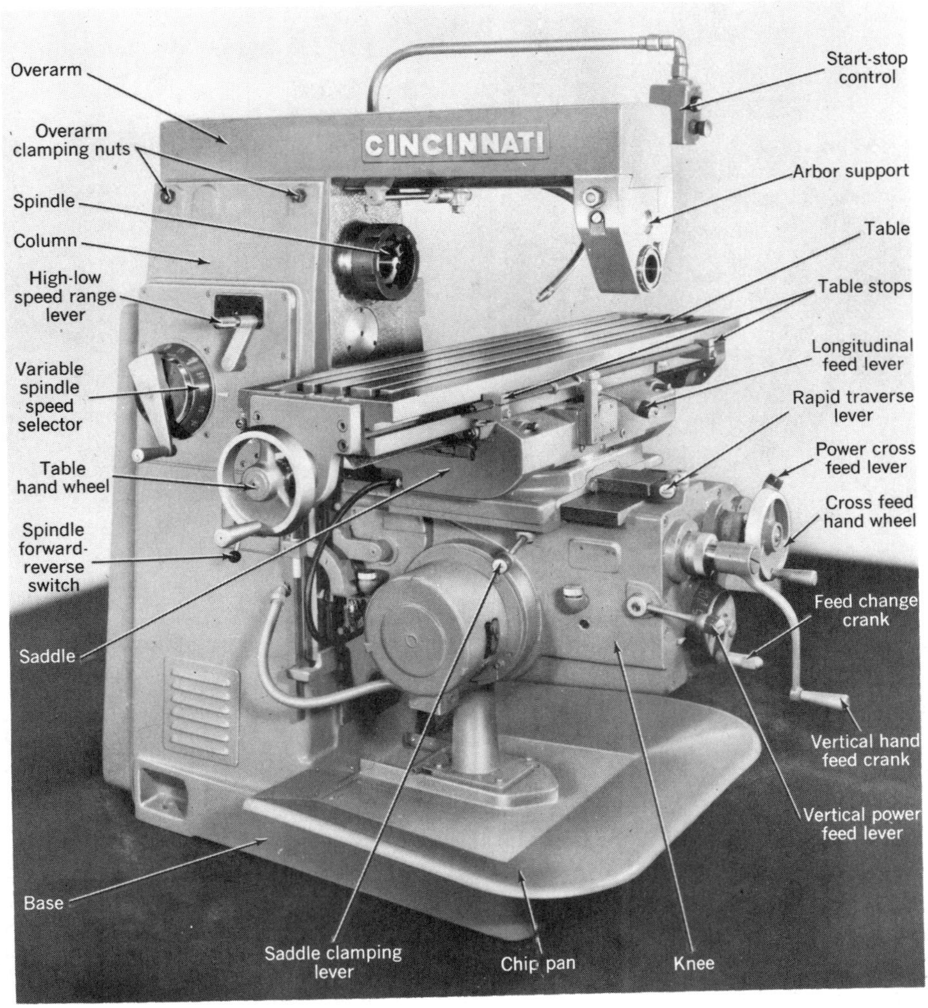

Figure 3.11 Horizontal Knee-and-column Milling Machine

into the work material while the drill rotates. There are several drilling machine types, each designed for a different environment. These types include:

1. Simple drill press.
2. Sensitive drill press.
3. Turret drill press.
4. Radial drill press.
5. Gang drill presses.

A simple drill press is the most common type and is found in nearly all shops.

Figure 3.12 Production Milling Machine Intended for High-Volume Repetitious Operations

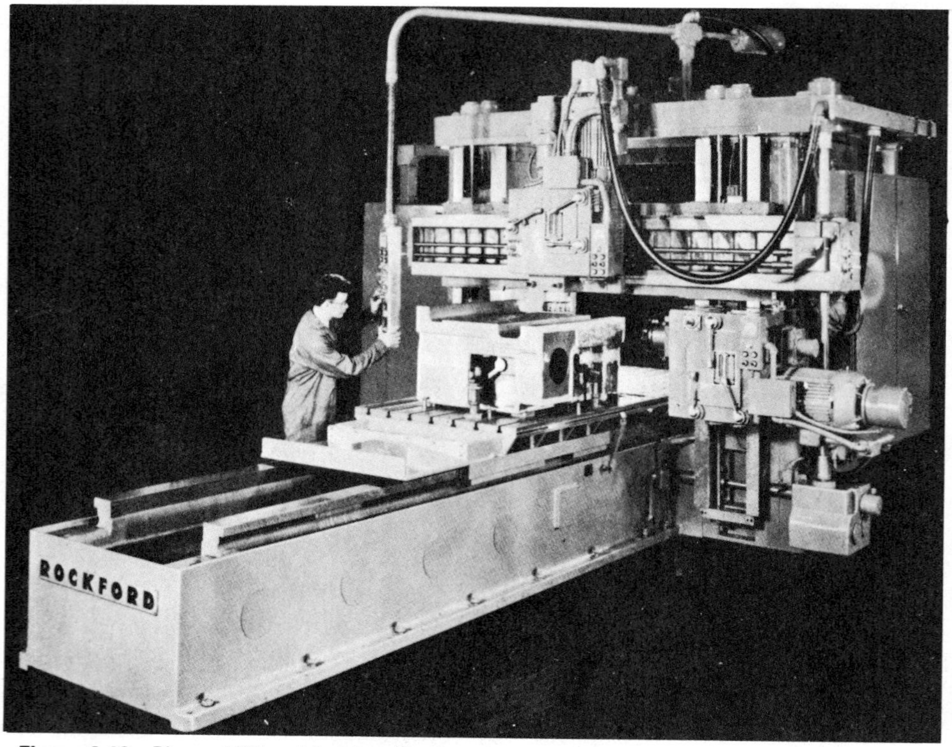

Figure 3.13 Planer Milling Machine (By Permission of McGraw-Hill)

Figure 3.14 An NC Machining Center

Figure 3.15 illustrates the common simple drill press. The machine is either table-mounted or pedestal-mounted with anywhere from a $\frac{1}{2}$-hp to 3-hp motor. A sensitive drill press is found in a toolroom where more precision is necessary. The term "sensitive" comes from the fact that the thrust lever that lowers the spindle and drill bit is geared so that the operator can have a "feel" for the thrust imparted to the drill. This is important when using small drill bits or when drilling hard work materials.

A turret drill press is designed with volume production in mind and is depicted in Figure 3.16. The turret is loaded with all the drills, taps, or reamers necessary to make the part. As each operation is completed and another, different tool is needed, the turret is rotated until the needed tool is in the downward position. The advantage of a turret over a simple drill press is that all operations can be performed on the part instead of completing the first operation on all parts before returning to the first part for the second operation, thereby eliminating material handling and promoting automation.

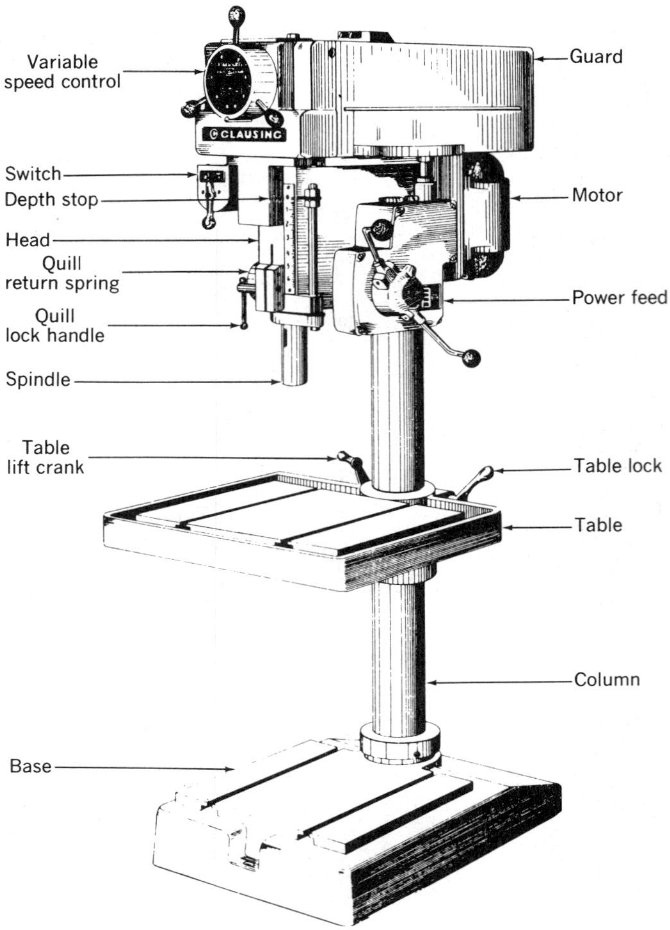

Figure 3.15 A Pedestal-Mounted Simple Drill Press Common to Most Job Shops

The radial drill press is physically the largest type and is shown in Figure 3.17. The term "radial" comes from its capability to swing the spindle radially over the work material. The radial drill press is the only type that is designed to move the spindle over the work material rather than move the work material under the spindle. Therefore it is primarily designed to drill holes in large or heavy parts that cannot be easily moved and positioned.

The gang drill press is nothing more than a series of simple drill presses attached to a single table. They are used for increasing automation similarly to the turret drill press. The difference is that the required tooling is not housed in a turret, but, rather, is fixed in a series of simple drill presses. The operator performs all the necessary operations on a part by moving from one simple drill press to the next.

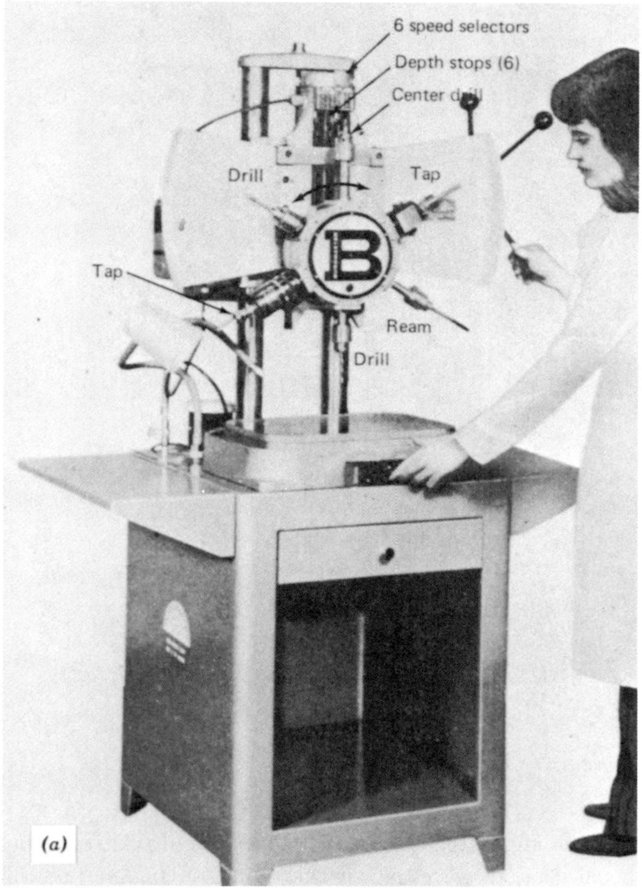

Figure 3.16 A Turret Drill Press Designed for Volume Production (By Permission of McGraw-Hill)

Other Chip-Removal Machinery

The previous three subsections have briefly described three major types of chip-removal manufacturing machinery. There are many more types and varieties of chip-removing machinery.

The saw family is a group of machines that are used to cut initial shapes from plate steel, bar stock, angle iron, and the like. A saw performs the cutting operation with a serated cutter that is either reciprocating or revolving. Typical examples of sawing machines are power hacksaws, band saws, and abrasive cutoff wheels. An example of a power hacksaw is shown in Figure 3.18.

Another family of chip-producing machines is the gear-cutting machinery. These are specialized machines designed to manufacture only gears. Typically, most specific

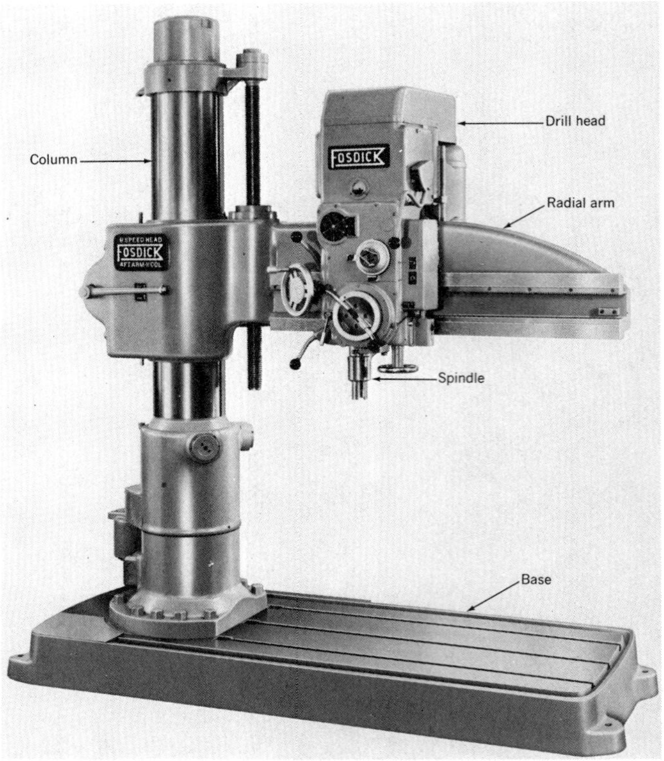

Figure 3.17 A Radial Drill Press

gear-cutting machines are designed to cut only one type of gear, be it an internal gear, spur gear, or bevel gear. Gear-cutting machines include hobbing machines, shapers, shavers, and Gleason cutters. It should be noted that nearly all gear cutters are designed for large-volume manufacturing. It also should be noted that low-volume production runs of some types of gears can be cut on knee-and-column milling machines. Figure 3.19 illustrates a hobbing machine cutting a spur gear.

Broaching machinery cuts specialized shapes such as internal splines, slots, and keyways. A broaching machine creates the desired shape with a single pass of a cutter. The cutter employed, a broach, is similar to a saw blade in style but has the required slot's thickness and shape. Because the broach cuts the slot in only one pass, each tooth of the broach exerts a large amount of force, thereby requiring the broaching machine to be very sturdy and rigid. Figure 3.20 describes a vertical broaching machine cutting a keyway in a spur gear.

A fourth machine family is the finishing machines. These machines remove very little material but impart a very smooth surface and allow for very close tolerances. These machines include lapping, honing, and polishing equipment. Figure 3.21 shows

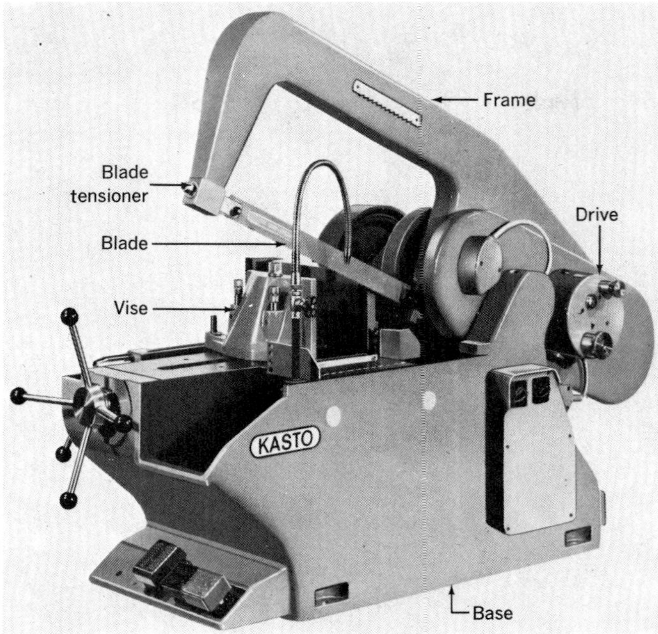

Figure 3.18 An Example of the Saw Family: A Power Hack Saw

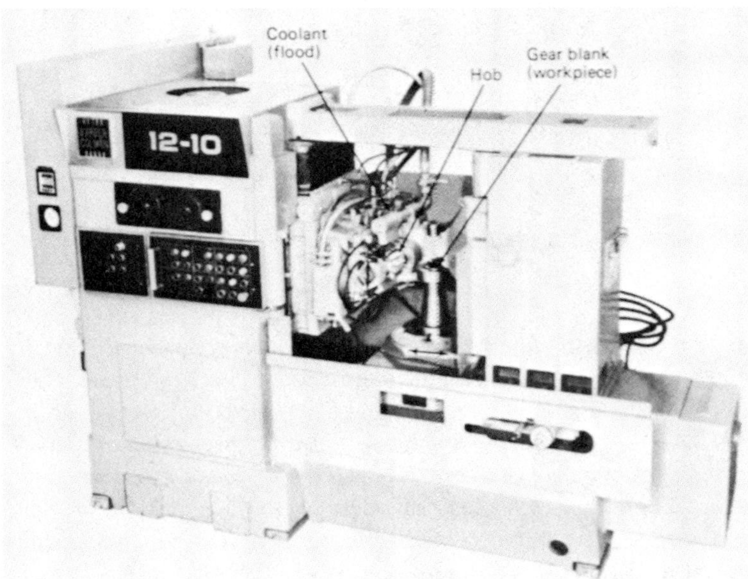

Figure 3.19 An Example of the Gear Cutting Family: A Hobbing Machine (By Permission of McGraw-Hill)

Figure 3.20 An Example of the Broach Family: A Vertical Broach (By Permission of McGraw-Hill)

the spur gear having the internal hole honed. A lapping machine typically addresses the exterior surfaces, whereas honing addresses the interior surfaces.

A fifth machine family is the grinding machines. Until the last decade, grinding machines have been considered within the finishing machine family. This is because grinders have historically been used to obtain tight tolerances and smooth surfaces. In the last decade, grinding technology has advanced to where grinders are doing roughing cuts, removing significant portions of material. Different grinding machine types include cylindrical grinders, rotary surface grinders, reciprocating surface grinders, centerless grinders, and internal grinders. A reciprocating surface grinder is shown in Figure 3.22.

Figure 3.21 An Example of the Finishing Family: A Honing Machine (By Permission of McGraw-Hill)

3.3.2 Metal Fabricating

Metal fabricating is the altering of the shape and size of plate, sheet, rod, or wire. These secondary processes are concerned with bending and cutting material. But unlike the chip-removal processes, these processes are not solely concerned with reducing the input product's size and shape by removing material in the form of chips. Rather, metal-fabricating processes attempt to remove a minimal amount of scrap material while forming the input product into a finished product. Figure 3.23 illustrates several products that require metal-fabricating processes.

As with chip-removal processes, an understanding of the machinery necessary to perform the processes will clarify these processes. Table 3.5 provides examples of metal-fabricating processes. Since it would be impossible to enumerate all the different metal-fabricating processes and associated machinery here, the following sections summarize the most important aspects.

It should be noted that metal-fabricating machinery is by far the most dangerous of all machinery types. Quite often the shape, size, and number of products to be manufactured make automated material handling impractical. Instead, it is necessary for an operator to manually load the product into the machine, placing a hand into the

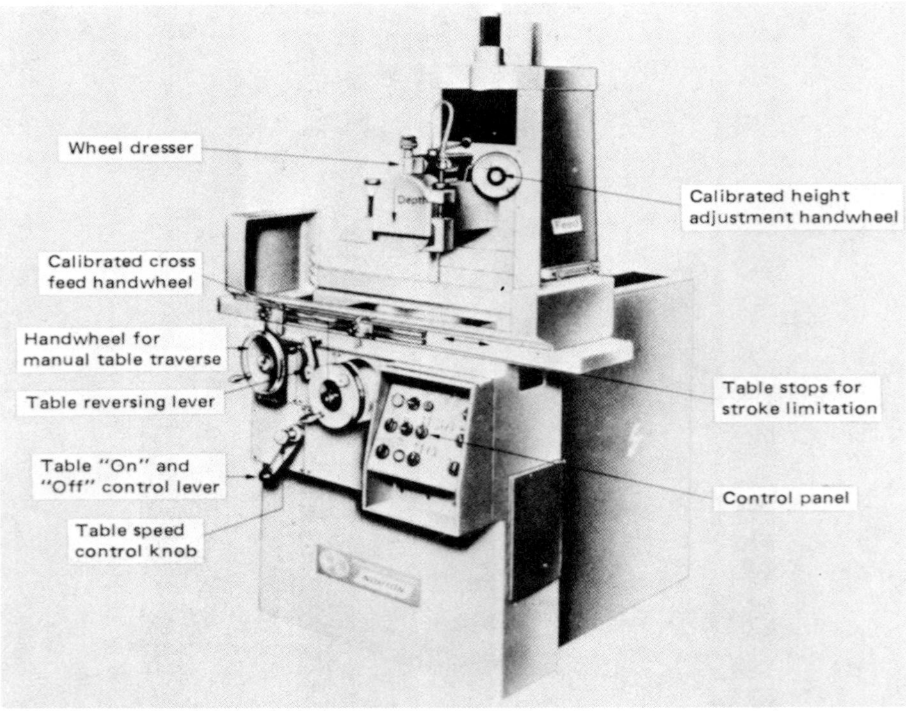

Figure 3.22 An Example of the Grinding Family: A Reciprocating Surface Grinder (By Permission of MacMillan)

point of operation. Because the product must be bent, formed, or cut, a large amount of force, applied very quickly, is necessary to overcome the material's inherent strength. If the hand is not removed by the operator before actuating the machine, significant injury can occur.

Because of regulations instituted under the Occupation Safety and Health Act

Table 3.5 Examples of Metal-Fabricating Processes

Process	Description	Machine
Punching	Creates a hole of nearly any shape	Punch press, numerical control press
Bending	Creates a single straight-line bend	Press brake
Shearing	Cuts the material along a straight line from one edge to another	Shear
Nibbling	Cuts the material by taking numerous small cuts	Punch press, numerical control press
Forming	Creates a complex bend with several angles along several axis	Punch press

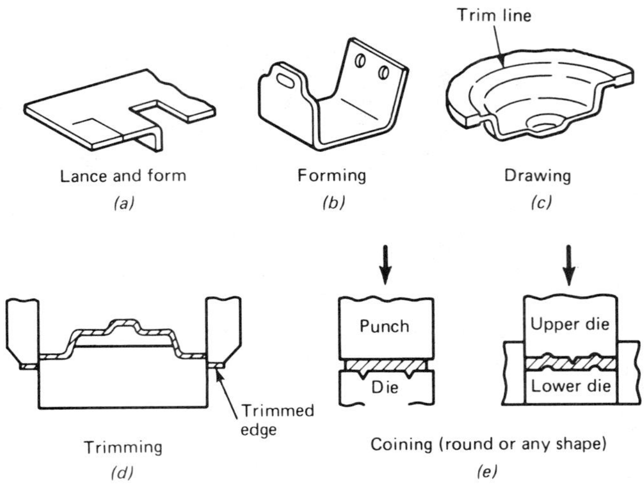

Figure 3.23 Several Products Requiring Metal-Fabricating Processes (By Permission of McGraw-Hill)

(OSHA) and the industry's concern, safety devices have been developed to absolutely minimize the chance of physical danger. These devices include:

1. Physical barriers prohibiting entrance to the point of operation during actuation.
2. Pull-back devices to pull the operator's hands away from the point of operation.
3. Limiting devices that deactivate the machinery if the operator remains in or near the point of operation.

The most important factor to remember about metal-fabricating processes and associated machinery is their potential for physical harm. Safety before productivity is a must.

Shears

A shear is designed to cut large sheets of material into smaller, more manageable pieces. Typically, input material for a shear is the finished product from a primary process. After the material has been processed by the shear, it usually will receive further secondary processes on punch presses or press brakes. Figure 3.24 illustrates a squaring shear with stock being fed into it.

The only process a shear can perform is cutting. A shear can only cut along a straight line from one side of the product to the other. It cannot cut any intricate designs, and it cannot hold tight tolerances. A shear's size and capacity are determined by the product's maximum width and thickness. Typical shear widths are 6, 8, 10, and 12 feet. Typical thicknesses are less than $\frac{1}{4}$ in., $\frac{1}{4}$ to $\frac{1}{2}$ in. and more than $\frac{1}{2}$ inch.

The shear performs its cutting process by forcing two cutting edges together in a scissor action. Therefore the cut progresses from one side of the product to the other. The force for the process is provided either by a mechanical flywheel or by a hydraulic mechanism.

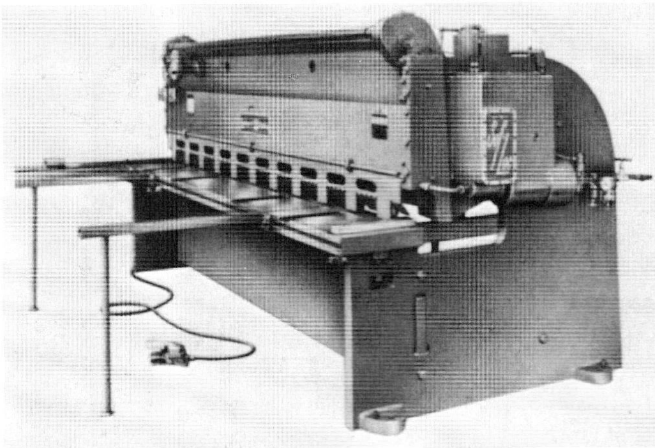

Figure 3.24 A Mechanically Operated Shear (By Permission of McGraw-Hill)

Press Brakes

A press brake is designed to apply a single straight-line bend to a sheet of material. The term "press" in press brake comes from the concept of pressing or forcing an angle or bend into the material. The term "brake" is based on the fact that the bending process is designed to occur very slowly to allow for safety and to control the amount of bend desired.

Similar to a shear, a press brake works along only a straight line and must create a bend from one side of the material to the other. A press brake's size is determined by its available force and the maximum width that can be utilized. The force is measured in tonnage, and press brakes range from 30 up to hundreds of tons. Widths vary from 4 to 12 ft, and they can be coupled together for greater widths. Figure 3.25 illustrates a 45-ton, 8-ft-wide unit. Like shears, the power for a press brake can come from mechanical or hydraulic power as well as pneumatic power.

Punch Presses

Punch presses are the most common types of metal-fabricating machinery and can perform more different types of processes than any other metal-fabricating machine type. From Table 3.5 this should become obvious. There are two major punch press types: C-frame and straight side. Figure 3.26 illustrates an open back inclinable (OBI) C-frame punch press with automated material feeder. A C-frame press is designed to perform processes requiring limited force (less than 60 tons). It is designed for easy feeding of materials from right to left without any obstructions. The machine has an open back and can be inclined backward to allow for the finished product to fall by gravity out the backside.

Figure 3.25 A Hydraulic Press Brake (By Permission of McGraw-Hill)

A straight-side press is designed for performing processes that require excessive forces. This press type is supported at all four corners to prevent the press from deflecting. A straight-side press usually requires loading from the front because of the supports. Typically they are not designed to produce as many parts per time unit as are the C-frame presses. Figure 3.27 illustrates a hydraulic straight-side press being manually loaded.

A C-frame press is usually powered by a mechanical mechanism with energy being supplied by a flywheel. A straight-side press can be powered mechanically or hydraulically. Hydraulics are employed for the larger units because a flywheel cannot generate the necessary power.

Other Metal-Fabricating Machinery

There are many other types of metal-fabricating machinery, many designed to perform a specific secondary process. A partial list includes:

1. Three-roll roll former.
2. Two-roll roll former.

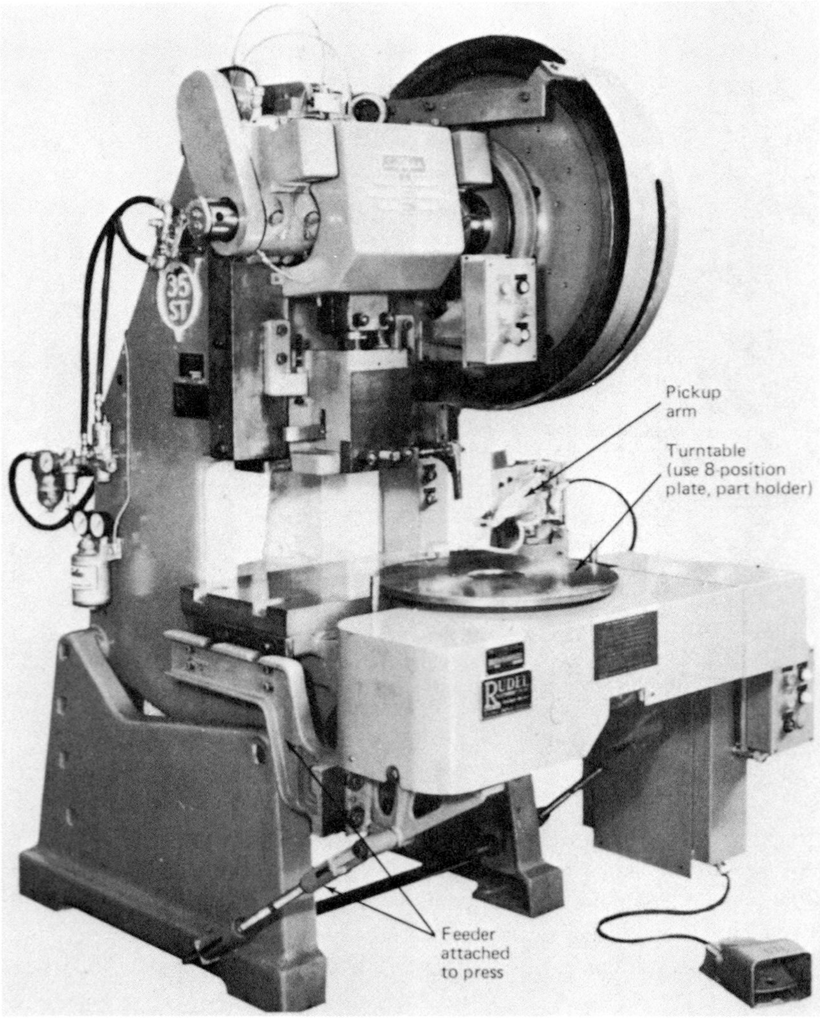

Figure 3.26 An OBI C-frame Punch Press (By Permission of McGraw-Hill)

3. Continuous-pass roll former.
4. Numerically controlled punch press.
5. Burning machinery.
6. Four-slides.

Figure 3.28 illustrates a two-roll roll former manufacturing a tube. Both the two-roll and three-roll roll formers are designed to perform processes creating cylindrical- or conical-shaped products. A continuous-pass roll former can manufacture products of

Figure 3.27 A Straight-Side Hydraulic Punch Press (By Permission of McGraw-Hill)

almost infinite length in nearly any continuous shape from cylindrical tubing to semi-rectangular house guttering.

Numerically controlled punch presses are designed to perform several different processes on a product. In addition, they can easily switch from manufacturing one product to another. These machines are controlled by digital electronic controls with the necessary information being stored in virtual memory, on paper tape or cassette. To orientate the product at the point of operation, these machines have devices to move the material automatically. Most numerically controlled punch presses have the ability to automatically change tooling, which allows for many different processes. Figure 3.29 illustrates a modern version of a numerically controlled punch press manufacturing a complex product requiring numerous different metal-fabricating processes.

Figure 3.28 A Two-Roll Roll Former

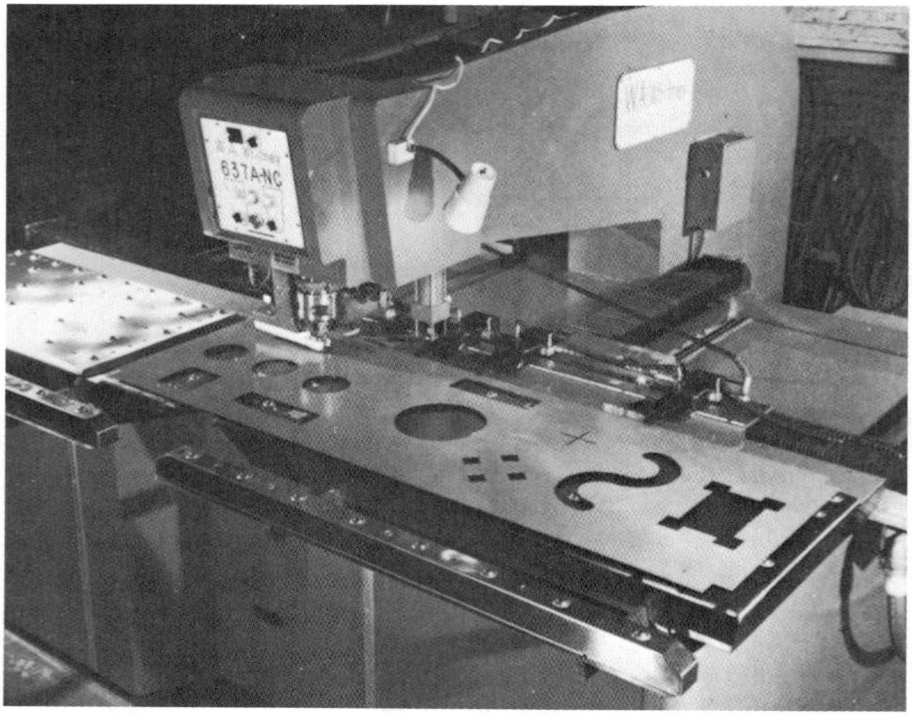

Figure 3.29 A Numerical Control Punch Press (By Permission of McGraw-Hill)

3.4 PROCESS PLANNING

Process planning is the determination of the appropriate type and order for the activities or operations (processes) that are necessary to manufacture the product to the design specifications. It is also the determination of the product's appropriate routing through the manufacturing facility. The routing consists of the specific machinery and associated tooling the product will visit to be altered from an input material to a finished product. Process planning's objective is to determine a set of operations and associated routing to minimize the manufacturing cost. This objective is bounded because there are numerous constraints associated with manufacturing processes. For example, a hole must be drilled before it can be tapped because the hole must be created before the thread can be put into the hole's sides. Additional constraints include the machinery's capability to perform a specific process, and the ability to clamp the part to a machine to perform the given process.

Process planning is both an art and a science. Although techniques and tools are available to facilitate and standardize process planning, a considerable amount of finesse based on experience is necessary. Unfortunately, that experience cannot be acquired in a short time span, but instead requires years of dedicated work. Because of the considerable experience necessary, a concentrated effort has been made totally to automate the process planning activity.

After the process plan has been developed, it is used by engineering, shop foremen, and hourly workers. Engineering uses it to estimate costs, to determine the amount of raw material to order, to determine when to release the material to the manufacturing area, to plan capacity, to determine system bottlenecks, and to perform other necessary control functions. The personnel in the manufacturing area use the process plan to sequence the products on the machines, physically route the products, and monitor the products' progress through the facility. Typically an abbreviated process plan is attached to a product to assist the manufacturing personnel in its tracking and control.

3.4.1 Manual Process Planning

For an understandable process plan, it is necessary to complete a minimum of two documents. The first is a product description that illustrates how the product should physically appear. It may be a three-view drafting, isometric drawing, picture, scale model, or sample. The product description is usually provided by the design engineers; however, it may not be complete enough for process planning and may require additional input by the industrial engineer.

A product description should include:

1. Material requirements.
2. Product usage.
3. Testing and quality limits.
4. Packing and packaging requirements.

5. Detailed shape and size definitions.
6. Unique identification.

Figure 3.30 illustrates a rather complete product description. This is used by the industrial engineer in determining the necessary processes to take the input raw material and convert it to a finished product. In addition, the product description is sometimes used by the hourly worker to visualize and determine the precise dimensions, composition, and/or surface finish desired. In many cases when the product description is a drafting, it is attached to the product as it moves through the facility and is used by succeeding workers to determine appropriate machine settings.

The second document is a process description, which can carry titles such as routing sheet, process plan, operations sheet, or instruction packet. The process description is the document that requires the most experience to construct. It sets down the detailed step-by-step processes (operations) to convert a raw material into a finished product. Critical here is the need to order the processes for maximum manufacturing efficiency.

The logic necessary to create the process description can be subdivided into five iterative steps.

1. Determine all necessary processes.
2. Group those processes by machine capability.
3. Order the processes within a group.
4. Order the groups of processes.
5. Iteratively check for binding constraints and nonfeasibility.

While performing these steps, the engineer must be aware of numerous inputs and constraints. Of course, the major input is the product description with the product's physical design constraints. Other inputs and constraints may include:

1. Manufacturing facility layout.
2. Machinery capabilities.
3. Tooling and holding device capabilities.
4. Clamping and reference surfaces.
5. Desired final surface finishes.
6. Safe manufacturing processes.
7. Manufacturing cost considerations.
8. Product delivery date (due date).

The last two points are major overriding considerations in most instances. It is necessary to design a process that is cost effective and will allow the product to be shipped to the consumer by its due date. If these two points are not consistently met, then the company will lose money and future customers.

Figure 3.31 illustrates a process description. Note that it uniquely defines the

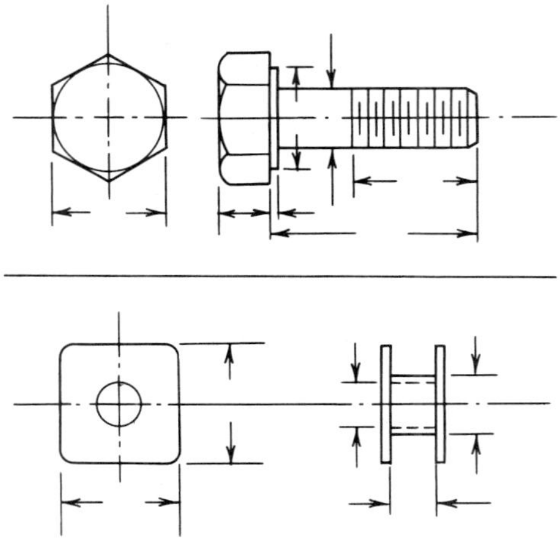

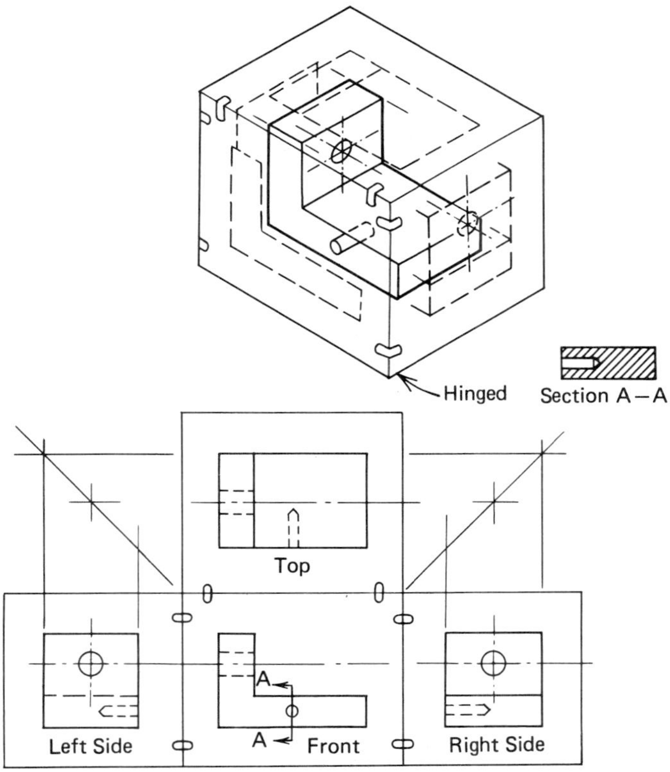

Figure 3.30 Examples of Product Descriptions (By Permission of the R.D. Irwin Company)

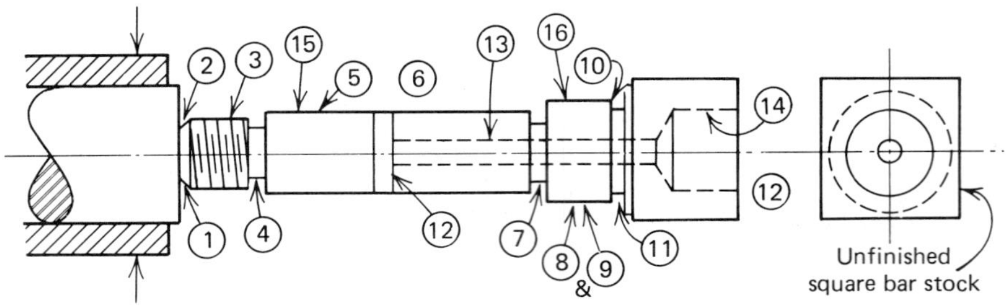

Operation Sheet

PRODUCT NAME Fuel Pin PRODUCT NO. 12-562
MATERIAL 1020 C.R.S. PRINT NO. 1-A-391
PLANNER T.I.M. DATE 6/5/-- REVISIONS R.G.

OPERATION NO.	OPERATION DESCRIPTION		OPERATION	TOOLS
10	Face and cut off, chamfer, neck, rough turn, finish turn, neck, rough turn, neck, washer face, drill large concentric hold, drill small concentric hole.	--	T.L.	For special tools only
15	Thread		T.L.	
20	Drill lateral hole	--	D.	
30	Grind small diameter	--	Gr.	
40	Grind large diameter	--	Gr.	

Figure 3.31 An Example of a Process Description (By Permission of the R.D. Irwin Company)

product with the same identification number found on the product description. Also, the planner, date, and revisions are clearly noted. The processes are listed in the necessary completion order along with the associated machinery and tooling. It should be apparent that an industrial engineering planner must have a thorough knowledge of the feasible processes, each machine's capabilities, and available tooling. To alleviate some of this required knowledge, automated process planning has been and is being developed.

3.4.2 Automated Process Planning

Before automated process planning can be addressed, *computer aided design* (CAD) must be defined. CAD is product design using the computer's capabilities to minimize repetitive and tedious tasks. With CAD, the design engineer works at a terminal that includes a cathode-ray tube (CRT), keyboard, and automatic printer instead of at a drafting table. On the CRT, the designer can create three-dimensional or three-view drawings by using specialized software. This software contains logic to create lines, planes, and shapes and to orientate them together correctly on the CRT screen. The software can also store and retrieve past product drawings. To analyze product design features, many software systems can apply theoretical loads or forces and analyze the product's strength and durability. In total, the CAD hardware and software can perform tasks required of a draftsman with greater speed and accuracy.

With a CAD-created design, it would be desirable to automatically determine a complete, optimal process plan. It would also be desirable to automatically determine the control commands for the computer numerical control machinery. With today's state-of-the-art technology these capabilities are not totally available, although development is ongoing. Specifically, the objective of automated process planning is to quickly and inexpensively determine the optimal process plan, where overall optimality is based on total cost. Figure 3.32 presents a schematic for an automated process planning system.

The process plan and associated manufacturing costs should be easily and quickly determined so that the design engineer, utilizing that information, can expediently design and redesign the product to minimize manufacturing costs (or to make manufacturing feasible) while still meeting product specifications. Currently with a manual system, a process planner may require hours to create a process plan, therefore negating expedient information feedback to the designer. And because process plans require a

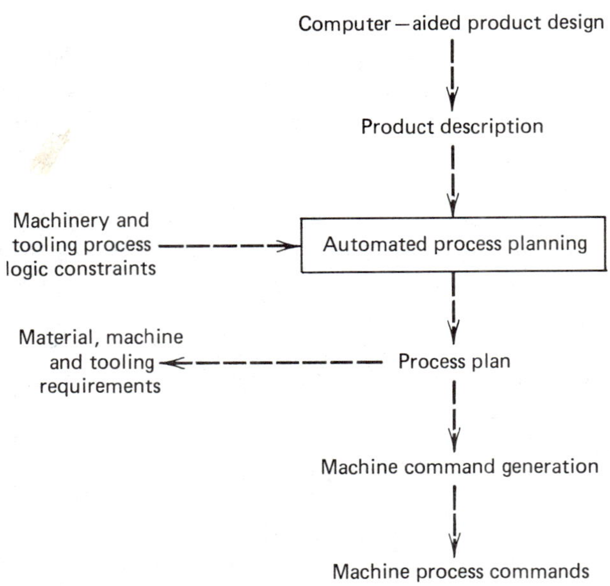

Figure 3.32 Automated Process Planning Schematic

significant amount of time and expense to create, it is not economically feasible to create numerous plans to analyze minor design variations, even though those variations may reduce manufacturing costs.

Automated process-planning system-logic development is extremely difficult because of the vast number of different product types, number of different processes available, feasible orderings of those processes, and implied manufacturing constraints. For example, assume one product that requires 10 processes. First, the automated process-planning system must determine the 10 correct processes from the thousands available, noting that in some instances there may not be a clear definitive choice between two processes (punch a hole with a OBI C-frame punch press, or drill a hole with a simple drill press). Then it has to determine the correct ordering of the 10 processes. For only 10 processes, the total number of possible different process-ordering combinations is 10!, or 3,628,000. Now, many of these combinations can be eliminated because they are logically infeasible, but the computer must contain the necessary logic to make those decisions. The remaining combinations must be searched based on known criteria and constraints to determine the optimal plan. The precise criteria definitions for the thousands of processes and their interactions are extremely difficult to convert into computer logic. From this example, it is hoped that the reader will have an appreciation for the complexity and difficulty associated with creating the software necessary for a complete automated process-planning system.

3.5 COMPUTER-INTEGRATED MANUFACTURING

During the mid-1970s the single-chip microprocessor was developed. This electronic unit consists of thousands of transistors, diodes, capacitors, and so on. It is the basis for all the personal computers and the more sophisticated hand calculators/computers. Microprocessors have also been integrated into many manufacturing systems to increase overall productivity.

Before the advent of the microprocessor, initial computer applications were grouped under the heading *computer-aided manufacturing* (CAM). Primarily, computers were used to control machinery directly or to manage the system's operations. The computers used to manage the system were large mainframes located in isolation rooms performing batch-data processing. These computers did not perform real-time shop floor control. Instead, they were primarily accountants, recording and documenting the manufacturing system's past activities. Computers were also used to perform analysis required by and to assist the staff engineers. As such, it was appropriate to consider these computers as aids to the manufacturing system.

Computer applications have greatly increased in the recent few years. Now, with microprocessors, they are becoming integrated into the manufacturing environment, hence the term *computer-integrated manufacturing* (CIM). Computer integration has occurred throughout the manufacturing facility including:

1. Machine control.
2. Automated material-handling equipment:
    a. Wire-guided vehicles
    b. Roller conveyor systems
    c. Carousels.
3. Automatic Storage and Retrieval Systems (ASRS).
4. Test-monitoring equipment.
5. Quality-assurance systems.
6. Shop floor control:
    a. Real-time data collection
    b. Real-time scheduling
    c. Real-time process monitoring.

3.5.1 Numerical Control Machinery

Even before the advent of the microprocessor, electronic digital controls were being incorporated into the design of chip producing, metal fabricating, robotic, and other manufacturing machinery. As early as the 1960s, controls were being attached to milling machines and drill presses, where the objective was to automate for multiple identical products and to eliminate human error. Since then, numerical control has been incorporated into high-volume production machines replacing the mechanical cams and gears that are expensive to create.

Numerical control is a system where activities or actions are monitored and controlled by direct insertion of numerical (digital) data. Numerical control is a very effective method to provide automation in the manufacturing facility. Typically this objective is accomplished by electronic controls. There are three types of numerical control:

1. Numerical control (NC).
2. Direct numerical control (DNC).
3. Computer numerical control (CNC).

NC is the oldest and simplest of the three types. With NC the logic (program) is coded and placed on a paper or Mylar 1-in.-wide tape that the NC machine reads. The logic is placed on the tape by punching holes into the tape in a cross-row by cross-row pattern. The logic contains sets of Cartesian coordinates that the machine will travel to in a sequential manner. The coordinates detail the desired tool cutting pattern that will result in a product with the desired shape and size.

DNC operates similarly, but the necessary information is transmitted from a host mainframe computer via hardwire instead of being read from a paper tape. DNC's advantage is that a product's program can be transmitted to any one of several machines or to a set of machines for concurrent manufacturing. DNC also eliminates the paper tape, which can wear or be destroyed, and it allows for increased intermachine control.

With DNC, the complete set of control commands is downloaded to a machine and stored in its memory. When the product has been completed, the machine requests the next program from the host computer.

Computer numerical control is the newest of the three types, and has gained prominence with the microprocessor. With CNC each machine has its own independent computer. This computer can store many programs simultaneously and can retrieve additional programs from floppy disks or cassette tapes. CNC does not require a hardwire link to a mainframe computer, nor does it require paper tape input. Rather, the program can be typed directly into the controller via a keyboard and CRT screen, often while the machine is performing another task. It should be noted that NC machines are considerably less expensive than CNC machines, and DNC machines provide better manufacturing facility control than NC or CNC machines.

When comparing NC, DNC, and CNC against other manual or mechanically automated machinery, several advantages and disadvantages become apparent. The advantages include:

1. Easy product design changes.
2. Simplified and reduced inspection.
3. Reduced scrap caused by human error.
4. Reduced space and costs required for tooling, fixtures, and controls.
5. Increased machine flexibility.
6. Reduced operator expertise.

The disadvantages include:

1. Increased machine costs.
2. Increased maintenance costs.
3. Increased number of skilled personnel to program and maintain.
4. Increased process planning necessary for efficient machining.

For numerical control machinery, be it NC, DNC, or CNC, the advantages far outweigh the disadvantages, and the number of these machines being manufactured is quickly increasing as can be seen in Figure 3.33.

3.5.2 Robotics

The Robot Institute of America defines a robot as a "programmable, multi-function manipulator designed to move material, parts, tools, or specialized devices through variable programmed motions for the performance of a variety of tasks." A robot is composed of a controller, mechanical arm, and power supplies. Figure 3.34 depicts a Cincinnati T3 robot, which is very common in manufacturing facilities. Robots are capable of lifting, placing, and assembling products. They can also perform welding,

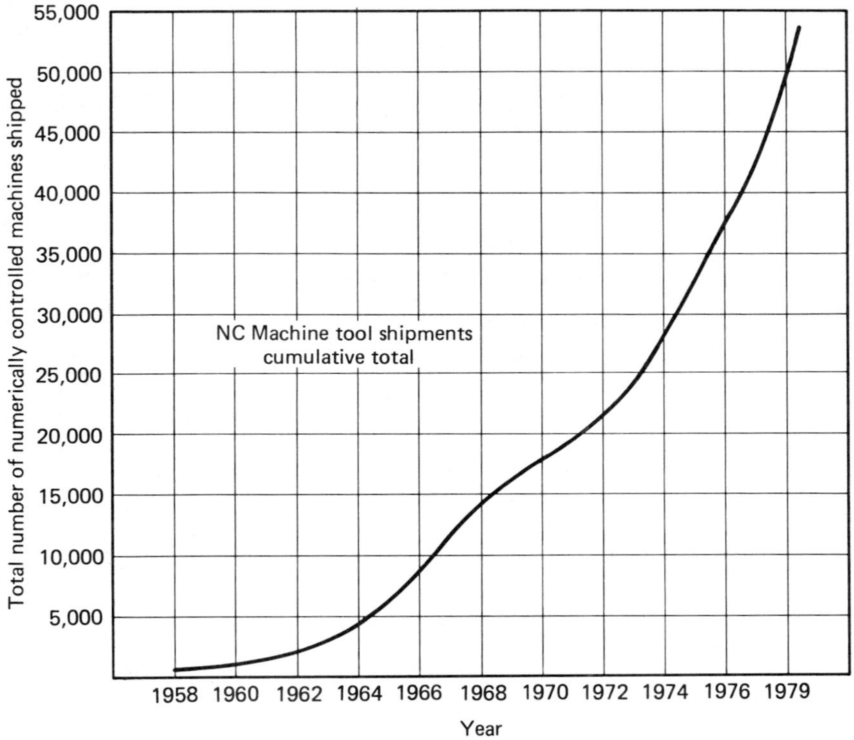

Figure 3.33 Increases in Numerical Control Machinery (From the U.S. Department of Commerce)

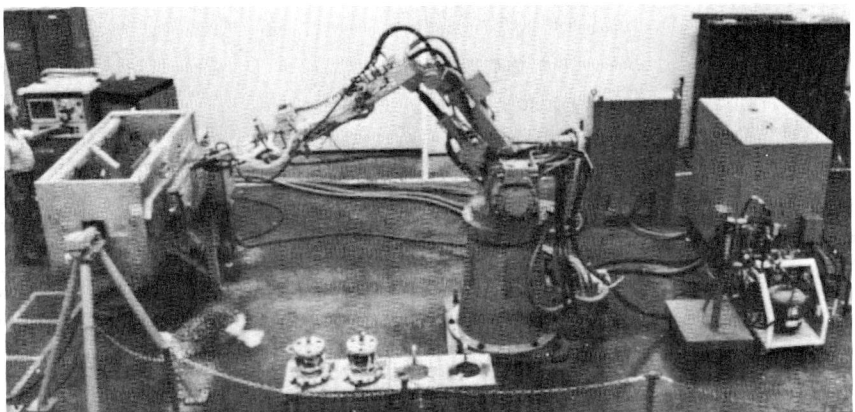

Figure 3.34 Cincinnati T3 Robot (By Permission of Industrial Press)

painting, deburring, and other manufacturing operations within specific limitations. Those limitations are primarily in regard to weight and positional accuracy. In most industrial applications, robots are used to replace or supplement humans, but through the robot's increased productivity, there is usually no need to fire the replaced employees. Instead, these people are often placed in other positions created by the robot's productivity.

Robots have several distinct advantages and disadvantages. The advantages include:

1. Capability to work three continuous shifts.
2. Lift loads beyond human capability.
3. Extended reach beyond human capability.
4. Content to perform repetitious, tedious activities.
5. Capability to work in a hazardous environment.
6. Eliminate potential human harm.
7. Minimum quality-control defects.
8. Can be depreciated.
9. Easy to train and retrain.
10. High motion flexibility.
11. Good return on investment.

Robots' disadvantages include:

1. Cannot think or reason.
2. No sight capability.
3. Cannot easily sort disorganized objects.
4. Cannot move beyond reach capability.
5. High initial capital cost.
6. Require skilled maintenance personnel.
7. Require spare parts (in some cases, spare robots).

Robots have become an accepted portion of the manufacturing facility. Several large companies (General Motors, IBM, General Electric) have begun to manufacture them; Cincinnati Milicron and Unimation have long been robot manufacturers. Robots first received wide acceptance in Japan. This, at first review, appears unusual because of the vast amount of available labor. But the Japanese have learned that because of the robot's productivity, they can manufacture a more cost-competitive product. With such a product, sales increase and there is a definite need to employ more people.

Robots are now being integrated into U.S. manufacturing facilities. It has been forecast that the number of robots will triple from 1982 to 1995. Therefore, the future industrial engineer will be heavily involved with robots.

3.6 MANUFACTURING FACILITIES AND MANAGEMENT ORGANIZATION

To fully comprehend the manufacturing processes function, it is necessary to have an understanding of the physical facility and the manufacturing management organization. Many industrial engineers begin their working careers as manufacturing supervisors and progress upward, finally reaching a position as manager of manufacturing operations or plant manager. In addition, during the careers of nearly all industrial engineers, they will come into contact with a manufacturing facility. Therefore it is necessary to understand the inherent differences between various manufacturing facilities' capabilities and constraints, as well as the manufacturing system's organization.

3.6.1 Manufacturing Facilities

It is important to understand the physical organization of a manufacturing facility. This organization is governed by numerous criteria, three of which follow:

1. The number of different types of products being processed in the facility.
2. The volume of products of each specific product type.
3. The change, over time, of the different product types.

Depending on these criteria and others, the nature and organization of a manufacturing facility can change. The manufacturing facility's orientation is called the plant layout or facility design. Techniques for precisely configuring a facility layout or plant design have been established, and these techniques are an integral portion of an industrial engineer's training (see Chapter 11). It is appropriate at this time for the reader to be aware of a manufacturing facility's many different general configurations.

A facility can operate as a continuous system, or it can manufacture discrete products. An example of a continuous system is an oil refinery, where the product flows continuously through the pipes, reactors, and other equipment in the system. In a continuous system there typically is a single path or a very limited number of paths through the system. Also, there are few if any times when the material is not being moved and processed through the system. An example of a discrete product-manufacturing system would be an automotive engine plant. In such a plant, each product or group of products (batch) moves from one machine to another, receiving individual processes while at each machine. The products will also probably have to wait at each machine, and their routings will probably be different.

A facility that manufactures discrete products can be categorized into five specific types.

1. Job shop.
2. Flow shop.
3. Production line.

4. Transfer line.
5. Assembly line.

Each facility type has been designed to satisfy specific manufacturing requirements. Each has inherent advantages and disadvantages that result in a cost-effective manufacturing operation.

A job shop is a manufacturing facility where the machinery is organized by its capabilities. A job shop is also called a functional layout or a process shop, as illustrated in Figure 3.35. In a job shop the lathes, mills, and drills are grouped separately. The job shop's advantage is that all the similar machines and associated technology are grouped together and the facility can handle many different product types. The disadvantage is that a product must be moved great distances within the shop to complete all of its operations.

In a flow shop, the machinery is not solely organized by function but rather is organized to allow for a logical product flow through the facility. The flow shop cannot handle as many different product types as can a job shop, but because of its organization, material handling is reduced. A production line is a further restriction of a flow shop. The number of product types is further reduced, and this allows for a marked increase in automation. A transfer line is a totally automated system designed to handle one and only one product. An assembly line is very similar to a production line but is intended for the manual or semiautomated assembly of many input products into one final product.

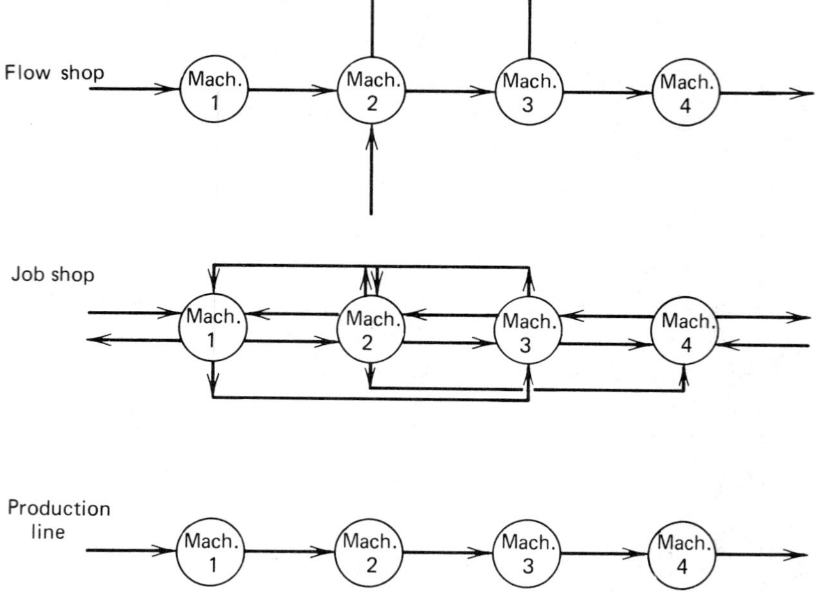

Figure 3.35 Flow Shop, Job Shop, and Production Line Facility Layouts

3.6.2 Manufacturing Management

Manufacturing management is directly concerned with the processes that the products are undergoing. As mentioned earlier, most industrial engineers working with manufacturing processes carry the title of ''manufacturing engineer'' and are considered line personnel. Again, a line person is typically responsible, in some way, for the daily or hourly process control.

Figure 3.36 illustrates a typical management structure of a large company possessing one or more manufacturing facilities. The senior person at each facility is the plant manager. This person may carry many different titles, such as plant superintendent, operations manager, vice president for operations, or even president. But his or her responsibilities are basically the same regardless of the title. These responsibilities include control of the overall manufacturing line function as well as the supporting staffs of design engineers, financiers and accountants, employee relations personnel, and marketing personnel.

Design engineers are typically electrical, aeronautical, and mechanical engineers supported by draftsmen. These people are responsible for product design, development, and testing. With them resides the computer-aided design (CAD) and computer graphics capability.

Personnel and employee relations staff are responsible for hiring and firing and for employee benefits. These people are responsible for coordination with any and all union activities. Finance and accounting control accounts receivable and accounts payable. Their major reports are the Balance Sheet and Income Statement. Marketing is the sales and advertising branch of a manufacturing facility. In addition, in many companies they are responsible for forecasting expected demand. The sum of sales and expected demand constitute the manufacturing requirements.

The senior management person most directly responsible for manufacturing process control is the manager of manufacturing operations. This person is the senior line employee and has responsibility for inventory levels, quality assurance, production engineering, and all manufacturing and assembly. The inventory manager, who reports to the manager of manufacturing operations, is responsible for the receiving, holding, moving, and shipping all products. This function is sometimes referred to as the trafficking activity. Material-handling equipment and operators as well as the warehouse and its personnel are managed by the inventory manager.

The inventory manager as well as all other personnel at this management level are senior engineers, quite often industrial engineers who started as foremen, supervisors, or basic industrial engineers and have been promoted. The quality control manager is responsible for all the inspectors and for determining if the products meet satisfactory quality levels. If the product does not meet standards, then it is this person's responsibility to determine the cause and make the necessary changes.

The production engineering manager supervises the engineers who are assigned directly to the manufacturing facility and its operations. These people can be industrial, mechanical, electrical, or chemical engineers who are seeking to make the manufacturing facility run more efficiently. The industrial engineer may be responsible for

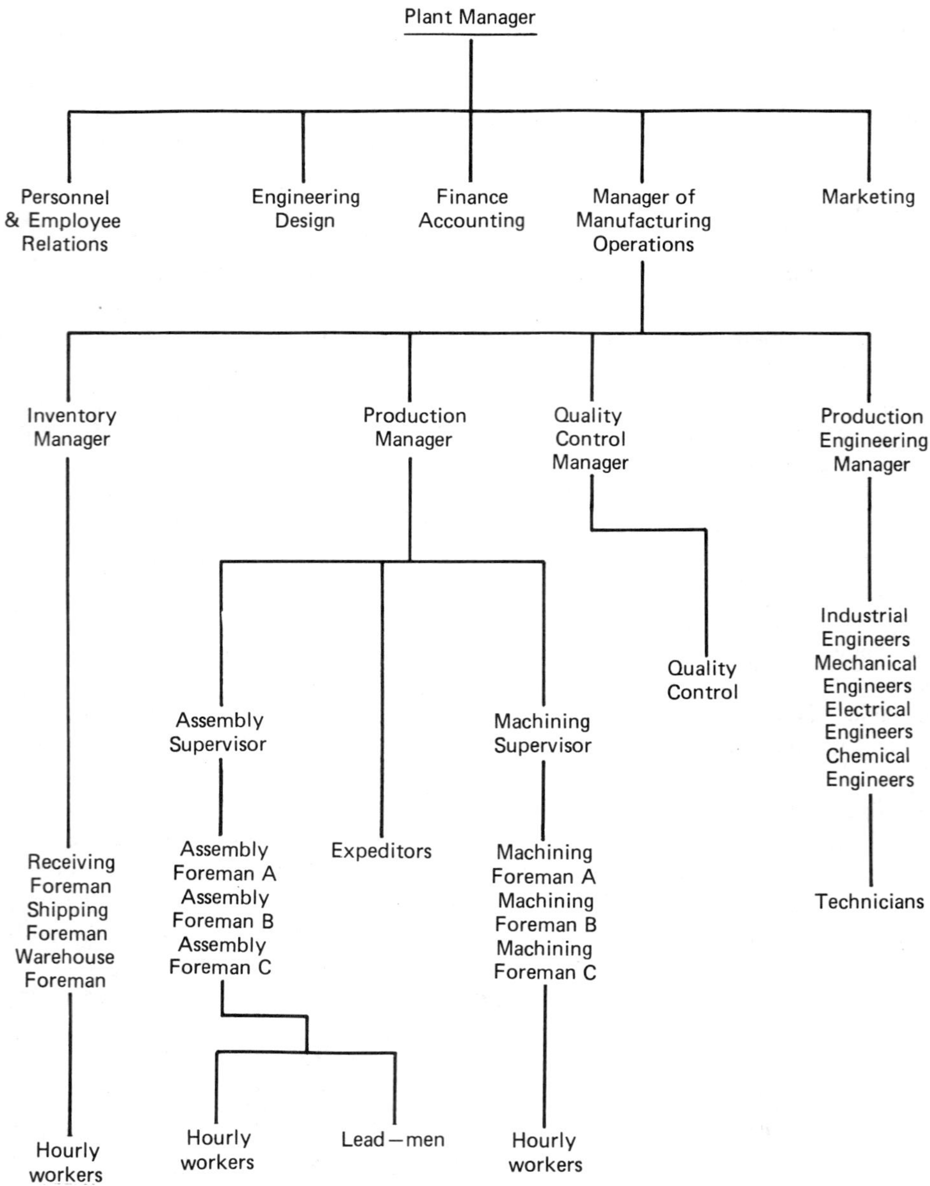

Figure 3.36 Typical Manufacturing Management Organization

such things as:

1. Plant layout or relayout.
2. Time and motion studies.
3. Scheduling.
4. Material requirements planning (MPR).
5. Capacity requirements planning (CPR).
6. Process planning.
7. Product flow determination.
8. Economic analysis and feasibility studies.
9. Tool and fixture design.
10. Numerical control, part programming.

These areas and others are the industrial engineer's domain and a typical starting point for a new graduate.

The production manager has responsibility for the actual manufacturing operations. Depending on the facility's size and the number of hourly workers there will be several levels of lower management below the production manager. These levels are also positions that can be filled by industrial engineers. Here, a supervisor will manage from two to six foremen. The foreman will manage from 6 to 20 hourly workers.

In addition, there may be one or more lead-men under a foreman. A lead-man performs hourly worker activities but also directs the work of people nearby. An expeditor is that person who is responsible for accelerating the processing of certain high-demand or late products. While expeditors are necessary, they are often an indication of poor overall manufacturing control and are to be avoided.

This overview of a manufacturing management organization has been very cursory. In a company with hundreds or even thousands of employees there are many more management levels and responsibility areas. In addition, the secondary personnel levels under all major functions except manufacturing operations have been omitted. Finally, no two companies are organized exactly alike; therefore, responsibilities may vary drastically from company to company.

3.7 SUMMARY

The manufacturing processes field contains topics too numerous to discuss completely. Rather, this chapter has attempted to highlight some of the major points. It should be remembered that many industrial engineers eventually work as manufacturing engineers dealing with manufacturing processes. In addition, nearly every product undergoes multiple primary and secondary processes which the industrial engineer is responsible for designing or controlling. Therefore it is encumbent on the IE to have a thorough knowledge of manufacturing processes.

This chapter has reviewed three primary processes: forming, casting, and forging.

Discussion of the secondary processes in the areas of chip producing and metal fabrication dwelt on the machinery necessary to perform the processes. Manual and automated process planning were introduced based on product and process descriptions. Computer-integrated manufacturing was presented along with discussions of numerical control machinery and robotics. Finally, manufacturing facility and management organizations were discussed.

REFERENCES AND SUGGESTED READING

1. *A Management Guide to Numerical Control Machine Tools,* The Institute of Production Engineers, London, 1978, p. 209.

2. E. B. Buffa, *Modern Production/Operations Management,* 6th ed., Wiley, New York, 1980, p. 661.

3. R. B. Chase and N. J. Aquilano, *Production and Operations Management,* Irwin, Homewood, IL, 1977, p. 703.

4. J. J. Childs, *Principles of Numerical Control,* Industrial Press, New York, 1982, p. 302.

5. E. P. DeGarmo, J. T. Black and R. A. Kohsher, *Materials and Processes in Manufacturing,* MacMillan, New York, 1984, p. 969.

6. J. F. Engelberger, *Robotics in Practice: Management and Application of Industrial Robots,* AMACON, New York, 1980, p. 291.

7. J. H. Greene, *Production and Inventory Control, Systems and Decisions,* Irwin, Homewood, IL, 1974, p. 703.

8. M. P. Groover, *Automation, Production Systems, and Computer-Aided Manufacturing,* Prentice-Hall, Englewood Cliffs, NJ, 1980, p. 601.

9. T. G. Gunn, *Computer Applications in Manufacturing,* Industrial Press, New York, 1981, p. 199.

10. W. R. Tanner, Ed., *Industrial Robots, Volume I, Fundamentals* (2nd ed.), Society of Manufacturing Engineers, Dearborn, Mich., 1981, p. 440.

11. H. C. Kazanas, G. E. Baker, and T. G. Gregor, *Basic Manufacturing Processes,* McGraw-Hill, New York, 1981, p. 378.

12. *Machinery's Handbook (21st Ed.),* Industrial Press, New York, 1980, p. 2439.

13. *Manufacturing Engineers Handbook,* (4th ed.), Society of Manufacturing Engineers, Dearborn, Mich., 1976, p. 1216.

14. F. W. Wilson, Ed., *Manufacturing Planning and Estimating,* Society of Manufacturing Engineers, McGraw-Hill, New York, 1963.

15. K. H. Moltrect, *Machine Shop Practice, Volume 1 and 2,* Industrial Press, New York, 1981, p. 516.

16. B. W. Niebel and A. B. Draper, *Production Design and Process Engineering,* McGraw-Hill, New York, 1974, p. 809.

17. R. E. Peterson and E. A. Silver, *Decision Systems for Inventory Management and Production Planning,* Wiley, New York, 1979, p. 787.

18. G. W. Plossl, *Manufacturing Control, The Last Frontier for Profits*, Reston, Reston, VA, 1973, p. 166.

19. V. E. Repp and W. J. McCarthy, *Machine Tool Technology*, McKnight, Bloomington, IL, 1979, p. 809.

20. A. D. Roberts and S. C. Lapidge, *Manufacturing Processes*, McGraw-Hill, New York, 1977, p. 418.

21. J. A. Schey, *Introduction to Manufacturing Processes*, McGraw-Hill, New York, 1977, p. 367.

22. H. W. Yankee, *Manufacturing Processes*, Prentice-Hall, Englewood Cliffs, NJ, 1979, p. 765.

REVIEW QUESTIONS AND PROBLEMS

3.1 What is a manufacturing process?

3.2 Why are industrial engineers concerned with manufacturing processes?

3.3 What is the difference between a primary and a secondary process?

3.4 Besides the processes listed in Table 3.3, list five more secondary processes with their input and finished products.

3.5 What type of lathes are designed for mass production?

3.6 List three types of milling machines.

3.7 What is an advantage of a numerical control machine?

3.8 What is the basic difference between a line IE and a staff IE?

3.9 What is a C-frame punch press? What processes can it perform?

3.10 What is metal fabrication?

3.11 Define the difference between a job shop and a production line.

3.12 If a transfer line is designed to produce only one type of product, approximately how many product types can be manufactured in a job shop?

3.13 What is the difference between a product description and a process description?

3.14 Why is automated process planning such a valuable technique?

3.15 Provide two reasons why manufacturing process knowledge is important to the industrial engineer?

3.16 What are two advantages of a robot?

3.17 Why are industrial engineers concerned with automation?

3.18 Define and briefly discuss CAM.

3.19 Illustrate why a product may undergo several primary processes.

3.20 As an industrial engineer starting work performing manufacturing engineering tasks, conceptualize what those tasks and activities might be.

Chapter 4

Work Methods and Measurement[1]

4.1 INTRODUCTION

Perhaps the most basic and most common function of traditional industrial engineering is the activity of measuring how long it should properly take to perform a job, and finding ways to allow a worker to perform his job more efficiently. These two areas of IE/OR are generally called *work measurement* and *work methods* improvement, respectively.

A work method is the way in which a given job or task is performed. The method includes the body motions used by the operator or person performing the job, as well as the tools, fixtures, and other work aids used to complete the job. A description of a work method normally includes not only what tools and equipment are used, but also how frequently they are employed, their locations, and the sequence in which use of the tools as well as other parts of the methods take place. Such a description is a key part of the subject area of work methods; however, the term also includes analyzing and improving the method being studied. The description, analysis, and improvement of work methods are described in the first part of this chapter.

The second half of this chapter deals with the techniques commonly used to determine the time a particular work task should require for completion. These techniques together with their application comprise the area of work measurement. Use of these measurement techniques is an integral part of work methods improvement, since they allow us to evaluate a key criterion of improvement—whether the time required to complete the job has actually been reduced by a proposed methods change. After first giving a brief history of the combined areas of work methods and measurement, these two subjects are treated as somewhat independent areas in the remainder of this chapter.

---

[1]This material was contributed by D. L. Kimbler and Robert Craig.

4.2 HISTORY OF WORK METHODS AND MEASUREMENT

The study of work methods and the measurement of work are as old as work itself. Early craftsmen undoubtedly studied the motions of their craft. More recent history includes the use of motion and time studies in the manufacture of pens in France and excavation in the United States, both in the late 1700s. Thomas Jefferson is reputed to have also used time study in his manufacture of nails near Monticello.

The modern documented use of work methods and measurements begins with Taylor and Gilbreth (see Chapter 1). Frederick W. Taylor formalized the principles of scientific management. Some of these dealt with matching tools to jobs and workers, and standardizing methods. Others dealt with the use of work incentives and close supervision. Frank Gilbreth took a more detailed approach in his studies of bricklaying using the concepts of micromotions and the ''therblig'' as a basic unit of motion. (Notice that ''therblig'' is an anagram of ''Gilbreth.'')

The mechanical view of work in the early part of the twentieth century was tempered by the introduction of behavioral science. Lillian Gilbreth, a psychologist, worked with her husband Frank in their industrial consulting firm and expanded his mechanical beginnings to include a more complete view of the worker. Among the best known examples of this behavioral approach are the studies of motor assembly by Western Electric Company, the Hawthorne studies. These studies, thought at the time to demonstrate the benefits of worker attitude resulting from corporate concern (the ''Hawthorne effect''), were in fact inconclusive [2]. Their immediate effect, however, was to spur additional behavioral research in the workplace. This research eventually grew into the formal study of human factors.

World War II, with its demands on production, provided a new impetus for study of work. The War Production Board set out detailed procedures in key industries, including the use of time study [8]. At the same time, the requirements for trained operators of complex new military equipment brought renewed interest in human factors. It is the view of many historians that conflict accelerates technical advances; in industrial engineering, war brought on the development of operations research and human factors engineering.

The time of development of work methods and measurement was also the time of development and expansion of the organized labor movement. These were times of change, distrust, and hostility in many instances. Companies demanded their right to manage, including prescribing methods and setting standards. Unions resisted these efforts, sometimes justifiably. So-called efficiency experts were sometimes used to implement unreasonable standards. Unions began to view efforts to improve production as simply a ''speed-up'' of the line. One result was a Federal law against some uses of time study. Many of these adversary attitudes were to color labor attitudes for years.

Today, work methods and measurement are still integral parts of industrial engineering. The strict principles of Taylor have been tempered by the humanistic views of the Gilbreths. Union attitudes have changed to the extent that some unions themselves use time study, although others still oppose it. Perhaps the most significant aspect of the current practice of work methods is that while the mechanical detail is important, the quality of working life is seen as having great potential impact on productivity. (See the discussion of quality circles in Chapter 15.)

Another aspect of the more recent practice of work methods and work measurement is the broadening scope of its application. Outside of the manufacturing environment, work study is applied in such diverse areas as services and white-collar jobs. The trend in recent years has been to a higher proportion of labor involved in services versus manufacturing. In this regard, the utilization of office automation systems can be viewed as simply another work system, subject to the same principles as manufacturing. The flow of office work products can be seen as a different kind of jobshop, but can be analyzed in the same way. In addition, the use of methods improvement and work measurement can now be found in Post Office operations, fire departments, and hospitals.

4.3 METHODS ANALYSIS

As mentioned in the previous section, methods improvement or methods analysis is the detailed study of work methods, including human motions, fixtures, work aids, and the function of a work station within a larger system. Its place in industrial engineering can be found in its relation to motion study and operations analysis. Motion study is a part of methods analysis, dealing specifically with work motions. Operations analysis is a broader term that includes methods as well as such topics as materials, work flow, tooling, and line balancing. In this textbook, methods analysis is limited primarily to within a work station; that is, it deals with a worker and his local workplace.

The study of work methods, especially when the goal is method improvement, is an iterative process of description of the method itself, development of alternative methods (perhaps by modifying the method given), and evaluation of the results. This cycle may be repeated several times until the improvement objectives are met. The development of a new method is similar, and both depend on the ability to analyze in detail the components of a job.

The objectives of this improvement process vary in practice from fulfilling a labor contract requirement to trying to make a job less fatiguing. However, the most common objective is simply to make a job more productive. This means that we increase the ratio of outputs from that job to the total input resource consumed in completing the job. That is, the objective is to enable the job to be completed more efficiently. This can be done by reducing the time required to perform the job, making the job less physically demanding, improving the tools used on the job, and a variety of other such steps.

The remainder of this section will be concerned with the four basic steps in the methods improvement process—description, development of alternatives, modification, and evaluation. To discuss this process, we first describe the basic human motions involved in most jobs and the key principles that are employed to use these motions in the most economical manner possible. Modifying a job to utilize these principles of motion economy is a fundamental part of the step of developing alternative work methods. After describing these principles, we present the basic charts and diagrams that are used in the description step. Then we summarize the key aspects of the overall process of analyzing a method for improvement possibilities—that is, the use of the principles of motion economy as well as other techniques to generate alternative methods through improvements in the existing method. The evaluation step in the improvement process involves the use of work measurement techniques. This is discussed in the latter sections of this chapter.

4.3.1 Motion Economy

The work initially developed by the Gilbreths has been modified and further developed, resulting in a set of tools and principles that, for the most part, have been very useful since the 1920s. Although sometimes condensed or expanded, the concept of the therblig and the principles of motion economy are probably the most durable tools of the IE/OR profession.

Therbligs can be divided into three groups. The first group contains productive therbligs: reach, move, grasp, position, disengage, release, inspect, and use. All of these involve effective motions that contribute directly to the execution of a task. The second group is less productive, and they are primarily mental. These are change direction, pre-position, search, select, and plan. The presence of these therbligs in a task usually indicates that the task design can be improved. The third group, consisting of hold, avoidable delay, unavoidable delay, and rest, are prime candidates to be eliminated from a task, since they do not directly contribute to the completion of a task.

The idea of the therblig led to two other Gilbreth contributions, the classification of motions and the principles of motion economy. Motions can be put into five groups, with increasing effort required for higher classification.

1. Finger motions.
2. Finger and wrist motions.
3. Finger, wrist, and lower arm.
4. Finger, wrist, and full arm.
5. Finger, wrist, arm, and upper body.

The principles of motion economy exist in several forms. Gilbreth [7] proposed 17 of them, and Ralph M. Barnes [1] listed 22. Whereas some of these principles are general, others are sometimes too specific or of limited application. The most applicable principles are restated here, some in a combined form. These principles are:

1. Begin and end each element simultaneously with both hands.
2. Make arm motions simultaneous and symmetrical with respect to the center of the body.
3. Keep hand motions at the lowest appropriate classification.
4. Avoid jerky motions and sharp changes in direction.

5. Develop rhythm to avoid fatigue.
6. Relieve arm motions by using foot pedals.
7. Use feeders, holding fixtures, and ejectors where possible.
8. Provide fixed and proper locations for tools and materials.
9. Take advantage of gravity and momentum, and provide power assist when working against them.
10. Locate controls for ease of operation.
11. Preserve workpiece orientation with sequential tasks.
12. Provide working conditions that do not contribute to fatigue or low productivity.

These principles are used as starting points in job analysis and design. However, these principles are in one sense misleading. It may be inferred that rigid observation of these principles, combined with exclusive use of group one therbligs, will result in the best possible method. In practice, this is not generally true. Motion economy is achieved by considering each therblig and its relationship with other therbligs within the task. A task using only move, reach, grasp, release, and use may be efficient, but there is still the possibility of improvement by using different fixtures, better tools, or shorter move distances. Each task must be considered as a whole, and these principles are simply guides for analysis.

4.3.2 Graphical Tools

A basic step in improving the method used to perform a task is to first describe that method adequately. This is necessary so that the analyst can readily identify areas of the method that have potential for improvement. For example, the description of a job may point out that unnecessary walking is taking place during a cycle of the job.

The tools used to describe work methods are primarily graphical. Two of the earliest methods are high-speed motion picture photography and the cyclegraph. The high-speed films are used to identify therbligs that cannot be detected by simple observation of the task. The film is analyzed and a simultaneous motion cycle chart (simo chart) is used to record therbligs and their timing. Each frame includes a clock or electronically generated time indicator that is used to provide time information and eliminate the effects of variation in camera speed.

The cyclegraph is a still-photography technique in which the operator has lights attached to hands or other parts of the body that are to be studied. The film is exposed during the operation, resulting in a continuous line being recorded tracing operator motions. A useful modification is the chronocyclegraph, in which the lights flash under control of a timer. The result is a dashed line that indicates both path and speed. (See Figure 4.1.)

Data may also be collected by other methods, such as videotape, time-lapse photography, computer logging, and direct observation. Data collection and analysis are aided by a variety of charting methods. The result is typically a textual-graphical model of an operation or manual segment of a task.

Figure 4.1 Illustration of a Chronocyclegraph

Three typical models are the Right- and Left-Hand Chart, the Process Chart, and the Multiple Activity Chart. The Right- and Left-Hand Chart (Figure 4.2) may be used with direct observation to record a sequence of motions by therbligs, indicating timing, simultaneous motions, and unproductive time. This chart is highly useful for analysis of manual operations.

The Process Chart is used at a less detailed level for more complicated operations, as shown in Figure 4.3. Each element is listed at the left, and the appropriate process symbol is marked. These symbols are used in other charts as well, and have the following meanings.

1. Circle—operation, machine cycle, or use of tools.
2. Arrow—transportation or movement.
3. Square—inspection or test.
4. Squared circle—delay.
5. Triangle—storage.

Other columns are used for distances, process times, quantities, and general notes. The process symbols may also be used to describe material flow and sequence of assembly.

The Multiple Activity Chart (Figure 4.4) is used when an operation is made up of two or more simultaneous sequences of activities. It is especially useful with a single operator tending several machines or with several operators whose tasks are

SUMMARY PER 1 PIECES

	PRESENT LH	PRESENT RH	PROPOSED LH	PROPOSED RH	DIFFERENCE LH	DIFFERENCE RH
○ OPERATIONS	4	4				
⇨ TRANSPORTS	5	3				
□ HOLDS	0	0				
▽ DELAYS	3	5				
TOTAL	12	12				

RIGHT AND LEFT HAND CHART

OPERATION CLOCK ASSEMBLY

PRESENT } METHOD. DATE 10/10/83

OPERATOR — ANALYST D. KIMBLER

LAYOUT

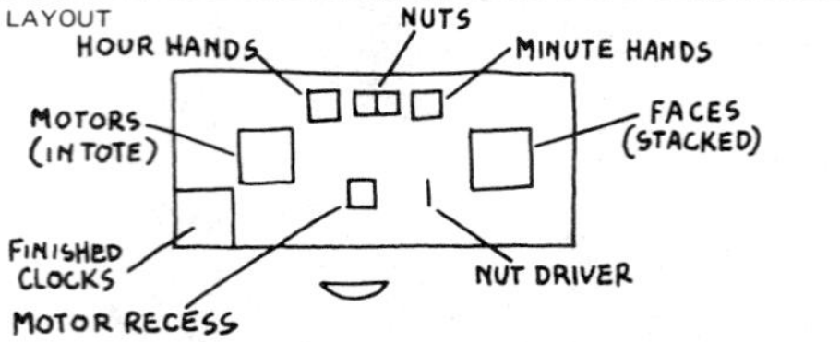

PARTS SKETCH

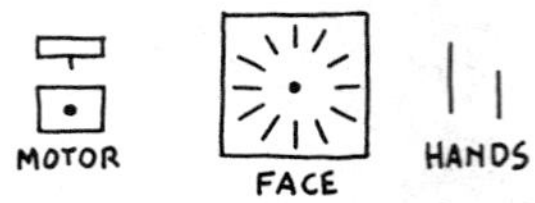

LEFT HAND		RIGHT HAND
GET MOTOR	1	GET FACE
INSERT IN RECESS	2	WAIT
WAIT	3	PLACE ON SHAFT
GET NUT	4	GET DRIVER
START ON THREADS	5	WAIT
WAIT	6	TIGHTEN NUT, DROP DRIVER
GET HOUR HAND	7	GET MINUTE HAND
PRESS ON HOUR HAND	8	WAIT
GET TOP NUT	9	PRESS ON MINUTE HAND
START ON THREADS	10	WAIT
WAIT	11	TIGHTEN NUT (FINGERS)
TO FINISH STACK	12	WAIT
	13	
	14	
	15	
	16	
	17	
	18	
	19	
	20	
	21	
	22	

Figure 4.2 A Right-Hand, Left-Hand Chart

FLOW PROCESS CHART

ANALYSIS	QUESTION EACH DETAIL
WHY?	
What? Where? When? Who? How?	

NO. ______
PAGE ___ OF ___

SUMMARY

	PRESENT NO.	PRESENT TIME	PROPOSED NO.	PROPOSED TIME	DIFFER'CE NO.	DIFFER'CE TIME
○ OPERATIONS						
⇨ TRANSPORTATIONS						
□ INSPECTIONS						
▽ DELAYS						
D STORAGES						
DISTANCE TRAVELED		FT.		FT.		FT.

JOB ______
□ MAN OR □ MATERIAL ______
CHART BEGINS ______
CHART ENDS ______
CHARTED BY ______ DATE ______

	DETAILS OF (PRESENT) METHOD	Time Minutes	Operation / Transport / Inspection / Delay / Storage	Distance in feet	Quantity	Time	ACTION: Eliminate	Combine	Change: Seque.	Place	Person	Improve	NOTES
1			○ ⇨ □ D ▽										
2			○ ⇨ □ D ▽										
3			○ ⇨ □ D ▽										
4			○ ⇨ □ D ▽										
5			○ ⇨ □ D ▽										
6			○ ⇨ □ D ▽										
7			○ ⇨ □ D ▽										
8			○ ⇨ □ D ▽										
9			○ ⇨ □ D ▽										
10			○ ⇨ □ D ▽										
11			○ ⇨ □ D ▽										
12			○ ⇨ □ D ▽										
13			○ ⇨ □ D ▽										
14			○ ⇨ □ D ▽										
15			○ ⇨ □ D ▽										
16			○ ⇨ □ D ▽										
17			○ ⇨ □ D ▽										
18			○ ⇨ □ D ▽										
19			○ ⇨ □ D ▽										
20			○ ⇨ □ D ▽										
21			○ ⇨ □ D ▽										
22			○ ⇨ □ D ▽										
23			○ ⇨ □ D ▽										
			○ ⇨ □ D ▽										

Figure 4.3 A Process Chart

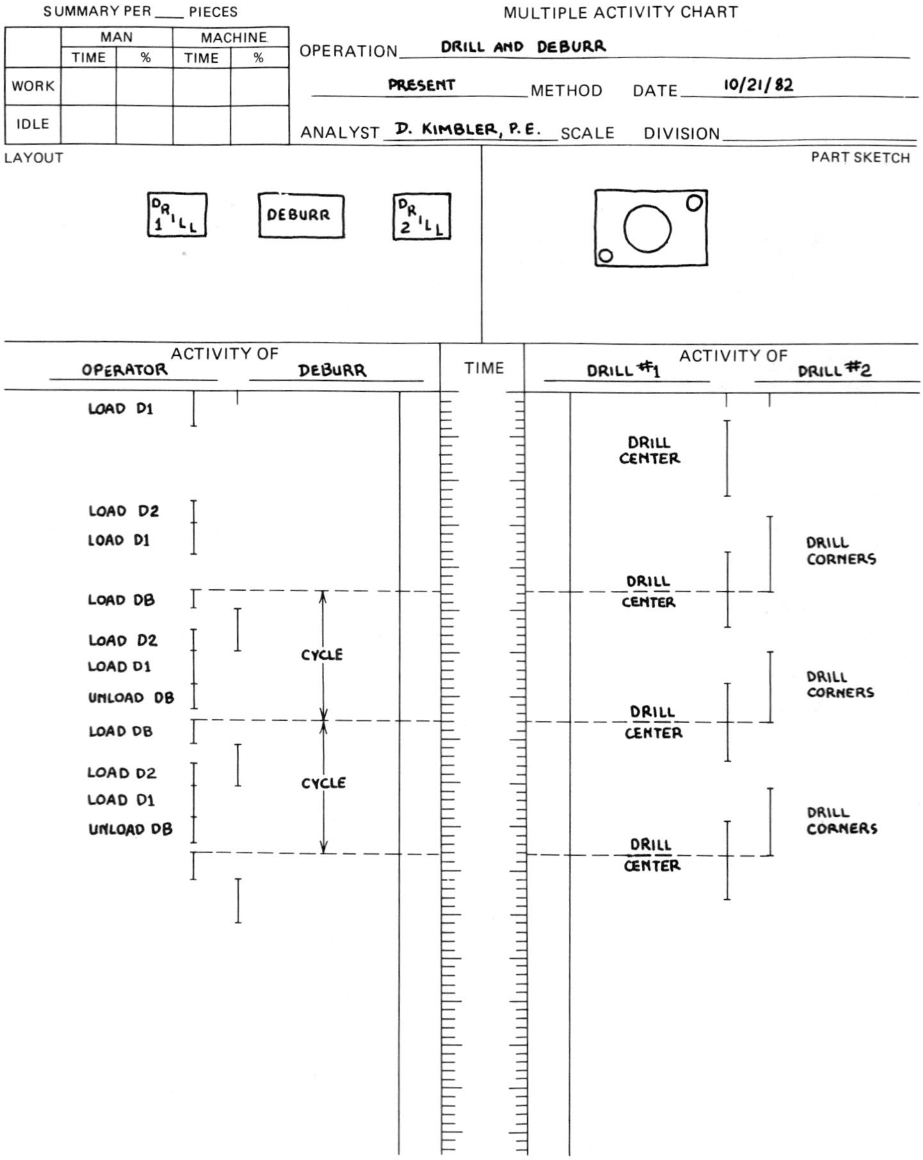

Figure 4.4 A Multiple Activity Chart

periodically synchronized. The chart vividly illustrates productive and idle times, and can be used to redesign operation sequences to increase productive time.

4.3.3 Improvement Analysis

After the job or task to be improved is described using the graphical models mentioned in the previous section, the next step in the improvement process is to analyze the method for inefficiencies that can be eliminated. This involves the use of the principles of motion economy and general principles of analysis. It is important at this stage to be methodical and to thoroughly document possible changes. Each improvement step should include an analysis of method and complete description of the new method. In this way possible improvements are less likely to be overlooked and potential problems in the new method will be detected.

Vital to improvement is the "questioning" attitude. Industrial organizations have a kind of "social momentum," usually manifested in continuing to do things as they have been done. This results in inefficient methods and obsolete tooling, as technological changes pass by the operational methods. The operators most closely involved with the process tend to overlook potential improvements. In many cases a brief examination by an "outsider," that is, someone with no vested interest in the past, will result in significant improvements.

At each step, improvements may be discovered by considering necessity, sequence, combination, and simplification. Every element should be examined with regard to its necessity, and unnecessary movements or operations should be eliminated. The sequence of operations may be altered into a more productive arrangement. In many cases, especially when operations have evolved with little design, operations may be effectively combined. Simplification will usually involve modifications to the workplace layout, fixturing, or tooling.

As an example, consider the assembly of two mating threaded parts. Unassisted assembly consists of holding one part in each hand, engaging the threads, and then turning. An obvious improvement is to develop a fixture to hold half of each assembly, so that two assemblies may be handled simultaneously. Production is not necessarily doubled, however, unless all motions are considered. The most effective fixture would allow the right hand to turn the top half while the left hand turns the inverted bottom half of the other assembly. In this way, hand motions are symmetrical and simultaneous; otherwise, imbalance of motions will reduce effectiveness and lead to fatigue.

The assembly of printed wire boards in the electronics industry provides several examples of effective motions. The insertion of components into the board is done in layers, building vertically. Components are delivered in standard orientations, and may have leads bent and cut to length. Board sizes may be standardized to facilitate handling and soldering. Assembly may be partially automated, with integrated circuits inserted by machine and odd-shaped components done by hand. High-volume electronics manufacturing is made possible by the use of refined and well-tuned work methods.

Several iterations of analyses and documentation will result in several alternatives for improvement. These alternatives may be ranked in order of preference, or each stage of analysis may result in an intermediate method, which is then subjected to analysis itself. In some cases, the final method proposal has evolved a step at a time from the existing methods. In others, each stage of analysis may contain some segment of the operation that, by itself, cannot be improved. These segments may be pieced together to form most of the proposed improved method.

Once the proposed method is developed, it is evaluated and then implemented. Ideally, each stage of improvement will include both evaluation and documentation sufficient to support implementation. In this way, efforts may be cut short and still result in an improved method. This constant focus on implementation also helps ensure that the final alternative will meet all objectives and requirements of the existing method.

4.4 EVOLUTION OF METHODS STUDY

As technology changes, so do the demands on work methods analysis. New products and processes require new methods, and new analytical tools allow fresh approaches to methods study. Many new projects in method design are being conducted in high-technology industries, with new problems and new tools.

As our manufacturing technology changes, the fundamental principles of work methods remain. What has changed in the practice of work methods is the application of new technology. Microcomputers now allow methods analysts to utilize their time more effectively in larger and more complex tasks. Pencil and paper exercises in developing alternative methods can now be done by computer simulation, and computerized predetermined systems (discussed later in this chapter) are used to blend human and automated operations.

4.4.1 Human Factors

The man–machine interface is of increasing importance as the technology of manufacturing becomes more sophisticated. Controls have evolved from mechanical linkages to computer input devices. The study of these tasks consists of not just control manipulation but operator thought processes and computer logic. Nuclear reactor control rooms, chemical process plants, and computer-integrated manufacturing systems all require a good man–machine interface design to be safe and effective.

The efficient operation of the modern industrial work station goes beyond simple motion economy. Operator effectiveness is a function of the entire work environment. Lighting, personal comfort, ergonomic tool design, and the use of work aids all contribute to the total station design. The early work in the study of human motions contributed to work methods and human factors engineering. These disciplines combine to best advantage in the design of tasks, work stations, and general work areas in modern industrial environments.

As the proportion of workers in service functions increases in relation to manufacturing, the focus of work methods is shifting. The techniques of work design can and are being applied in office environments, along with behavioral aspects of groups and individuals. The development of ergonomically sound and well-designed work methods and work spaces can have a significant effect on productivity at the individual level.

4.4.2 Automated Systems

The introduction of automation into the workplace has a considerable impact on both work methods design and task analysis. Many systems now have both manual and automated components, requiring that some tasks must be considered as combined systems. In other cases, the presence of a cell of automation requires redesign of manual operations that precede or follow it.

The manual aspects of partially automated systems are subject to the same improvement methods as purely manual operations. The difference lies in the reduced manufacturing tolerances due to automation. Operating rates for manual tasks must be consistent with the automated system. Automation may require special tooling, fixtures, or part orientation that will affect the manual task. Effective design will require larger system boundaries, and individual tasks must be designed with regard to the larger system.

The use of robots brings a new area into work methods design. Robotic tasks are frequently analogous to manual tasks, and anthropomorphic arms must be programmed to move effectively. The motion of these arms, and thus the development of their programs, is subject to the principles of motion economy just as are human motions.

The traditional tools and models of work methods may be applied to robotic systems. Process charts are used to map out the operation sequence, and left- and right-hand charts may be modified to design detailed operations. This is shown in Figure 4.5, a process chart for a robotic assembly task. The time-study methods described in the following sections have also been effectively applied to robotic systems.

As in combined systems, however, robotic programming requires more thorough analysis. The ability of a robot to learn and adapt is severely limited, requiring a much higher level of detail in robot task design.

4.5 WORK MEASUREMENT

An essential element of information needed by all managers to carry out their responsibilities is the knowledge of how long organization tasks should take to complete. This information, which provides management with quantitative measures of work performance, is basic to many of the decisions that must be made and the control to be exercised. The time required for the completion of tasks is determined in many situations through the techniques of work measurement. Work measurement may be defined as a methodology for determining the amount of time required under certain

FLOW PROCESS CHART

ANALYSIS
WHY?
What? Where? When? Who? How?

QUESTION EACH DETAIL

NO. 82126
PAGE 1 OF 1

SUMMARY

	PRESENT NO.	PRESENT TIME	PROPOSED NO.	PROPOSED TIME	DIFFER'CE NO.	DIFFER'CE TIME
◯ OPERATIONS	7					
⇨ TRANSPORTATIONS	6					
☐ INSPECTIONS						
▽ DELAYS	4					
D STORAGES						
DISTANCE TRAVELED		FT.		FT.		FT.

JOB PUMA 500 BLOCK ASSEMBLY
☑ MAN OR ☐ MATERIAL
CHART BEGINS
CHART ENDS
CHARTED BY D. KIMBLER, P.E. DATE 10/21/82

	DETAILS OF (PRESENT) METHOD	Time Minutes	Operation / Transport / Inspection / Delay / Storage	Marked symbol	Distance in feet	Quantity	Time	ACTION: Eliminate	Combine	Change: Seque.	Place	Person	Improve	NOTES
1	HAND TO FEEDER		◯ ⇨ ☐ D ▽	Delay										HAND EMPTY
2	GRASP BLOCK		◯ ⇨ ☐ D ▽	Operation										
3	TO FIXTURE		◯ ⇨ ☐ D ▽	Transport										
4	TO FEEDER		◯ ⇨ ☐ D ▽	Delay										HAND EMPTY
5	GRASP BLOCK		◯ ⇨ ☐ D ▽	Operation										
6	TO FIXTURE		◯ ⇨ ☐ D ▽	Transport										
7	TO BOLT FEEDER		◯ ⇨ ☐ D ▽	Delay										HAND EMPTY
8	GRASP BOLT		◯ ⇨ ☐ D ▽	Operation										
9	TO BLOCK CENTER		◯ ⇨ ☐ D ▽	Transport										
10	INSERT BOLT		◯ ⇨ ☐ D ▽	Operation										
11	TO NUT FEEDER		◯ ⇨ ☐ D ▽	Delay										HAND EMPTY
12	GRASP NUT (ROTATE HAND)		◯ ⇨ ☐ D ▽	Operation										AUXILIARY FINGERS
13	TO BOLT END		◯ ⇨ ☐ D ▽	Transport										
14	ENGAGE 1 THREAD		◯ ⇨ ☐ D ▽	Operation										
15	TO TIGHTEN FIXTURE		◯ ⇨ ☐ D ▽	Transport										
16	ENGAGE FIXTURE		◯ ⇨ ☐ D ▽	Operation										CONTROLLER SIGNAL TIGHTENS NUT
17	TO CONVEYOR		◯ ⇨ ☐ D ▽	Transport										REPEAT CYCLE
18			◯ ⇨ ☐ D ▽											
19			◯ ⇨ ☐ D ▽											
20			◯ ⇨ ☐ D ▽											
21			◯ ⇨ ☐ D ▽											
22			◯ ⇨ ☐ D ▽											
23			◯ ⇨ ☐ D ▽											
			◯ ⇨ ☐ D ▽											

Figure 4.5 Robotic Task Description Using Process Charts

standard conditions for tasks involving some human activity. The amount of time required as determined by this methodology is known as the *standard time* for that task. In the remainder of this chapter, we shall describe the various techniques commonly used in industry to determine these standard times and other related measures of work.

4.5.1 Standard Times

Standard times are used in carrying out many functions in industrial and service organizations. Their general purpose is to provide a means for measuring work—the length of time it normally takes to complete a particular task. The result is used in a variety of ways. For example, by knowing how many products can be produced per hour, we can determine the number of workers and the number of machines needed to meet production requirements. Similarly, standard times are used to balance or distribute required work among operators or employees (as in "assembly line balancing"), to evaluate and compare alternative production methods, to serve as a goal or yardstick in wage incentive programs, as the basis for cost estimation in setting selling prices, and as the basic data used in setting production schedules.

4.5.2 Calculating Standard Times

The commonly accepted definition [4] of a standard time is:

> (1) The time which is determined to be necessary for a qualified workman, working at a pace which is ordinarily used under capable supervision and experiencing normal fatigue and delays, to do a defined amount of work of specified quality when following the prescribed method; (2) the normal or leveled time (to accomplish a task) plus allowances for fatigue and delays.

As can be seen from this definition, a standard time is not just an observed or "clocked" time to do a job; it accounts for the fact that the observed time may not have been recorded under usual or normal conditions. Therefore, a standard time includes compensating time for delays caused by operator fatigue resulting over a long period of repetition of the job, equipment failures, and so on, as well as compensating time to reflect the pace or skill level of the average or normal worker.

There are a variety of alternative ways in which standard times are set or determined. Often in labor negotiations, standard times are bargained for or negotiated. In other situations, the average job completion time experienced in the past is used as the standard. More objective, scientific approaches include stopwatch time studies, work sampling, predetermined time standards for small elements of a job, and the use of standard data. Each of these methods is described in the remainder of this chapter.

4.6 STOPWATCH TIME STUDY

This method of developing time standards is based on the results of directly observing and timing a job as it is being performed on the shop floor. The *observed time* obtained in this manner is then adjusted to reflect the pace and skill of an average or normal worker rather than the one performing the job when it was actually observed. The resulting adjusted time is called the *normal time*. Finally, the normal time is then modified to reflect any extra time required to complete the job over a long time span rather than the relatively brief time during which it was actually observed. Addition of these *allowances* converts the normal time into the final *standard time*.

Certain steps are usually followed in carrying out the stopwatch time study. Companies usually establish a policy on how these steps should be conducted. This helps ensure uniformity and accuracy of time studies. The basic steps are as follows:

1. Ascertain the need for the study. This information may become known in one of several ways. Reports developed by the production planning function may show new items to be produced, all of which may require studies. Engineering change notices may indicate the need for rate changes. Method changes or production for which standards do not exist may be reported by the production foreman. In some situations, the request for a study may be received from the operator who wants full coverage on his or her wage incentive rates.
2. Notify the foreman of your intentions.
3. Select an operator for study. If the operation is being performed by several operators, one should be selected who is performing at close to a normal effort and skill level.
4. Check to see if the work method being used can be improved. The guidelines and procedures discussed in the previous sections of this chapter should be followed to improve and standardize the method.
5. The next step is one of the most important steps but, unfortunately, one that frequently is done poorly. This step is the recording of complete information on the operation. Included should be such items as date, name of operation, name of foreman, name of analyst, department name and location, tools, machines, attachments, jigs, fixtures, assembly or part numbers, drawing numbers, materials used, tools, specification numbers, speeds, feeds, and distances. This information should be supplemented with a sketch of the work area showing locations and distances involved in the work area. Time study forms by many organizations have blank spaces for much of this information along with a gridded area to facilitate sketching the workplace.
6. Break the job being studied into separate tasks or elements. For example, an assembly job might have elements identified such as "position subcomponent *A* on the table" and "tighten all 5 screws." In general, machine-paced tasks should be separated from operator-paced elements, and elements should be of relatively short duration with distinct starting and stopping points—such as dropping a part in a tote bin with its associated noise.
7. Record the times associated with several repetitions of the job. This step is the basic "time study" itself. Times are obtained through the use of a decimal stopwatch such as that pictured in Figure 4.6, a "quick click" set of watches as shown in Figure 4.7, or a more sophisticated timing device like that shown

Figure 4.6 Decimal Minute Snapback Timeout Stopwatch (Courtesy of Meylan Stopwatch Corporation)

Figure 4.7 "Quick Click" Time Study Board Provides for Direct Readings without the Necessity for Calculations Required by the Continuous Method of Time Study (Courtesy of Meylan Stopwatch Corporation)

in Figure 4.8, which can be connected directly to a computer without the need for manual computations. Observed times and other pertinent data are recorded on a form such as the one shown in Figure 4.9.

There are two basic methods generally used in operating the stopwatch or similar timing device—the *snapback method* and the *continuous method*. In the snapback approach, the analyst reads the watch at the end of each successive element and immediately snaps the watch counter or hand back to zero. The hands move from zero as the next element continues. A direct reading is taken and recorded just prior to returning the hands to zero again. Figure 4.9 illustrates this mode of time study.

In the continuous method, the watch is started at the beginning of the job and is not stopped until the study is completed. When the end of an element is observed, the analyst reads the time under the moving watch hand and records it. Individual element times are obtained at the end of the study by subtracting readings from that for predecessor element readings. An illustration of a continuous stopwatch study is provided in Figure 4.10.

While the observations are being made and recorded, the analyst also observes and evaluates the effort and skill of the operator. Based on this evaluation, a decision is made as to what leveling factor or *performance rating* is appropriate for each element. The performance rating (which is covered in detail later in this chapter) is a percentage correction factor used to adjust actual observed times to normal times. The performance rating is the most subjective aspect of the entire process and the most frequent basis of controversy with the operators. However, skilled time study analysts should be expected to come within ±5 percent of each other on their performance ratings.

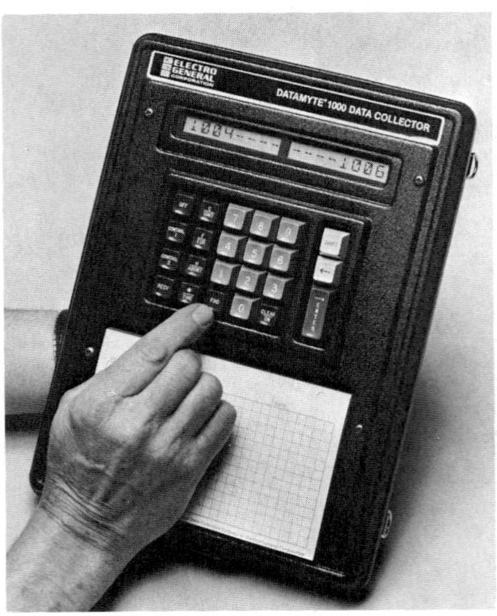

Figure 4.8 "DATAMYTE 1000 Data Collector" (Courtesy of Electro General Corporation)

Part Name DRAWER 1024 CHEST Part No. 487A Sheet 1 of 1 Sheets

START W10:05 STOP W10:13 TIME 8

EL No.	ELEMENTS	1	2	3	4	5	6	7	8	9	10	Total Time	No. Obs.	Avg. Time	Perf. Rating	Normal Time	Freq.	Allowed Time
1	GET DRAWER FROM SKID &	5	6	6	5	7	5	5	5	5	6	108	20	.054	110	.049	1	.049
	POSITION TO MACH. BED	5	6	6	5	5	5	5	6	5	5							
2	SAND ONE SIDE FRONT & BACK	8	8	9	8	8	8	9	7	10	8	161	19	.085	110	.077	1	.077
	DOVETAIL JOINTS	9	9	8	8	(17)	9	8	8	11	8							
3	TURN DRAWER END FOR	5	5	5	5	6	4	4	4	5	6	93	19	.049	100	.049	1	.049
	END	5	4	5	5	5	5	5	5	5	A							
4	REPEAT (2)	8	8	8	9	8	8	8	9	8	9	168	20	.084	105	.080	1	.080
		8	8	8	10	10	8	8	8	9	8							
5	ASIDE ONE DRAWER FROM	6	7	6	6	6	7	6	6	7	6	133	20	.066	95	.069	1	.069
	MACHINE BED TO SKID	7	7	8	8	7	7	7	7	6	6							
6	MOVE SKID OF DRAWERS ASIDE 10′–15′ (13 DRAWERS)	30	32									62	2	.310	100	.310	1/13	.024
7	MOVE EMPTY SKID TO POSITION (13 DRAWERS)	8	9									17	2	.085	95	.085	1/13	.069
8	MOVE SKID OF DRAWERS TO POSITION AT MACHINE	20	22									42	2	.210	100	.210	1/13	.016

FOREIGN ELEMENTS

A	20	TALK TO SUPERVISOR	F		
B			G		
C			H		
D			I		
E			J		

Total Allowed Time		.433
PFD Allow.	%	1.15
Std. Min. per Pc.		.498
Std. Pcs. per Min.		2.008

Figure 4.9 Completed Time Study Illustrating the Snapback Method of Stopwatch Reading

4.6.1 Time Study Calculations

The first step in actually calculating a standard time using the results of the study is to determine the exact observed time to be used for each element (in view of the many repetitions observed during the study). Several methods have been used for this purpose. However, the most frequently used procedure is simply to take the average of the good observations for each element as recorded on the time study form.

The next step is to calculate the normal time T_n as follows:

$$T_n = (T_0)\left(\frac{P}{100}\right) \tag{4.1}$$

where P is the percentage performance rating or leveling factor based on 100 percent being normal. Note that a rating factor such as 110 percent says that the observed worker was working 10 percent faster than the normal worker, so that the observed time T_0 is slowed down by 10 percent in deriving the normal time.

Next the allowances are applied. The three categories of allowances normally used are:

1. *Personal.* Included in this category are times for trips to the water fountain, rest rooms, and similar needs. These needs will vary with the location, con-

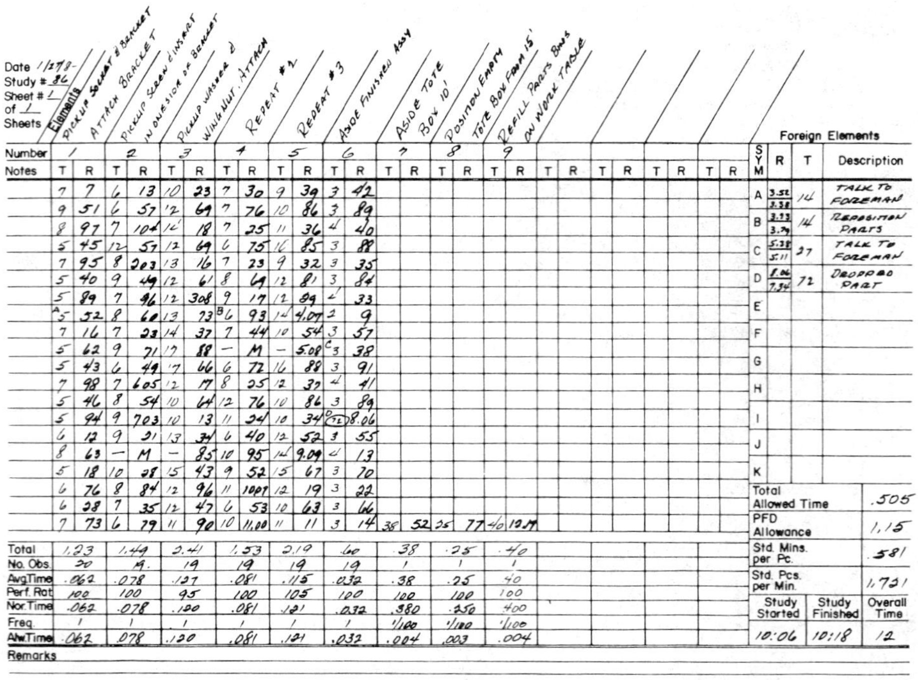

Date 1/27/8- Study # 86 Sheet # 1 of 1 Sheets

Elements	Pickup socket & bracket, attach bracket		Pickup screen & insert in one side of bracket		Pickup washer & wingnut, attach		Repeat #2		Repeat #3		Aside finished assy		Aside tote box 10'		Position empty tote box from 15'		Refill parts bins on work table	
Number	1		2		3		4		5		6		7		8		9	
Notes	T	R	T	R	T	R	T	R	T	R	T	R	T	R	T	R	T	R
	7	7	6	13	10	23	7	30	9	39	3	42						
	9	51	6	57	12	69	7	76	10	86	3	89						
	8	97	7	104	14	18	7	25	11	36	4	40						
	5	45	12	57	12	69	6	75	10	85	3	88						
	7	95	8	203	13	16	7	23	9	32	3	35						
	5	40	9	49	12	61	8	69	12	81	3	84						
	5	89	7	96	12	308	9	17	12	29	4	33						
	A 5	52	8	60	13	73	B 6	93	14	4.07	2	9						
	7	16	7	23	14	37	7	44	10	54	3	57						
	5	62	9	71	17	88	—	M	—	5.08	C 3	38						
	5	43	6	49	17	66	6	72	16	88	3	91						
	7	98	7	605	12	17	8	25	12	37	4	41						
	5	46	8	54	10	64	12	76	10	86	3	89						
	5	94	9	703	10	13	11	24	10	34	D (72)	8.06						
	6	12	9	21	13	34	6	40	12	52	3	55						
	8	63	—	M	—	85	10	95	14	9.09	4	13						
	5	18	10	28	15	43	9	52	15	67	3	70						
	6	76	8	84	12	96	11	1007	12	19	3	22						
	6	28	7	35	12	47	6	53	10	63	3	66						
	7	73	6	79	11	90	10	11.00	11	11	3	14	38	52	25	77	40	12.17

	1	2	3	4	5	6	7	8	9
Total	1.23	1.49	2.41	1.53	2.19	.60	.38	.25	.40
No. Obs.	20	19	19	19	19	19	1	1	1
Avg.Time	.062	.078	.127	.081	.115	.032	.38	.25	.40
Perf. Rat	100	100	95	100	105	100	100	100	100
Nor.Time	.062	.078	.120	.081	.121	.032	.380	.250	.400
Freq.	1	1	1	1	1	1	1/100	1/100	1/100
Alw.Time	.062	.078	.120	.081	.121	.032	.004	.003	.004

Remarks

Foreign Elements

SYM	R	T	Description
A	3.52 / 3.38	14	Talk to foreman
B	3.93 / 3.79	14	Reposition parts
C	5.38 / 5.11	27	Talk to foreman
D	8.06 / 7.34	72	Dropped part
E			
F			
G			
H			
I			
J			
K			

Total Allowed Time	.505
PFD Allowance	1.15
Std. Mins. per Pc.	.581
Std. Pcs. per Min.	1.721

Study Started	Study Finished	Overall Time
10:06	10:18	12

Figure 4.10 Completed Time Study Illustrating the Continuous Method of Stopwatch Reading (Adapted from S. M. Lowry, H. B. Maynard, and G. J. Stegemerten, *Time and Motion Study and Formulas for Wage Incentives*, 2nd ed., McGraw-Hill, New York, 1932, p. 135)

ditions, and kind of tasks performed. For many production work situations, 5 percent allowance is a figure frequently applied.

2. *Delay*. Included in this category are production delays beyond the control of the operator. This allowance has provisions for machine breakdowns, scheduling difficulties, raw material stockouts, supervisory errors, interruptions, and the like. Avoidable delays are usually not included in the authorized allowance.

3. *Fatigue*. This includes allowances made to provide for the operator's need to rest periodically, particularly if the work involves heavy lifting or exposure to an unpleasant environment including high heat, humidity, noise, vibration, and so on. The operator working in a strenuous job will be unable to maintain production at the calculated normal rate for long periods of sustained production. The fatigue allowance provides for this need.

The allowance factors are applied in the following manner:

$$T_s = (T_n)\left(1 + \frac{P + F + D}{100}\right) \tag{4.2}$$

where

T_s = Standard time (decimal minutes or hours)
T_n = Normal time (decimal minutes or hours)
P = Personal allowance (percent)
F = Fatigue allowance (percent)
D = Delay allowance (percent)

Example 4.1

An example will help to illustrate these relationships. Let us assume that in a particular plant the personal allowance is 5 percent, fatigue 7 percent, and delay 5 percent. For one of the elements in a stopwatch time study the recorded observations were 0.17, 0.16, 0.17, 0.16, 0.14, 0.18, 0.19, 0.20, 0.17, 0.16 minutes and the performance rating was 95 percent. Determine the standard time for this element.

$$T_0 = \frac{\text{Total Time}}{\text{Number of Observations}}$$

$$= \frac{.17 + .16 + .17 + .16 + .14 + .18 + .19 + .20 + .17 + .16}{10}$$

$$= 0.170 \text{ min}$$

$$T_n = (0.170 \text{ min})\left(\frac{95}{100}\right) = 0.162 \text{ min}$$

$$T_s = (T_n)\left(1 + \frac{P + F + D}{100}\right) = 0.162\left(1 + \frac{5 + 7 + 5}{100}\right)$$

$$= 0.162(1.17) = 0.190 \text{ min}$$

The standard time for the entire operation can be obtained by adding T_s for all elements in that operation. ■

The allowances used vary considerably from one organization to another. They also vary in the method of application. Instead of applying the allowance by multiplying the percentage factor times the normal time, some organizations state the personal allowance in minutes per day. Thus, for example, the allowance would not be expressed as 4 percent but as 19 minutes in an 8-hour work day (480 min × 4 percent = 19.2 min). Personal allowance information presented in this format is easier for the workers to understand and accept. In other plants, fatigue allowance is sometimes handled by setting up required rest periods. This approach is generally less acceptable because some stronger workers may not need to rest and would prefer the increased earnings possible by working continuously while on an operation with an incentive rate.

Examples of allowances that are sometimes used are as indicated below.

	Men	*Women*
Personal allowance	4%	6%
Fatigue allowance	8%	10%
Delay allowance	2%	2%

The fatigue allowance in heavy manual work may run as high as 15 percent in some situations. A substantial amount of research under the label of "human factors" has been conducted to establish theoretical values for the allowances under various conditions. For further discussion in this area, see Ref. [2].

4.6.2 Sample Size Required

There are a number of possible sources of error in conducting a time study. One of these sources is in the size of the sample of observations. Obviously, a wide range of variation in the element readings would require a larger number of observations to maintain the same degree of accuracy. If an initial group of observations taken indicates a large variation among the watch readings, there is a high probability of a bias in the sample. That is, the number of observations is too small to provide a good representation of the true time to complete the job. The question then becomes one of deciding what degree of accuracy is desired and then how many observations are needed to meet that desired accuracy.

In many situations, the analyst simply uses his past experience to judge the number of readings required. A more precise approach that is often used is based on establishing the statistical validation of the accuracy of the study. The procedure is normally applied to one of the job elements. The first step is to establish policy standards for the confidence level and acceptable accuracy of results required. A confidence level of 95 percent with a ±5-percent confidence interval is typically used in time study accuracy standards. This means that in 95 out of 100 times we could expect the error to be no greater than 5-percent of the true element time. For this degree of accuracy, the following formula may be used to test the adequacy of a sample of N time readings initially taken.

$$N' = \left[\frac{40N}{\sum X} \sqrt{\frac{\sum X^2 - \left(\sum X\right)^2 / N}{N - 1}} \right]^2 \qquad (4.3)$$

where N' = Number of observations needed for required accuracy
X = Element readings
N = Number of observations actually taken

If N' turns out to be less than or equal to N, then the initial set of N readings is adequate to ensure that the average reading obtained so far is within 5 percent of the true time to complete that element. If N' is greater than N, additional observations are made and N' is recalculated using all values of X obtained. In some sets of data, N' may change significantly when additional observations are made. It may be necessary to conduct these calculations for each element since the value of N' will vary for each element. The element with the largest value of N' is often used in the computations.

In order to ensure more accurate sample results, we have to use a tighter or more restrictive confidence interval or confidence level. In terms of the sample size calculation given in Eq. (4.3), a tighter confidence statement means that the numerical parameter would increase in value. For example, a 95-percent confidence level within a 2-percent confidence interval would require a value of 100 rather than the value of 40 shown in Eq. (4.3). For a 5 percent confidence interval having a 99-percent confidence level, a value of 60 would be used, whereas a value of 150 would be used for a 2-percent confidence interval having a 99-percent confidence level. Further discussion of determining the proper sample size can be found in references such as [1] or [8].

Example 4.2

This problem illustrates the method used for calculating an adequate sample size. Assume that a time study has been made that has several elements, all of which have what appears to be similar variability. However, the third element has the largest time and is considered the most important element; therefore, we concentrate on it. This element had the following recorded observations at the end of the initial time study (expressed in hundredths of a minute): 10, 9, 10, 9, 10, 10, 11, 10, 10, 11. An accuracy of ±5-percent confidence interval at a 95-percent confidence level is desired. Therefore, applying Eq. (4.3):

$$N' = \left[\frac{40N}{\sum X} \sqrt{\frac{\sum X^2 - \left(\sum X\right)^2 / N}{N - 1}} \right]^2$$

$$= \left[\frac{40 \times 10}{100} \sqrt{\frac{1004 - 10000/10}{9}} \right]^2 = 7.1 \text{ observations}$$

The value of N' is 7.1, well below the actual number of readings, 10. Therefore, within our acceptable limits of accuracy, the initial study results would stand with no further observations needed. ■

4.6.3 Performance Rating

As noted earlier, an important yet very difficult part of determining a standard time is deriving the proper performance or leveling factor to convert the observed time [T_0 in Eq. (4.1)] into a normalized time (T_n). The need for this conversion arises from the fact that there are usually wide variations of skill, motivation, pace, and the like, displayed by employees doing similar jobs. The performance of the particular employee doing the job while the time observations are being recorded may not be very representative of the "normal" performance. The observed employee may be highly skilled, or may be working at a speed significantly faster than that of the average operator. The observed times associated with such untypical performance must be corrected before they can represent the time that should be expected of a *normal* job operator or performance.

There is no universally accepted procedure for establishing the performance rating, and a number of different methods have been espoused by various practitioners. However, the heart of any performance rating scheme is the analyst's concept of "normal" skill and effort. The accuracy of the entire process is dependent upon this concept, which is the criterion by which the performance is measured. This criterion may not always be the same from one analyst to another. Similarly, the individual analyst's criterion may vary from one job to another and from one operator to another. Many companies use "rating films" as a means of training time study analysts in understanding the company's *definition* of normal performance. However, because of its subjective nature, performance rating, which is extremely important to the practice of time study, is the weakest link in the process.

Several of the performance rating methods in use attempt to overcome this weakness by employing formalized procedures to ensure consistency and reliability. As an illustration, consider the plan developed by Lowry, Maynard, and Stegemerten [3]. This is one of the oldest yet most frequently used performance rating systems. Four factors—skill, effort, condition, and consistency—are evaluated in this plan to arrive at a performance rating for the operator. The first two factors are considered by far the most important. Each of them carries somewhat special meanings that must be understood by the analyst before applying them. For example, skill is defined as "proficiency at a given method." The analyst evaluates the skill level by noting such factors as coordination, rhythm, hesitation, precision of movements, and lost motions. Variation from normal or average skill by a particular operator is indicated by grades as shown in Figure 4.11. As listed, there are six classes of skill, which are broken down into 16 percentage values.

To determine the performance rating or leveling factor, the appropriate rating for each of the four factors is noted, and equivalent numerical values are taken from the table and added algebraically. An example will illustrate how the method is applied. Suppose that a time study is made in which the factors for particular element are rated as B2C2AC. The performance rating for that particular element would be determined as follows:

			Percentage	
B2	Skill	(Excellent)	+0.08	
C2	Effort	(Good)	+0.02	
A	Conditions	(Ideal)	+0.06	
C	Consistency	(Good)	+0.01	
			+1.00	
			1.17	= 117 percent

Skill			Effort		
+0.15 +0.13	*A*1 *A*2	Superskill	+0.13 +0.12	*A*1 *A*2	Excessive
+0.11 +0.08	*B*1 *B*2	Excellent	+0.10 +0.08	*B*1 *B*2	Excellent
+0.06 +0.03	*C*1 *C*2	Good	+0.05 +0.02	*C*1 *C*2	Good
0.00	*D*	Average	0.00	*D*	Average
−0.05 −0.10	*E*1 *E*2	Fair	−0.04 −0.08	*E*1 *E*2	Fair
−0.16 −0.22	*F*1 *F*2	Poor	−0.12 −0.17	*F*1 *F*2	Poor
Conditions			**Consistency**		
+0.06	*A*	Ideal	+0.04	*A*	Perfect
+0.04	*B*	Excellent	+0.03	*B*	Excellent
+0.02	*C*	Good	+0.01	*C*	Good
0.00	*D*	Average	0.00	*D*	Average
−0.03	*E*	Fair	−0.02	*E*	Fair
−0.07	*F*	Poor	−0.04	*F*	Poor

Figure 4.11 Performance Rating Table (Adapted from S. M. Lowry, H. B. Maynard, and G. J. Stegemerten, *Time and Motion Study and Formulas for Wage Incentives*, 2nd ed., McGraw-Hill, New York, 1932, p. 135)

On the other hand, a lower performance rating would result from an operation evaluated E2F1DE as follows:

			Percentage
E2	Skill	(Fair)	−0.10
F1	Effort	(Poor)	−0.12
D	Conditions	(Average)	0.00
E	Consistency	(Fair)	−0.02
			+1.00
			0.76 = 76 percent

Other performance rating plans include "objective rating" and "synthetic rating," the latter based on comparisons of element times with known standards such as available from predetermined time systems or previous time studies. Attempts have also been made to develop rating methods based on experimental results of physiological changes that occur in the body as work is being performed [1].

4.7 WORK SAMPLING

Random sampling procedures based on the laws of probability have been used for numerous applications in industry, such as statistical quality control. Another very useful sampling application was developed that permitted a quick determination of the percentage of downtime on production machinery. Later this technique, called "work sampling," was refined to include all types of activity, both human and machine. Work sampling may be defined as a technique for developing a representative sample of a human activity or machine operation by the use of random sampling procedures. Work sampling complements stopwatch time study and can be very useful in analyzing some situations that would be difficult to handle with a stopwatch. It is especially good for nonrepetitive or long-cycle operations and situations where it is desired to observe several operations at one time.

Work sampling is used in a variety of applications. One use involves determining the percentage of an individual's time that is spent in certain categories, such as unavoidable delays, productive time, idle time, inspection time, or time handling material. Similarly, it is used to ascertain equipment utilization in categories such as operating time, setup time, or downtime. In addition, work sampling is used in establishing personal and delay allowances.

Work sampling also may be used to establish production standards directly. The significant difference in this method and other sampling studies is that the observer, in addition to observing and recording the appropriate category, also evaluates the performance of the operator and records that rating. The performance ratings are averaged to get an overall rating for the period of the study. The following example will show how this method works.

Example 4.3

It was desired to set a standard for the time required to drill a hole in a low-volume, special-order subassembly. A work sampling study was set up over a 25-day period. During this time, the time study analyst made random observations of the drilling operation, recording the operator's performance rating as well as indicating how the operator is spending his time when the observation is made (specifically, "working" or "idle").

A schedule of exactly when these random observations were to take place was first prepared. Based on the use of a random number table, a sample of 400 observations was scheduled at times like 8:15 A.M. on day 1, 8:42 A.M. on day 1, 9:15 A.M. on day 2, and so forth. The total number of observations is based on considerations similar to those used to set the sample size in a stopwatch time study.

At the end of the 25-day study, 240 total hours had been clocked or logged in by the drill press operator, and 547 subassembly pieces had been drilled. The work sample's results showed that out of the 400 observations, the operator had actually been "working" when 360 of the observations were made—or 90 percent of the time. In addition, the company has a policy of allowances for personal, fatigue, and delay requirements of 15 percent. The operator was rated by the analyst as performing 5 percent better than a normal operator, so that a rating factor of 1.05 was used.

The time standard for this study was calculated as follows:

$$T_N = \frac{\begin{pmatrix}\text{Total time worked} \\ \text{in minutes}\end{pmatrix}\begin{pmatrix}\%\text{ work} \\ \text{occurrence}\end{pmatrix}\begin{pmatrix}\text{Rating} \\ \text{factor}\end{pmatrix}(1 + \text{allowances})}{\text{Number of pieces produced}}$$

or

$$T_N = \frac{(240 \text{ hrs})(60 \text{ min/hr})(0.90)(1.05)(1.15)}{547} = 28.6 \text{ min/piece}$$

■

4.8 PREDETERMINED TIME STANDARDS

In the early days of formal industrial work measurement, Frederick W. Taylor refined the procedures for establishing work standards by dividing operations into individual work elements, describing each element, timing each element, and adding appropriate allowances. Later Frank Gilbreth developed a technique for studying the basic motions involved in carrying out a task by the use of motion picture film. A natural extension of their efforts was the assignment of standard time values to these basic motions. The motions themselves were carefully classified and named. Thus, it became feasible to use these data to establish time standards for current or future operations without the use of a stopwatch. Each operation could be analyzed by breaking it down into its basic motions and assigning time values for these elements. This general procedure is called a predetermined time system. This approach avoids many of the psychological problems inherent in the face-to-face encounter involved in "direct" time study methods like work sampling or stopwatch studies.

One of the better known predetermined time systems is methods-time measurement (MTM) [5]. The founders of this technique relied heavily on the use of motion picture film in making their analyses. The MTM data are the result of frame-by-frame analysis of motion picture films in many diversified areas of work. A large number of experienced time study analysts were used to determine performance ratings using the Lowry, Maynard, and Stegemerten method mentioned earlier. These ratings were then used to normalize element times. A new time unit was developed and assigned 0.00001 hours as the value of one "time measurement unit." As a result, the following time conversions are appropriate:

1 TMU = 0.00001 hr	1 hr = 100,000 TMU
1 TMU = 0.0006 min	1 min = 1,667 TMU
1 TMU = 0.036 sec	1 sec = 27.8 TMU

The MTM-1 data card, shown in Figures 4.12 and 4.13, has 14 tables. Thirteen of these tables give predetermined time values for the various types of basic motions. One indicates the difficulty that can be expected when performing motions simultaneously. Use of MTM data for setting a time standard is illustrated in the following example.

Example 4.4

Assume that an operator reaches with his right hand to a single, $\frac{1}{2}$-in. O.D. flat washer that has $\frac{5}{16}$-in. I.D. The washer is located 12 in. away on the work table. His left hand remains idle. He picks up the washer, returns it to a bolt that is positioned upright in a fixture, places the washer on the bolt, and releases. For this activity, the work elements and times taken from the charts in Figure 4.12 are:

Left Hand	TMU	Right Hand	Description
	9.6	R12A	Reach 12 in. to object in fixed position.
	3.5	G1B	Grasp object lying close against flat surface.
	15.2	M12C	Move 12 in. to exact location.
	5.6	PISE	Position loose fit, symmetrical, easy to handle.
	2.0	RL1	Normal release by opening fingers as independent motion.
Total	35.9		

35.9 TMUs = 0.000359 hr = 0.0215 min

■

Extensive research has been conducted in various parts of the world to extend the capabilities and functions of the MTM systems. Through these efforts a whole family of MTM systems has been developed extending the basic MTM-1 scheme. This includes MTM-2, a system developed by establishing motion combinations. It contains a more limited number of distance ranges and cases of control than MTM-1. Operations suited to this system are those with cycle times greater than 1 min, those in which the cycle is not highly repetitive, and those where motions are neither complex nor contain simultaneous hand motions. Other MTM systems include MTM-3, MTM-GPD, MTM-V, MTM-C, MTM-B, MTM-M, and 4M. This latter version of MTM is often used by directly observing an operation and recording the element data directly into a computer for calculations. The data entry device is a portable microprocessor like the one shown in Figure 4.8.

TABLE IX – BODY, LEG, AND FOOT MOTIONS

TYPE		SYMBOL	TMU	DISTANCE	DESCRIPTION
LEG–FOOT MOTION		FM	8.5	To 4"	Hinged at ankle.
		FMP	19.1	To 4"	With heavy pressure.
		LM__	7.1	To 6"	Hinged at knee or hip in any direction.
			1.2	Ea. add'l inch	
HORIZONTAL MOTION	SIDE STEP	SS__C1	*	<12"	Use Reach or Move time when less than 12". Complete when leading leg contacts floor.
			17.0	12"	
			0.6	Ea. add'l inch	
		SS__C2	34.1	12"	Lagging leg must contact floor before next motion can be made.
			1.1	Ea. add'l inch	
	TURN BODY	TBC1	18.6	—	Complete when leading leg contacts floor.
		TBC2	37.2	—	Lagging leg must contact floor before next motion can be made
	WALK	W__FT	5.3	Per Foot	Unobstructed.
		W__P	15.0	Per Pace	Unobstructed.
		W__PO	17.0	Per Pace	When obstructed or with weight.
VERTICAL MOTION		SIT	34.7	—	From standing position.
		STD	43.4	—	From sitting position.
		B,S,KOK	29.0	—	Bend, Stoop, Kneel on One Knee.
		AB,AS,AKOK	31.9	—	Arise from Bend, Stoop, Kneel on One Knee
		KBK	69.4	—	Kneel on Both Knees.
		AKBK	76.7	—	Arise from Kneel on Both Knees.

TABLE X – SIMULTANEOUS MOTIONS

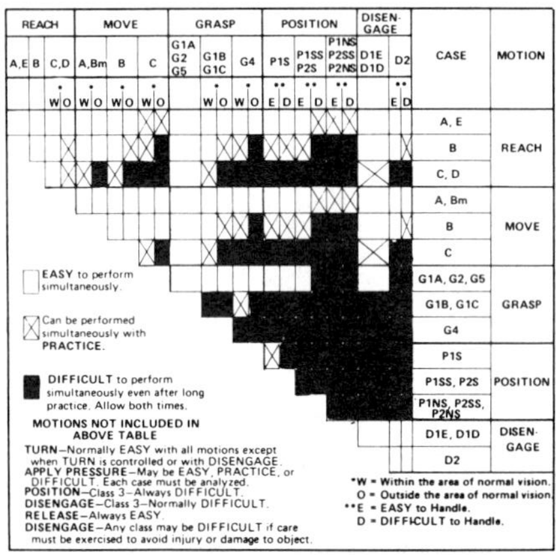

SUPPLEMENTARY MTM DATA

TABLE 1 – POSITION – P

Class of Fit and Clearance	Case of † Symmetry	Align Only	Depth of Insertion (per ¼") 0 >0≤1/8"	2 >1/8≤¾	4 >¾≤1¼	6 >1¼≤1¾
21 .150" – .350"	S	3.0	3.4	6.6	7.7	8.8
	SS	3.0	10.3	13.5	14.6	15.7
	NS	4.8	15.5	18.7	19.8	20.9
22 .025" – .149"	S	7.2	7.2	11.9	13.0	14.2
	SS	8.0	14.9	19.6	20.7	21.9
	NS	9.5	20.2	24.9	26.0	27.2
23* .005" – .024"	S	9.5	9.5	16.3	18.7	21.0
	SS	10.4	17.3	24.1	26.5	28.8
	NS	12.2	22.9	29.7	32.1	34.4

*BINDING—Add observed number of Apply Pressures.
DIFFICULT HANDLING—Add observed number of G2's.

†Determine symmetry by geometric properties, except use S case when object is oriented prior to preceding Move.

TABLE 1A – SECONDARY ENGAGE – E2

CLASS OF FIT	DEPTH OF INSERTION (PER 1/4") 2	4	6
21	3.2	4.3	5.4
22	4.7	5.8	7.0
23	6.8	9.2	11.5

TABLE 2 – CRANK (LIGHT RESISTANCE) – C

DIAMETER OF CRANKING (INCHES)	TMU (T) PER REVOLUTION	DIAMETER OF CRANKING (INCHES)	TMU (T) PER REVOLUTION
1	8.5	9	14.0
2	9.7	10	14.4
3	10.6	11	14.7
4	11.4	12	15.0
5	12.1	14	15.5
6	12.7	16	16.0
7	13.2	18	16.4
8	13.6	20	16.7

FORMULAS:

A. CONTINUOUS CRANKING (Start at beginning and stop at end of cycle only)

$$TMU = [\,(N \times T) + 5.2\,] \cdot F + C$$

B. INTERMITTENT CRANKING (Start at beginning and stop at end of each revolution

$$TMU = [\,(T + 5.2)\,F + C\,] \cdot N$$

C = Static component TMU weight allowance constant from move table
F = Dynamic component weight allowance factor from move table
N = Number of revolutions
T = TMU per revolution (Type III Motion)
5.2 = TMU for start and stop

METHODS-TIME MEASUREMENT
MTM-I APPLICATION DATA

1 TMU = .00001 hour		1 hour = 100,000.0 TMU
= .0006 minute		1 minute = 1,666.7 TMU
= .036 seconds		1 second = 27.8 TMU

Do not attempt to use this chart or apply Methods-Time Measurement in any way unless you understand the proper application of the data. This statement is included as a word of caution to prevent difficulties resulting from misapplication of the data.

MTM ASSOCIATION
FOR STANDARDS
AND RESEARCH
16-01 Broadway
Fair Lawn, N.J. 07410

MTMA 101
PRINTED IN U S A

Figure 4.12 MTM-1 Data Card showing Tables 1, 1A, 2, IX and X. (Used with permission of MTM Association)

TABLE I – REACH – R

Distance Moved Inches	Time TMU A	Time TMU B	Time TMU C or D	Time TMU E	Hand In Motion A	Hand In Motion B	CASE AND DESCRIPTION
3/4 or less	2.0	2.0	2.0	2.0	1.6	1.6	A Reach to object in fixed location, or to object in other hand or on which other hand rests.
1	2.5	2.5	3.6	2.4	2.3	2.3	
2	4.0	4.0	5.9	3.8	3.5	2.7	
3	5.3	5.3	7.3	5.3	4.5	3.6	B Reach to single object in location which may vary slightly from cycle to cycle.
4	6.1	6.4	8.4	6.8	4.9	4.3	
5	6.5	7.8	9.4	7.4	5.3	5.0	
6	7.0	8.6	10.1	8.0	5.7	5.7	
7	7.4	9.3	10.8	8.7	6.1	6.5	C Reach to object jumbled with other objects in a group so that search and select occur.
8	7.9	10.1	11.5	9.3	6.5	7.2	
9	8.3	10.8	12.2	9.9	6.9	7.9	
10	8.7	11.5	12.9	10.5	7.3	8.6	
12	9.6	12.9	14.2	11.8	8.1	10.1	
14	10.5	14.4	15.6	13.0	8.9	11.5	D Reach to a very small object or where accurate grasp is required.
16	11.4	15.8	17.0	14.2	9.7	12.9	
18	12.3	17.2	18.4	15.5	10.5	14.4	
20	13.1	18.6	19.8	16.7	11.3	15.8	
22	14.0	20.1	21.2	18.0	12.1	17.3	E Reach to indefinite location to get hand in position for body balance or next motion or out of way.
24	14.9	21.5	22.5	19.2	12.9	18.8	
26	15.8	22.9	23.9	20.4	13.7	20.2	
28	16.7	24.4	25.3	21.7	14.5	21.7	
30	17.5	25.8	26.7	22.9	15.3	23.2	
Additional	0.4	0.7	0.7	0.6			TMU per inch over 30 inches

TABLE II – MOVE – M

Distance Moved Inches	Time TMU A	Time TMU B	Time TMU C	Hand In Motion B	Wt. (lb.) Up to	Dynamic Factor	Static Constant TMU	CASE AND DESCRIPTION
3/4 or less	2.0	2.0	2.0	1.7				
1	2.5	2.9	3.4	2.3	2.5	1.00	0	
2	3.6	4.6	5.2	2.9				A Move object to other hand or against stop.
3	4.9	5.7	6.7	3.6	7.5	1.06	2.2	
4	6.1	6.9	8.0	4.3				
5	7.3	8.0	9.2	5.0	12.5	1.11	3.9	
6	8.1	8.9	10.3	5.7				
7	8.9	9.7	11.1	6.5	17.5	1.17	5.6	
8	9.7	10.6	11.8	7.2				
9	10.5	11.5	12.7	7.9	22.5	1.22	7.4	B Move object to approximate or indefinite location.
10	11.3	12.2	13.5	8.6				
12	12.9	13.4	15.2	10.0	27.5	1.28	9.1	
14	14.4	14.6	16.9	11.4				
16	16.0	15.8	18.7	12.8	32.5	1.33	10.8	
18	17.6	17.0	20.4	14.2				
20	19.2	18.2	22.1	15.6	37.5	1.39	12.5	
22	20.8	19.4	23.8	17.0				
24	22.4	20.6	25.5	18.4	42.5	1.44	14.3	C Move object to exact location.
26	24.0	21.8	27.3	19.8				
28	25.5	23.1	29.0	21.2	47.5	1.50	16.0	
30	27.1	24.3	30.7	22.7				
Additional	0.8	0.6	0.85		TMU per inch over 30 inches			

TABLE III A – TURN – T

Weight	30°	45°	60°	75°	90°	105°	120°	135°	150°	165°	180°
	Time TMU for Degrees Turned										
Small – 0 to 2 Pounds	2.8	3.5	4.1	4.8	5.4	6.1	6.8	7.4	8.1	8.7	9.4
Medium – 2.1 to 10 Pounds	4.4	5.5	6.5	7.5	8.5	9.6	10.6	11.6	12.7	13.7	14.8
Large – 10.1 to 35 Pounds	8.4	10.5	12.3	14.4	16.2	18.3	20.4	22.2	24.3	26.1	28.2

TABLE III B – APPLY PRESSURE – AP

FULL CYCLE			COMPONENTS		
SYMBOL	TMU	DESCRIPTION	SYMBOL	TMU	DESCRIPTION
APA	10.6	AF + DM + RLF	AF	3.4	Apply Force
			DM	4.2	Dwell, Minimum
APB	16.2	APA + G2	RLF	3.0	Release Force

TABLE IV – GRASP – G

TYPE OF GRASP	Case	Time TMU	DESCRIPTION	
PICK-UP	1A	2.0	Any size object by itself, easily grasped	
	1B	3.5	Object very small or lying close against a flat surface	
	1C1	7.3	Diameter larger than 1/2"	Interference with Grasp
	1C2	8.7	Diameter 1/4" to 1/2"	on bottom and one side of
	1C3	10.8	Diameter less than 1/4"	nearly cylindrical object.
REGRASP	2	5.6	Change grasp without relinquishing control	
TRANSFER	3	5.6	Control transferred from one hand to the other.	
SELECT	4A	7.3	Larger than 1" x 1" x 1"	Object jumbled with other
	4B	9.1	1/4" x 1/4" x 1/8" to 1" x 1" x 1"	objects so that search
	4C	12.9	Smaller than 1/4" x 1/4" x 1/8"	and select occur.
CONTACT	5	0	Contact, Sliding, or Hook Grasp.	

EFFECTIVE NET WEIGHT			
Effective Net Weight (ENW)	No. of Hands	Spatial	Sliding
	1	W	$W \times F_c$
	2	W/2	$W/2 \times F_c$

W = Weight in pounds
F_c = Coefficient of Friction

TABLE V – POSITION* – P

CLASS OF FIT		Symmetry	Easy To Handle	Difficult To Handle
1–Loose	No pressure required	S	5.6	11.2
		SS	9.1	14.7
		NS	10.4	16.0
2–Close	Light pressure required	S	16.2	21.8
		SS	19.7	25.3
		NS	21.0	26.6
3–Exact	Heavy pressure required.	S	43.0	48.6
		SS	46.5	52.1
		NS	47.8	53.4

SUPPLEMENTARY RULE FOR SURFACE ALIGNMENT	
P1SE per alignment: >1/16 ≤1/4"	P2SE per alignment: ≤1/16"

*Distance moved to engage—1" or less.

TABLE VI – RELEASE – RL

Case	Time TMU	DESCRIPTION
1	2.0	Normal release performed by opening fingers as independent motion.
2	0	Contact Release

TABLE VII – DISENGAGE – D

CLASS OF FIT	HEIGHT OF RECOIL	EASY TO HANDLE	DIFFICULT TO HANDLE
1–LOOSE–Very slight effort, blends with subsequent move.	Up to 1"	4.0	5.7
2–CLOSE–Normal effort, slight recoil.	Over 1" to 5"	7.5	11.8
3–TIGHT–Considerable effort, hand recoils markedly.	Over 5" to 12"	22.9	34.7

SUPPLEMENTARY

CLASS OF FIT	CARE IN HANDLING	BINDING
1– LOOSE	Allow Class 2	———
2– CLOSE	Allow Class 3	One G2 per Bind
3– TIGHT	Change Method	One APB per Bind

TABLE VIII – EYE TRAVEL AND EYE FOCUS – ET AND EF

Eye Travel Time = $15.2 \times \frac{T}{D}$ TMU, with a maximum value of 20 TMU.

where T = the distance between points from and to which the eye travels.
D = the perpendicular distance from the eye to the line of travel T.

Eye Focus Time = 7.3 TMU.

SUPPLEMENTARY INFORMATION

– Area of Normal Vision = Circle 4" in Diameter 16" from Eyes

– Reading Formula = 5.05 N Where N = The Number of Words.

Figure 4.13 MTM Tables (Continued) (Used with permission of MTM Association)

4.9 OTHER APPROACHES

In the previous sections of this chapter, we have described the primary methods for determining standard times. These are stopwatch studies, work sampling, and predetermined times. There are several variants of these as well as other approaches that are often used in certain situations. For example, stopwatch studies are often applied to motion pictures of a job rather than directly to the job. This has the advantage of avoiding face-to-face contact between the employee or operator and the stopwatch-carrying analyst.

A variant of predetermined times is a method called "standard data." One version of standard data entails use of a data file of historical standards for elements that are common to a variety of tasks, much like the MTM system. Another version is to develop an equation that relates the desired job standard (as the dependent variable) to one or more characteristics of that job (as the independent variables). The numerical version of the equation is often developed through statistical techniques such as regression analysis. For example, the Tomlinson Company, a furniture manufacturer in High Point, North Carolina, has an operation that involves assembly of chair backs. They have developed a standard data model or equation that relates the normal time required to assemble and clamp a chair back to variables such as the number of holes involved (H), the number of rails (R), and the number of nails required (N). The equation, applicable to "Loose Dowel Joint" type backs, is as follows (where T_N is in units of hours per 100 backs).

$$T_N = 0.459 + 0.050H + 0.082R + 0.050N \tag{4.4}$$

Therefore, if a particular customer's order called for assembling 1,000 Loose Dowel Joint chairs and each chair was to have 2 rails, 1 nail, and 4 holes, then the normal time for this job would be $T_N = 0.459 + (0.050)(4) + (0.082)(2) + (0.050)(1) = 0.873$ hr per 100 chair backs, or $10\ T_N = 8.73$ hr for the entire job.

4.10 SUMMARY

Work methods and measurement are among the most durable of IE/OR tools. The study of methods and attempts to measure performance preceded the formal development of industrial engineering by centuries. As the profession has evolved, so has the study of work. This study is still an important part of our profession because it deals with the heart of the matter, the effective use of people and materials.

This brief description of work methods and measurement has barely scratched the surface of the topic; more detailed descriptions may be found in the suggested reading. Yet the important points are here, and they are fundamental. This branch of IE/OR has three purposes—description, analysis, and measurement—all related to man at work. The improvement of productivity ultimately depends on the effectiveness of individual operations. These operations must be adequately described before any progress can be made. Analysis proceeds by attempting to modify the basic operation. And measurement allows a directly quantifiable comparison to be made.

Although work methods and measurement are generally considered to be classical industrial engineering tools, they are also relevant to applications of operations research. Operating rates, labor requirements, and special tooling are all inputs or coefficients in quantitative models. They provide the details that allow a model to yield its results and be an accurate descriptor or prescriptor of the manufacturing operation. Even large models with aggregated data must begin with details of operations. Within this context, work measurement is the starting point for quantitative analysis.

The future of this branch of IE/OR is expected to be as rich as its past as it continues to evolve. Robotics does not eliminate the need for measurement; rather, it creates the need for new approaches to an old problem in a new situation. The use of computers is having a significant impact through microprocessor-based time study systems, automated data collection, standard data filing systems, computer-aided standard generation, and computer-based synthetic motion times. The shifts of labor from blue- to white-collar and from manufacturing to service industries do not eliminate the need for the study of work; they merely change the context and location of the study. The human operator is still part of the system. The study of work merely adapts itself to new kinds of work as it has in the past, and continues.

REFERENCES AND SUGGESTED READING

1. Ralph M. Barnes, *Motion and Time Study: Design and Measurement of Work*, Wiley, New York, 1968.
2. James A. Lee, "The Hawthorne Relay Studies: A Faith Impervious to Iconoclasm," Proceedings of the AIIE Spring Conference, Atlanta, 1980.
3. S. M. Lowry, H. B. Maynard, and G. J. Stegemerten, *Time and Motion Study and Formulas for Wage Incentives*, McGraw-Hill, New York, 1940.
4. H. B. Maynard (Ed.), *Industrial Engineering Handbook*, McGraw-Hill, New York, 1971.
5. H. B. Maynard, G. J. Stegemerten, and J. L. Schwab, *Methods Time Measurement*, McGraw-Hill, New York, 1948, p. 12.
6. R. L. Morrow, *Time and Motion Economy*, Ronald Press, New York, 1946.
7. Marvin E. Mundel, *Motion and Time Study*, Prentice-Hall, 1978.
8. Benjamin W. Niebel, *Motion and Time Study*, Irwin, Homewood, IL, 1976.

REVIEW QUESTIONS AND PROBLEMS

4.1 For the following common tasks, list all therbligs contained in each task:

(a) Opening and pouring a canned drink
(b) Unlocking and opening a door

(c) Setting a thermostat
(d) Signing a letter with a fountain pen that is in a desk drawer

4.2 Develop a right-hand–left-hand chart for the tasks in Problem 4.1.

4.3 Three machines have cycle times of 0.24, 0.13, and 0.44 min, respectively. Each machine must be manually started. Develop a Multiple Activity Chart for one operator controlling all three machines. Rearrange the machine order in an attempt to maximize operator idle time and reduce machine idle time.

4.4 Which therbligs are most applicable to robots? Which are not applicable at all?

4.5 What is the most difficult part of stopwatch time study? Why?

4.6 In a production line setup there is a high-volume, highly repetitive activity for which time standards are needed. Discuss the advantages and disadvantages of the various work measurement systems for this particular requirement.

4.7 A mutual trust and confidence between the work measurement analyst and the production operator are essential to a smooth working organization. Discuss the problems that you might expect in this relationship and what can be done about them.

4.8 What are the qualifications needed in a good time study analyst?

4.9 Why is it necessary to record the exact information on the work area layout, machine speeds, feeds, and so on in work measurement data when collected?

4.10 Compare the advantages and disadvantages of the continuous and snapback methods of time study.

4.11 It has been said that production workers will protect inaccurate production standards by producing only the quantities expected. Is this true? What are the implications?

4.12 Calculate the equivalent in TMUs of 0.0060 hr/piece, 12 pieces/min, 33 sec/piece, and 0.411 min/piece.

4.13 Would you recommend that the production worker be allowed to examine the time study form if he or she should so request? Why?

4.14 Why are there so many different concepts of what constitutes normal performance?

4.15 What are some of the means whereby the operator can deceive the time study analyst during a study?

4.16 What is the single most important problem causing production standards to become obsolete?

4.17 In a particular time study, it is desired that an accurate time be obtained for an individual element of short duration. To do this, the analysis decides to use cycle timing, which gives the data below:

1st reading: $t_a + t_b + t_c = T_1 = 0.42$ min

2nd reading: $t_a + t_b = T_2 = 0.17$ min

3rd reading: $t_a + t_c = T_3 = 0.40$ min

Determine the time for each element.

4.18 In a time study, the stopwatch observations were as follows for the most variable element of the study:

0.19, 0.18, 0.17, 0.18, 0.15, 0.19, 0.15, 0.18, 0.16, 0.20

Are there enough observations for a ±5-percent accuracy with a 95-percent confidence level?

4.19 The following data are a summary of a stopwatch time study taken on an assembly operation performed manually:

Element Number	Average Time	Frequency of Occurrence
1	0.75	1/unit
2	0.21	1/unit
3	0.34	1/unit
4	1.15	1/unit
5	10.00	1/(12 units)

Performance was rated using the Lowry, Maynard, Stegemerten system as follows:

Element Number	Skill	Rating Effort	Conditions	Consistency
1	C2	B2	B	C
2	C1	C2	B	C
3	C2	C2	B	C
4	C2	C1	B	D
5	D	B2	E	D

Allowances for this particular operation are as follows:

Personal	5 percent
Fatigue	6 percent
Delay	4 percent

Calculate the standard time for this operation.

4.20 The following is a sketch of a work area for an assembly operation:

The bolts are $\frac{1}{4}$ in. diameter × 3 in. L machine bolts. The washers are flat $\frac{5}{16}$-in. I.D. × $\frac{1}{2}$ in. O.D. The sequence of motions used in the assembly are:

(a) LH reaches for bolt jumbled in bin, picks up, returns to assembly area. Simultaneously, RH reaches for washer jumbled in bin, picks up, and returns to assembly area.
(b) LH holds bolt; RH places washer on bolt.
(c) LH holds subassembly; RH procures nut from bin, picks up, returns to assembly area.
(d) LH holds subassembly; RH starts nut and then threads with three, one-half turns (finger, hand, and wrist motion).
(e) LH discards assembly to tote box 24 in. to left.

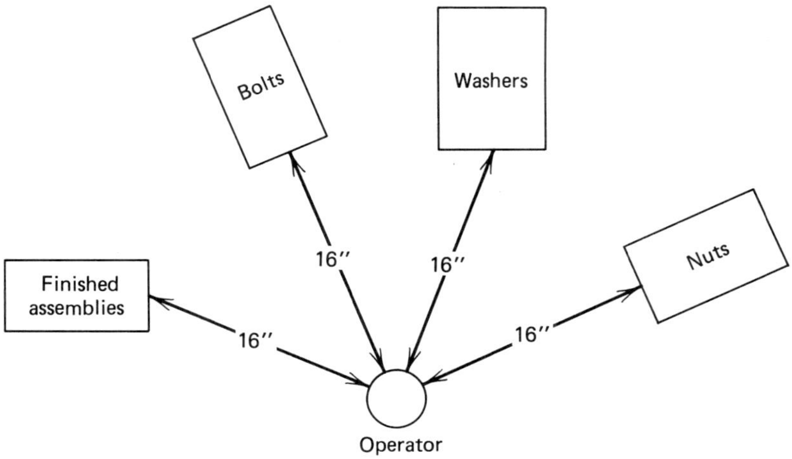

The materials handler keeps the gravity feed bins filled. However the operator moves the finished assembly tote box to a work table 10 ft away. The tote box will hold 50 assemblies. In moving the tote box, the operator uses the following sequence of motions.

> Stand up from chair, turn body 90 degrees to tote box, bend, pick up tote box (about 15 lb), rise, sidestep 12 in., and walk 10 ft to table. Place tote box on work table. Side step 18 in., reach and procure empty tote box located 30 in. away. Return tote box to body, turn 180 degrees, walk to finished assembly station. Turn 90 degrees, bend, place empty tote box. Rise, turn 90 degrees, side step 15 in., turn 45 degrees, sit down in chair.

After the work table becomes full of tote boxes, the materials handler removes them. Determine the production standard using MTM-1, assuming that personal, fatigue, and delay allowances total 16 percent.

Chapter 5

Job Evaluation and Wage Incentives

An important component in the structure and operations of any firm is its wage and salary administration system. The design of such systems can significantly influence the costs and productivity of the operations of the organization. There are many facets to a wage payment system, such as fringe benefits, base rates for direct workers, and overtime compensation allowances. Two such areas in which industrial engineering analysts have been traditionally involved are (1) the design and conduct of plans to determine the relative worth of the different jobs in the organization (i.e., "job evaluation") and (2) the design of monetary incentive systems to compensate increased worker productivity (i.e., "wage incentive systems"). Both of these design activities are discussed in this chapter.

5.1 JOB EVALUATION

One of the first and most important issues that confronts the management of a firm is what should be the base wage rate or base salary for each job in the organization. There are many factors that influence this decision, such as the availability of certain types of workers and the wage demands of unions. However, the most basic consideration is perhpas the relative worth of each job in terms of those criteria deemed important by the firm. The process of determining this job worth is called *job evaluation*. It is important to note that it is the *job* that is being evaluated, not the worker who is performing that job.

The job evaluation study begins with an investigation of the characteristics of a job, and ends with a determination of the worth of the job *relative to* other jobs in the organization. The relative worth of the jobs studied can be expressed on a simple ranking basis or on a quantitative point scale (e.g., job A is worth 350 points whereas job B is worth 400 points). However, regardless of the intermediate value comparison scheme used, the final evaluation of jobs is normally expressed in terms of base wages, or ranges of base wages.

The worth or value of a job is dependent on factors inherent in the job itself such as the amount of physical effort required to properly execute the tasks required in the job. Job evaluation attempts to assess the extent to which such factors are present in a job, and to assign a relative value to the job based on this assessment. The job's base rate is then set in accordance with its assessed value.

As an illustration of the utility of job evaluation, consider the jobs in a typical fire department. These include an engine pump operator, a dispatcher, an engine crewman, and a fire prevention inspector. How much should the city pay an engine crewman, say, who risks his life and expends a great deal of physical effort in fighting a fire, versus an inspector, who inspects public and commercial buildings for violations of fire prevention codes? The engine crewman has the more physically demanding job, but the inspector has a great deal of responsibility to ensure the safety of building occupants. The same types of comparative factors must be considered when the city sets the base rates for policemen, sanitation workers, building inspectors, and so on. Many cities have faced strikes because sanitation workers claim that their jobs are worth as much as those of policemen or because firemen demand equal pay rates with policemen. In such instances, job evaluation would be a useful tool to determine the merits of these claims and demands.

5.1.1 Alternative Evaluation Methods

There are four basic schemes or methods that are commonly used to evaluate the jobs being examined. Three of these are briefly described below. The first two methods—job ranking and job classification—are usually referred to as qualitative approaches because they do not involve any effort to quantify the demands and responsibilities of the jobs, or the relative value of those demands and responsibilities. In contrast, the point system and the factor comparison method are called quantitative approaches since they make a direct attempt at quantifying the value of the characteristics of the jobs.

5.1.2 Job Ranking

This method involves simply developing an ordered list of the jobs being examined. For example, if jobs A, B, and C are being evaluated, we might decide that B is the most important, followed by A and then C. Therefore, the results of the evaluation would be the list: B, A, C.

A committee is often used to establish the job rankings. The committee collectively determines the rankings through discussions of what the various jobs consist of, and how the company values the characteristics inherent in the jobs. For example, if job A requires a great deal of physical strength as compared to job B, then in terms of this one factor, job A would be ranked ahead of B. However, if job B demands a high degree of mental effort as compared to A, and if the company deems mental effort more important than physical effort, then the committee would rank job B over job A. Generally, when job ranking is used, there is no attempt to compare individual job factors or characteristics. Rather, the jobs are treated as a whole. In this way, many ''unidentifiable'' aspects of the jobs can enter into the evaluation.

Job ranking is a simple method and is easily understood by labor. Therefore, it is both inexpensive to administer and is readily accepted by labor and their unions. However, when jobs are evaluated or appraised as a whole, the rankings are often influenced by the biases of the evaluation committee members. Therefore, the accuracy of the rankings is often questionable.

5.1.3 Job Classification

A second qualitative evaluation method that is commonly used is job classification. This method is very similar to job ranking, with the basic difference being that in the job classification approach jobs are assigned to a hierarchy of basic job classes or grades with no ranking of jobs within the same class. Each job class should have a job description to illustrate the types of jobs that fall into that category. A committee usually is formed to develop these descriptions, along with determining the overall classification scheme—that is, the number of job classes to have and what kinds of jobs are ''equal'' and should therefore be placed together in the same class.

The determination of the types of jobs to group together and the overall ranking of job classes represents a decision as to what job characteristics or factors are important in the evaluation process as well as a trade-off among the relative merits of these factors. For instance, a particular job class may be defined so that jobs falling in that grade or class involve a good deal of physical activity but very little responsibility for the activites of other workers. However, the next highest job class may involve jobs that do have a fair amount of responsibility but not as much physical activity required.

Regardless of the final definition of job classes, the specific jobs to be evaluated are examined (through a combination of direct observation and job descriptions) and then assigned to a particular job class. For example, a local police department using a job classifiction method may have 15 unique jobs and 5 job classes or grades. The beat patrolman's job and the job of dispatcher may be assigned to the third highest class, while the jobs of detective and desk sergeant may be assigned to the second highest grade.

One of the first job classification schemes developed was the federal government's Civil Service grading system using a ''General Schedule'' (G.S.) for professional and administrative jobs, and a ''Crafts, Protective and Custodial Schedule'' (C.P.C.) for other jobs. The General Schedule has 18 different job classes, whereas the C.P.C. has 10 classes. In the General Schedule, the first or lowest four classes generally do not require a college education, while grades G.S.-5 through G.S.-10 are considered lower-level management positions. Specialized scientific jobs and higher-level management positions fall in grades G.S.-11 and above [3]. Several of the G.S. grades are described in Figure 5.1.

Grade-General Schedule 1 includes all classes of positions the duties of which are to be performed, under immediate supervision, with little or no latitude for the exercise of independent judgment, (1) the simplest routine work in office, business, or fiscal operations, or (2) elementary work of a subordinate technical character in a professional, scientific, or technical field.

Grade-General Schedule 5 includes all classes of positions the duties of which are (1) to perform, under general supervision, difficult and responsible work in office, business, or fiscal administration, or comparable subordinate technical work in a professional, scientific, or technical field, requiring in either case (A) considerable training and supervisory or other experience, (B) broad working knowledge of a special subject matter or of office, laboratory, engineering, scientific, or other procedure and practice, and (C) the exercise of independent judgment in a limited field; (2) to perform, under immediate supervision, and with little opportunity for the exercise of independent judgment, simple and elementary work requiring professional, scientific, or technical training equivalent to that represented by graduation from a college or university of recognized standing but requiring little or no experience; or (3) to perform other work of equal importance, difficulty, and responsibility, and requiring comparable qualifications.

Grade-General Schedule 9 includes all classes of positions the duties of which are (1) to perform, under general supervision, very difficult and responsible work along special technical, supervisory, or administrative experience which has (A) demonstrated capacity for sound independent work, (B) thorough and fundamental knowledge of a special and complex subject matter, or of the profession, art, or science involved, and (C) considerable latitude for the exercise of independent judgment; (2) with considerable latitude for the exercise of independent judgment, to perform moderately difficult and responsible work, requiring (A) professional, scientific or technical training equivalent to that represented by graduation from a college or university of recognized standing, and (B) considerable additional professional, scientific, or technical training or experience which has demonstrated capacity for sound independent work; or (3) to perform other work of equal importance, difficulty, and responsibility, and requiring comparable qualifications.

Grade-General Schedule 13 includes all classes of positions the duties of which are (1) to perform, under administrative direction, with wide latitude for the exercise of independent judgment work of unusual difficulty and responsibility along special technical, supervisory, or administrative lines, requiring extended specialized, supervisory, or administrative training and experience which has demonstrated leadership and marked attainments; (2) to serve as assistant head of a major organization involving work of comparable level within a bureau; (3) to perform, under administrative direction, with wide latitude for the exercise of independent judgment, work of unusual difficulty and responsibility requiring extended professional, scientific, or technical training and experience which has demonstrated leadership and marked attainments in professional, scientific, or technical research, practice, or administration; or (4) to perform other work of equal importance, difficulty, and responsibility, and requiring comparable qualifications.

Figure 5.1 Examples of Grades in the Civil Service Job Classification System

5.1.4 Point Method

In this evaluation approach, specific characteristics of jobs are identified that are deemed to distinguish the worth of one job from another. These characteristic factors are assigned numerical values or points that reflect their relative worth. A job to be evaluated is awarded points based on the degree of each factor present in the job, and the job's base wage rate is then set according to the total points awarded. The assignment of points to assess the value of traits or characteristics of jobs is the component of point methods that distinguishes them as being quantitative evaluation methods.

Point methods are perhaps the most common job-evaluation approach. In one survery of companies using job evaluation, more than 80 percent of the firms used a point system, while in a second survey 123 of 322 companies used point methods [3]. One reason for this popularity is the quantitative nature of point methods. This allows the logic and results of the job-evaluation process to be critiqued by both management and labor. In fact, the job evaluation plan is a major bargaining item in many labor union negotiations. Another advantage of the quantitative approach is the reduction of judgmental bias that can enter into the evaluation process. However, it is often difficult to accurately assess the degree of job factors inherent in some jobs, which is not a problem in the qualitative evaluation methods.

Point methods involve five basic steps. First, the factors or characteristics that the firm wishes to use to distinguish between jobs need to be selected. For instance, ABC Steel Company might decide that the environmental conditions within which employees are required to work should be reflected in the employees' pay, whereas XYZ Computer Sales Company might not value working conditions for its employees and hence not reflect this factor in its pay rates. A list of factors typically used in factory situations is given in Figure 5.2. Notice that some of these factors would not be relevant for other situations, such as in evaluating clerical or administrative jobs.

The second step in a point method is to clearly define each job factor selected in step one. This aids in consistent evaluations regardless of who is actually carrying out the evaluation. The third step is to divide each job factor into degrees or levels. This allows a specific job being evaluated to be identified with the degree or amount of each factor present in that job. An example of the degrees used for the job factor "Complexity of Duties" along with the definition used for each factor level is shown in Figure 5.3.

The fourth step is to assign a numerical value to each job factor that will reflect the importance of that factor relative to the other factors being used. This value is normally expressed as a percentage. For example, the factor "Experience and Training" might be worth 40 percent of the total value of a job in a particular firm, whereas "Physical Effort" might be assigned 15 percent.

The final step in a point method is to develop a job numerical rating scale so that a given job can be assigned a number or value that reflects its worth to the firm. This scale is formed by converting the factor percentage values determined is step four into intervals based on levels or degrees defined for each factor. An illustration of the points assigned to all factors in one particular plan is given in Figure 5.4.

Skill

Education
Education or mental development
Trade knowledge
Schooling
Experience
Previous experience
Experience and training
Training time
Training required
Time required to learn trade
Time required to adapt skill
Job knowledge
Knowledge of machinery and dexterity with tools
Knowledge of materials and processes
Mentality
Mental capability
Accuracy
Ingenuity
Initiative and ingenuity
Judgment and initiative
Intelligence
Resourcefulness
Versatility
Job skill
Manual dexterity
Manual accuracy and quickness
Dexterity
Degree of skill and accuracy
Physical skill
Ability to do detailed work
Social skill

Effort

Mental effort
Mental application
Mental or visual demand
Concentration
Visual application
Fatigue due to eye strain
Physical effort
Physical application
Physical demand
Physical or mental fatigue
Muscular or nerve strain
Fatigue
Monotony of work
Monotony and comfort

Responsibility (for)

Safety of others
Material or product
Material and equipment
Equipment or process
Equipment
Product
Machinery and equipment
Work of others
Supervision of others
Supervision exercised
Cost of errors
Necessary accuracy in checking, counting, and weighing
Effect on other operations
Spoilage of materials
Protection of materials
Physical property
Plant and services
Cooperation and personality
Coordination
Details to master
Quality

Working Conditions

Unavoidable hazards
Hazards involved
Exposure to health hazard
Exposure to accident hazard
Occupational hazard disease
Danger—accident from machinery or equipment
Danger—from lifting
Surroundings
Dirtiness of working conditions
Environment
Job conditions
Disagreeableness

Figure 5.2 List of Factors Typically Used for Factory-Type Jobs

The magnitude of the points assigned in this fifth step are not important. However, the relative values of the points are important. That is, if factor A is twice as important as factor B, then the points assigned to factor A should be twice the number of points assigned to B. Similarly, the points associated with the degrees within a factor should reflect the relative distinction among the degrees.

Degree	Definition	Typical Jobs
1	Simple routine duties, requiring the use of only a few definite procedures and little individual judgment, the work either being performed under immediate supervision or involving little choice as to methods of performance	Keypunch operator, messenger
2	Duties are clearly prescribed by standard practice but require the use of several procedures and the making of minor decisions requiring some judgment	Clerk C: stenographer A, typist A
3	Duties involve an intensive knowledge of a restricted field and require the use of a wide range of procedures and the analysis of facts to determine what action, within the limits of standard practice, should be taken	Detail engineer, aircraft design; junior engineer, aircraft design; methods engineer B; major estimator A; major estimator B; executive secretary; placement man, personnel; librarian A
4	Duties involve general knowledge of company policies and procedures and their application to cases not previously covered. Duties require working independently toward general results, devising new methods, and modifying or adapting standard procedures to meet new conditions. Decisions, however, are based on precedent and company policy	Major engineer, layout, aircraft design; engineer, layout, aircraft design; lead technician A, motion picture laboratory; senior auditor A, internal auditing
5	Difficult work on highly technical or involved projects, presenting new or constantly changing problems. Duties require outstanding ability to deal with complex factors not easily evaluated, or the making of decisions based on conclusions for which there is little precedent	Assistant group engineer, aircraft design; lead engineer, aircraft design

Figure 5.3 Typical Definition and Degrees of the Job Factor "Complexity of Duties"

Factor	Degree					
	1	2	3	4	5	6
1. Work experience	21	38	55	71	88	105
2. Essential knowledge and training	17	31	44	58	71	85
3. Dexterity	4	9	15	20		
4. Character of supervision	10	20	30	40	50	
5. Character of supervision given	11	22	33	44	55	
6. Number supervised	7	13	18	24	29	35
7. Responsibilities for funds, securities, and other valuables	6	11	16	20	25	30
8. Responsibility for confidential matters	6	18	30			
9. Responsibility for getting along with others	6	14	22	30		
10. Responsibility for accuracy—effect of errors	6	12	18	24	30	
11. Pressure of work	4	9	15	40		
12. Unusual working conditions	2	6	10			

Figure 5.4 Illustration of Assignment of Points to Job Factors and Their Degrees

Once the points have been determined, job evaluation using a point system proceeds as in the case of job ranking or job classification. That is, the jobs being studied are evaluated and a job evaluation manual is constructed. In the case of point systems, the evaluation or rating of a job involves matching it up with a specific degree of *each* factor and assigning it the number of points associated with those factor degrees. Then, the total value of the job is obtained by adding all degree points that are assigned to the job. For example, suppose the point plan shown in Figure 5.4 were being used to evaluate jobs in a local Post Office. The analyst conducting the evaluation might, after studying job descriptions and observing the jobs while they are being performed, decide that the job of "counter clerk" has associated with it degree 2 of factor 1, degree 3 of factor 2, degree 1 of factor 3, and so forth on down to degree 1 of factor 12. Therefore, this job would have a value of 38 + 44 + 4 + . . . + 2, or, say, a total of 275 points. On the other hand, the job of "route delivery man" might be awarded 71 points for factor 1, 71 points for factor 2, 9 points for factor 3, and so forth, for a total of, say, 425 points. The route delivery man job is therefore evaluated to be about one and a half times as important as the counter clerk job.

There are other quantitative job evaluation approaches besides the point method described above. For example, the "factor comparison" method combines the quantitative features of a point system with the relative ranking judgment of a job-ranking method. A quantitative rating scale is developed for each job factor based on a comparison of selected "key" jobs, and each job being studied is placed on these scales by comparing it to the previously placed key jobs. The scales are defined in units of base pay rate. The factor comparison method as well as the other evaluation approaches mentioned earlier is discussed in a number of textbooks, such as Belcher [1], Brennan [2], and Lytel [5].

5.2 WAGE INCENTIVE SYSTEMS

Job evaluation is a tool to establish proper wage base rates. However, many firms use wages beyond base rates to help motivate and compensate increases in direct and indirect worker productivity or achievements. Such payments are called *wage incentives.*

The purpose of wage incentive is to motivate employees to do a "better" job. The merits of the performance on a job can be measured in several ways, such as the number of pieces or units produced per hour or per day, the time to produce one unit, the unit cost, the yield of finished products from raw material, or the level of equipment utilization. Regardless of the criteria used to measure job performance, the wage incentive system should inspire an employee to achieve greater performance levels than he otherwise would.

As a motivational tool, wage incentives are an alternative to other, indirect financial incentives, as well as to nonmonetary motivators. The use of relatively high base rates, guaranteed annual wages and large fringe benefits are all examples of indirect financial incentives to encourage greater than normal performance. Typical nonmonetary incentives are company policies such as job enlargement, provision of ideal working conditions, proper employee placement, and liberal use of recognition and status symbols. Each type of incentive can have a different motivational effect on different workers, depending on their financial situation, home life, and so on. However, direct wage incentives are perhaps the most common incentive used in industry.[1]

As an indication of the motivational power of wage incentives, Niebel [7, p. 629] cites the experience of a particular firm that implemented an incentive system where none had been previously used. Before implementation, the average performance of the plant was 17 percent below the standard level for normal performance. After implementation, workers performed up to 125 percent of the standard.

Wage incentive systems can be classified as either individual plans or group plans. In an individual-type plan, a worker's incentive pay is based solely on his own performance on the jobs to which he is assigned. Under a group plan, a collection or group of workers is treated together, and all members of the group are paid the same incentive pay regardless of individual contribution. The group's incentive pay may be based on the collective performance of the group (as in the case of measuring the tons of iron produced per day in a foundry), or on an indirect performance measure (such as the average output of other shop workers whose efforts are in some way affected by the performance of the group).

Both the individual and the group approaches have their relative advantages and disadvantages. However, the basis of any wage incentive plan—that is, the relationship between performance or output and pay—can be applied to an individual plan or a

[1]Although wage incentives are often applicable in government jobs, indirect financial incentives are more common in many governmental agencies. Early retirement and "automatic" promotion are examples of such incentives.

group plan. For example, if it is decided to use as the relationship between performance and pay:

$$E = 0.05p \tag{5.1}$$

(where E = dollars earned per hour and p = number of units produced per hour), then an individual plan could be followed in which performance p would be interpreted as the output of a single worker and earnings E as his hourly take-home pay. On the other hand, if the relationship is used on a group basis, the performance might be interpreted as the combined output of the group, and each worker in the group would receive E as his take-home pay.

5.2.1 Payment Schemes

As stated earlier, the essence of any wage incentive plan is the functional relationship between performance and pay. This relationship or payment scheme is the primary mechanism by which the motivational philosophy of the firm is reflected in the incentive system. Through the payment scheme, the firm can fix the percentage (if any) of earnings created by extra productivity that goes to the company rather than directly to the worker (this is called "gain sharing" or "participation"). Also, the form and magnitude of the payment scheme can provide high motivation for certain types of worker (e.g., those capable of high productivity) while at the same time negative motivation for other workers. In addition, the direct labor cost of producing a unit has a functional relationship with output and is affected by the design of the payment scheme. Therefore, through the proper construction of the payment scheme the firm can implement the type of motivational strategy it deems most suitable.

Following is a discussion of several commonly used payment schemes. In presenting the functional relationships involved, the following notation will be used:

E = the hourly take-home pay (i.e., base rate plus incentive pay).

p_s = the standard level of productivity (i.e., the output rate that is expected of a normal worker under normal conditions.[2]

p_a = the actual level of productivity attained by a specific worker (or group) on a specific job.

p = the hourly productivity, expressed as an index of standard productivity (i.e., $p = p_a/p_s$).

α = the percentage of increased earnings to be received by the worker (so that $1 - \alpha$ is the percentage of earnings retained by the firm; i.e., the "gain shared").

b = the hourly base rate of the worker.

c = the direct labor cost contribution in making one unit or piece.

[2]See Chapter 4 for a discussion of the meaning and determination of work standards.

Daywork[3]

Daywork involves no incentive pay. It is simply the base rate earnings of an employee that is paid regardless of productivity. The payment scheme then can be represented graphically in Figure 5.5, or mathematically in Eq. (5.2):

$$E = b \tag{5.2}$$

Note that the direct labor contribution to the unit cost of pieces being produced is

$$c = \frac{\text{Earnings}}{\text{output}} = \frac{b}{p}$$

This labor cost is shown with a dashed line in Figure 5.5.

Piece Rate

Under the straight piece-rate payment scheme, the worker's earnings increase in a one-for-one fashion with his productivity. There is a 1-percent increase in earnings for each 1-percent increase in output. The base rate is not guaranteed as it is under the daywork plan. The straight piece-rate plan is graphically represented in Figure 5.6 and mathematically in Eq. (5.3):

$$E = bp = b\left(\frac{p_a}{p_s}\right) \tag{5.3}$$

When standard productivity is measured in time units, such as hours to complete the job, this scheme is often referred to as a "time piecework" or "standard hour" plan. For example, the standard to complete a job requiring complex welding might be 4 hours, whereas the actual time taken to do the job by a worker might be 5 hours.

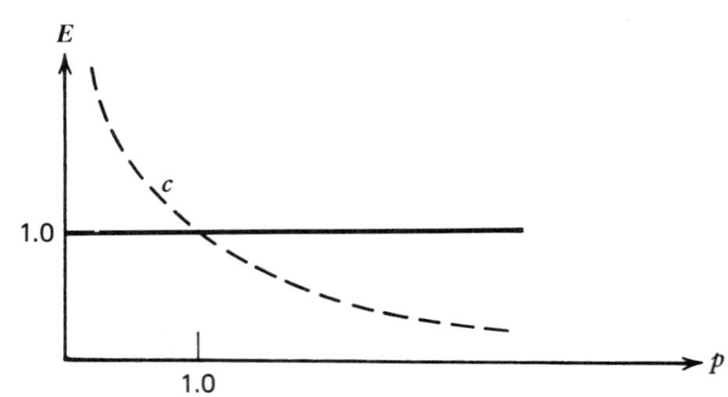

Figure 5.5 Daywork Payment Scheme

[3]The payment schemes are discussed as if they are applied to an individual incentive plan rather than to a group plan.

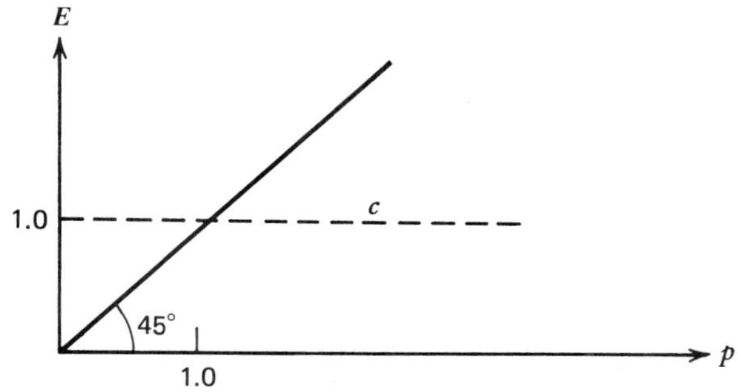

Figure 5.6 Straight Piece-Rate Payment Scheme

In this case $p_a = (1/5)$ "units" per hour, and $p_s = (1/4)$ "units" per hour, so that $E = (4/5)b = 0.80b$.

The straight piece-rate scheme is one of the oldest wage incentive plans, in use prior to the Industrial Revolution. However, it has the distinct disadvantage that there are no guarantees built in to prevent loss of a worker's earnings due to factors beyond his control, such as machine breakdowns, a shortage of raw material, or ineffective machine scheduling. To overcome such weaknesses, a guaranteed base earning is often included in a piece-rate scheme as in the next plan to be discussed.

Piece Rate with a Base

This scheme is identical to the straight piece-rate plan for jobs involving productivity beyond the standard level. However, it provides a guaranteed earnings if the worker's productivity is below the standard. This payment scheme is shown graphically in Figure 5.7 and mathematically in the following model.

$$E = \begin{cases} b & \text{if } p_a \le p_s \\ b(p_a/p_s) & \text{if } p_a > p_s \end{cases} \tag{5.4}$$

To illustrate this payment scheme, consider the following example of the operator of a wire-drawing machine in a mill producing steel wire that is under this type of plan.

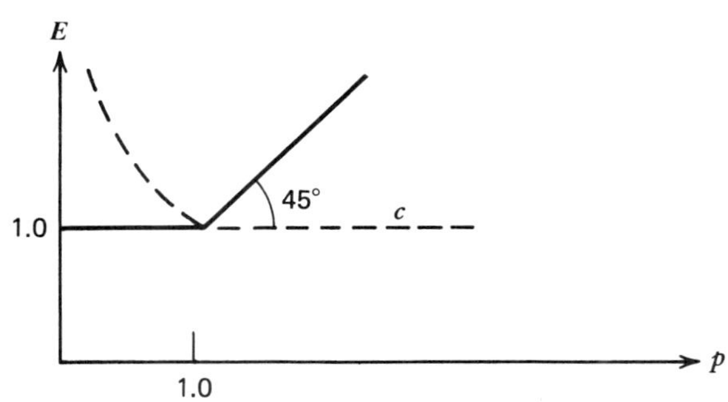

Figure 5.7 Piece-Rate with a Base Payment Scheme

The standard set for drawing $\frac{1}{4}$-in. wire down to $\frac{1}{16}$-in. wire is 1500 lb/hr. The operator is paid a base rate of \$5/hr. If, on Wednesday, the operator draws 24,000 lb of this type of wire in his 8-hr shift, his earnings for the shift would be

$$E_{\text{shift}} = 5 \times 8 \times \left(\frac{24{,}000}{1500 \times 8}\right) = \$80$$

because his average productivity, $p_a = 24{,}000/8$, is greater than the standard productivity, $p_s = 1500$. However, if the operator had drawn only 10,000 lb during the shift, his earnings would be \$40.

The piece-rate with a base payment scheme has a number of advantages and disadvantages. It is simple and easily comprehended, fair and equitable for workers, and usually acceptable to labor unions. In addition, there is often a reduction in unit cost resulting from above-standard production. On the other hand, the plan cannot be used very successfully where one operation or job depends upon another. In addition, because there is no upper limit on production, workers tend to overwork themselves, causing an increase in the chance of injury or a loss in product quality.

Piece Rate with a Shortened Base

Under this payment scheme, a worker receives the same guaranteed base earnings as the piece-rate with a base scheme, but the incentive pay (i.e., piece-rate earnings) starts before the standard productivity level is reached. A graphical representation of this scheme is given in Figure 5.8, and its mathematical model is as follows:

$$E = \begin{cases} b & \text{if } p_a \leq p_b \\ b(p_a/p_b) & \text{if } p_a > p_b \end{cases} \tag{5.6}$$

where p_b is the productivity level at which the guaranteed base earnings end and the incentive begins. The benefit of having the shortened base (i.e., incentive pay beginning at, say, 80 percent of standard productivity rather than 100 percent) is that workers

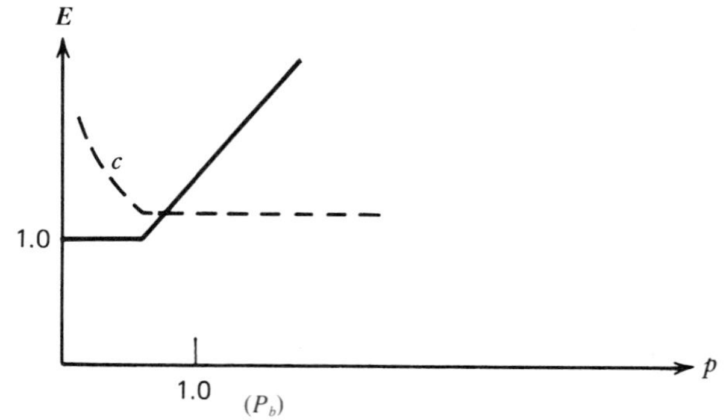

Figure 5.8 Piece-Rate with a Shortened Base Payment Scheme

who have the skills and so on to enable them to produce at a level slightly below standard will be motivated to attain their maximum productivity. Without the shortened base, those workers would react with the same motivation as if they were under a daywork plan, since they would have little chance of ever exceeding the 100 percent "jumping-off" point so that they could earn incentive pay. The choice of the base point p_b is obviously an important design parameter in this payment scheme.[4]

Gain Sharing

Figure 5.9 illustrates the type of payment scheme known as gain-sharing in which the monetary benefits of production beyond the standard level are shared between the worker and the firm. A percentage α of incentive earnings goes to the worker while the remaining incentive earning, $1 - \alpha$ percentage, goes to management. Gain-sharing can be applied to any type of piece-rate plan, such as the piece-rate with a base scheme or the shortened-base scheme. For example, the function relationship for the gain-sharing version of the piece-rate with a base scheme is:

$$E = \begin{cases} b & \text{if } p_a \leq p_b \\ b(p_a/p_b)\alpha & \text{if } p_a > p_b \end{cases} \tag{5.7}$$

The rationale of gain-sharing is often difficult to convey to workers. They tend to resent sharing the results of their additional productivity with management. Also, as indicated in Figure 5.9, gain-sharing schemes result in a variable unit cost that is an undesirable feature in operating a business.[5] These two features have diminished the utility of gain-sharing payment schemes so that they are not commonly used today.

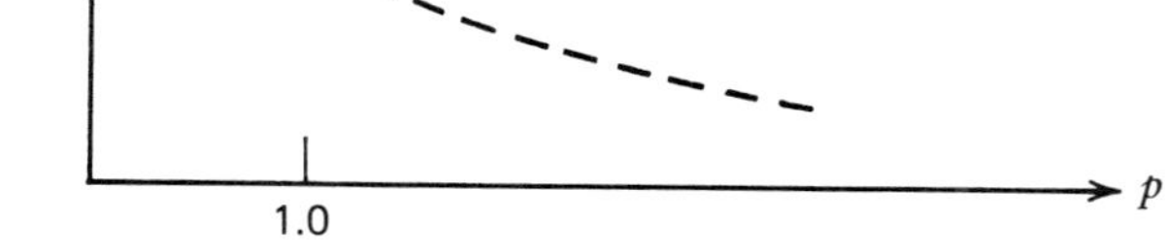

Figure 5.9 Gain-Sharing in a Piece-Rate with a Base Payment Scheme

---

[4]Note that as the base point is lowered, the unit cost for the same production level is increased, so that a trade-off between cost and motivational considerations is necessary in selecting p_b.

[5]Unit cost, which is the sum of direct labor cost and direct material cost and indirect labor or overhead cost, varies under any payment scheme, whenever the production quantity varies. Although direct labor unit cost is a constant in some schemes (see Figures 5.5–5.9), indirect cost on a unit basis varies as the production level varies.

5.2.2 Traditional Individual Wage Incentive Plans

The payment schemes presented previously are basic earnings versus productivity relationships that can be adopted by any firm and tailor-made to fit its specific situation and needs. Such an incentive plan would result when the firm ''parameterized'' the payment scheme model—that is, chose specific values for factors involved such as the productivity base level or the magnitude of a step increase. In addition, the firm might add qualifications or other provisions to the payment scheme, such as which jobs are covered by the plan or how any incentive earnings shared with the firm are to be used.

While a firm's total incentive plan is likely to be unique to that firm, there are a number of specific plans that have served as guides or ''model plans.'' Although most of these classic or traditional plans were developed prior to 1930, their logic and underlying motivational philosophy are still relevant today. For example, Fredrick W. Taylor's ''Differential Piece-Rate'' plan is a modified piece-rate scheme in which a relatively low rate (α) is paid for productivity below the standard level, while a relatively high rate (β) is paid for productivity beyond the standard. In addition, there is a step bonus (f)paid when a worker achieves standard productivity. The plan is shown graphically in Figure 5.10.

The Taylor plan is an example of a ''selective'' type incentive plan. That is, it was intentionally designed to encourage efficient workers and, at the same time, to discourage inefficient workers. This motivational effect results from the use of (1) the two significantly different piece rates and (2) the step increase at standard production, in conjunction with (3) a ''tight'' or high standard. A modification of the Taylor plan, called the Merrick Multiple Piece-Rate plan, uses three different piece rates in an attempt to overcome the negative motivational aspects of Taylor's plan for beginning and average workers.

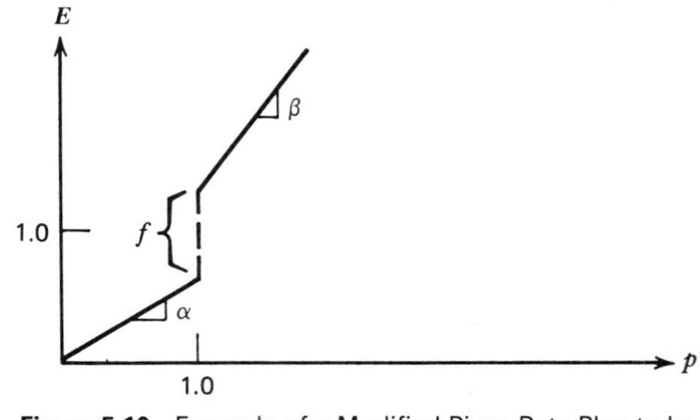

Figure 5.10 Example of a Modified Piece-Rate Plan to Include an Incentive Step

5.2.3 Group Incentive Plans

The Taylor plan is an example of a wage incentive plan designed specifically to be applied on an individual worker basis. However, it can also be applied to a group or team of workers by treating the productivity of the team as if it were the result of the efforts of a single worker, and applying the resulting earnings to each member of the group. An additional set of incentive plans that are specifically designed for group participation is the ''profit-sharing'' plans or ''personnel economic productivity'' plans [3, p. 254]. These plans generally have two basic features: a payment scheme whereby the group receives a bonus directly proportional to some measure of the firm's profitability for the time period in question, and a conscious attempt at promoting worker–management cooperation toward improving productivity.[6] The group in question can be just a small team of workers or all workers in a particular department. However, the more common application is on a plant-wide basis in which all direct and most indirect workers in the entire firm are treated as the group.

As in the case of individual incentive plans, there have been several classical or traditional group plans that have served as the basis or model for tailor-made plans for specific firms. These traditional plans include the Scanlon plan [3], the Rucker plan [1], the Lincoln plan [2], and the Kaiser plan [3]. These plans were initially developed for application on a plant-wide basis and are all similar in structure. Their primary differences are in the specific manner in which the bonus, standard, and so on, are calcualted, and the type of cooperative devices used.

5.2.4 Designing a Wage Incentive System

In engineering or designing a wage incentive system for a specific situation, a number of decisions must be made. Some of the considerations that should be taken into account in this design process are described below. To the extent possible, the decisions involved should be made in conjunction with the labor force or its union.

Individual versus Group Plan

In most situations, the incentive system can be applied either on an individual worker basis or on a group basis. Individual plans tend to have greater motivational value, but are more expensive to maintain and administer owing to the use of a specific standard for each individual worker's job. A group plan is usually best when the work is such that it is difficult to measure the contribution of individuals to the completion of the work, and when there is interdependence among work operations required to complete jobs. Other advantages usually cited [4] for group plans are that (1) less

[6] ''Profitability'' is being used very loosely here, since group incentive plans have general applicability in such service or nonprofit situations as hospitals [3], banks, maintenance crews, garbage collection, and police and fire protection. In such cases, ''profitability'' might be measured as cost savings, or improved service.

supervision is required, (2) new employee training time is reduced, (3) indirect labor can be included as part of the group and therefore readily covered by an incentive plan, (4) paperwork involved in administering the plan is reduced, (5) product quality is enhanced, (6) absenteeism is reduced, and (7) wages do not fluctuate from period to period as much. Some of the disadvantages associated with group plans are that (1) there is no check for individual efficiency, (2) high individual productivity is not encouraged, (3) good management is required, and (4) the possibility of personnel conflicts is greater.

An assembly line is a good example of a situation that is well suited to a group plan (not necessarily on a plant-wide basis). A job shop with small, varied jobs is an example of a good situation in which to use an individual plan. There are, moreover, many situations in which a combination of an individual and group plan is appropriate.

Choice of a Payment Scheme

The most basic component of an incentive plan is the wage payment scheme. Regardless of whether an individual or group basis is utilized, a payment scheme such as one of those represented in Figures 5.5 through 5.10 must be selected. The various types of schemes have different motivational effects, different unit cost effects, and different degrees of complexities. A worker is not likely to expend extra effort to be productive under a payment scheme that is too complex for him to understand and hence believe in. However, highly motivating aspects of a scheme usually add to the complexity of the plan. The designer of the plan must find a balance between motivational power and complexity.

Selection of a payment scheme involves not only the choice of the form or type of scheme, but also the specification of specific numerical values for the parameters of the scheme. For example, if the step scheme represented by Eq. (5.6) is chosen as the type of scheme to use, then a value must be specified for the productivity level P_b. There are many factors to consider in specific values for the parameters of a scheme, such as resulting motivational value and likely cash flow due to the expected earnings paid. One important consideration related to motivational effect is the relation of the scheme to the prevailing wage rate being paid by other companies in the vicinity. For instance, consider the numerical payment scheme shown in Figure 5.11. Because the incentive step is at 140 percent of standard, then only the most efficient workers will be able to meet this step. And since the earnings below the step are less than are being paid elsewhere, all but the very best workers will be discouraged and likely to seek work elsewhere.

Specification of a Standard

Perhaps the most critical parameter in a payment scheme is the standard level of productivity. Standards are not absolute values to be ''discovered.'' They are intended to be the productivity obtained by a *normal* worker under normal working conditions.

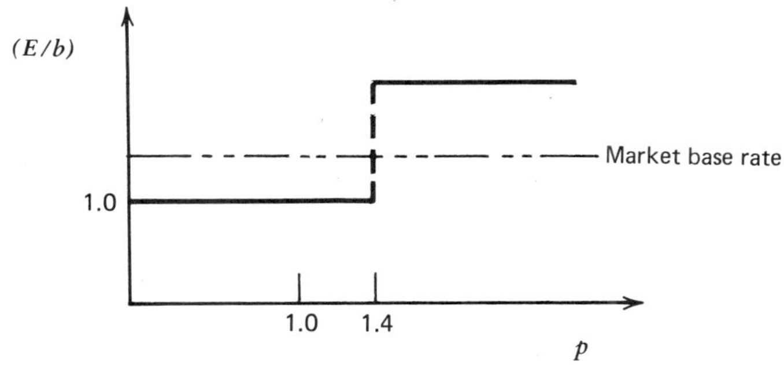

Figure 5.11 Payment Scheme in Relation to Prevailing Market Wage Rate

However, the definition of "normal" is up to the firm (often in conjunction with a union), and is a mechanism through which the firm can influence the value of the standard. Tight or high standards can be set by defining normal performance to be such that only few workers can attain that productivity. Loose or low standards can be set by using a liberal definition of normal such that, say, 90 percent of the workers assigned to the job can attain the resulting standard productivity. Other aspects of setting work standards are discussed in Chapter 4.

5.3 SUMMARY

Job evaluation is concerned with establishing the relative worth of a set of jobs. There are four basic approaches that can be used as alternative methods of evaluating jobs. They are (1) job ranking, (2) job classification, (3) the point method, and (4) factor comparison. The first two are considered to be qualitative approaches since they are based primarily on subjective input or data. The last two attempt to quantify the subjective factors involved in evaluating jobs and are thus called quantitative approaches. Regardless of the approach used, it is important to remember that it is the job, not the employee, that is being evaluated.

Wage incentive plans are intended to be a motivational device to encourage greater worker productivity. There are several components or aspects of an incentive system or plan, such as the type of payment scheme, the numerical values used for parameters of the scheme, suggestion systems, monitoring mechanisms, and employee performance feedback devices. There are alternatives that can be used for each of the components, such as the alternative types of payment schemes represented in Figures 5.5 through 5.10. The costs and benefits of each alternative should be examined before a specific component is included in the plan. The overall plan should be carefully engineered or designed to meet the objectives for the plan as established by the firm.

REFERENCES AND SUGGESTED READING

1. David W., Belcher, *Wage and Salary Administration*, Prentice-Hall, Englewood Cliffs, NJ, 1962.
2. Charles W., Brennan, *Wage Administration, Plans, Practices, and Principles*, Irwin, Homewood, IL, 1959.
3. J. D., Dunn and F. M., Rachel, *Wage and Salary Administration: Total Compensation Systems*, McGraw-Hill, New York, 1971.
4. M. R., Lott, "Wage Scales and Job Evaluation," Management and Administration, May 1925.
5. C. W., Lytle, *Job Evaluation Methods*, 2nd ed., Ronald Press, New York, 1954.
6. H. B., Maynard, *Industrial Engineering Handbook*, McGraw-Hill, New York, 1956.
7. Benjamin W., Niebal, *Motion and Time Study*, Irwin, Homewood, IL, 1972.
8. Carla S., O'Dell, *Gainsharing: Involvement, Incentives, and Productivity*, AMACOM, New York, 1981.
9. Robert E., Sibson, *Compensation*, AMACOM, New York, 1981.

REVIEW QUESTIONS AND PROBLEMS

5.1 What types of "models" are used in designing job evaluation plans? In designing wage incentive systems?

5.2 Discuss how the models mentioned in Question 5.1 can be validated in an actual application.

5.3 How is the activity of "work measurement" used in designing job evaluation plans? In designing wage incentive systems?

5.4 What is the purpose of job evaluation?

5.5 Define each of the following terms:

(a) Job analysis
(b) Job description
(c) Job evaluation manual
(d) Job ranking
(e) Job classification
(f) Point method
(g) Factor comparison method
(h) A job factor

5.6 What are the advantages and disadvantages of the job-ranking approach to job evaluation as compared to the point-method approach?

5.7 Develop a job evaluation plan for staff jobs in a university (these jobs include office secretaries, janitors, grounds keepers, campus police, and administrative staff).

5.8 A certain job factor has been assigned a relative importance of 30 percent in a particular point method of job evaluation. If 6 degrees are defined for this factor and the maximum number of points that can be awarded for the factor is 60, how many points would be assigned to each degree if (a) an arithmetic progression, (b) a percentage progression, and (c) a geometric progression were used? (See reference [6].)

5.9 The town of Blacksburg, Virginia, is developing a job evaluation point plan to determine equitable salaries for the town's police patrolmen, firemen, and garbage collectors.

(a) Construct a list of 10 or so job factors to be used in the plan.
(b) Assign a relative worth or value to each factor.
(c) Determine the appropriate number of degrees to be used for each separate factor.
(d) Assign points to each degree using a proportioning scheme that implies that the difference between any two degrees is 20 percent of the value of that factor.

1. Job Title LARGE HEADER SETUP OPERATOR 2. Department ______

3. No. on Job ______ 4. No. in Dept. ______ 5. Date ______

Statement of Job

The LARGE HEADER SETUP OPERATOR sets up and operates a battery of cold bolt headers using wire 3/8" diameter and over. Header operations include open and solid die work on one, two, or three blow headers. The LARGE HEADER SETUP OPERATOR must set up and adjust his machines and be able to handle a variety of materials and work to close tolerances. Occasionally tolerances of one one-thousandth of an inch must be held. Works from simple blueprints and usually has supervision on call when needed.

Duties

1. Sets up and operates a battery of cold bolt headers, using one, two, or three blows working on wire 3/8" diameter and over. Places and adjusts placement of hammers, solid or spring upset punches or hammers, and dies, solid or open. Occasionally must hold tolerances as small as one one-thousandth of an inch. Usual tolerance is two one-thousandths.
2. May have to touch up dies or hammers or change their shape or size slightly by polishing or scraping to achieve desired results. Frequently has to fit dies and hammers and adjust for wear of machine and errors in tool-

10. Contacts his supervisor for approval of the job and for further instructions on special and experimental products requiring more than usual supervision and instructions.
11. Checks production of previous shift to determine amount still to be run.

Worker Specifications

Normally two and one-half years of experience required to become familiar with the functions of the machine and to operate it competently. Tenth grade education or its equivalent is desirable. Must be able to understand simple drawings of product, use measuring instruments, and be able to understand all written instructions. Previous experience as a HEADUP JUNIOR SETUP OPERATOR is usually mandatory.

Figure 5.12 Example of a Job Description for a Factory Job (By Permission of McGraw-Hill)

5.10 Using the job description given in Figure 5.12, rate the job of "Large Header Setup Operator" using the point plan given in Figure 5.4.

5.13 Define each of the following terms.

(a) A wage incentive payment scheme
(b) Gain-sharing
(c) Daywork
(d) Piece-rate earnings
(e) Time bonus earnings
(f) Step incentive
(g) Group incentive plans.

5.14 Explain footnote number 4.

5.15 Explain the meaning of: The essence of any wage incentive plan is the functional relationship between performance and pay."

5.16 Develop a descriptive mathematical model of Taylor's Differential Piece-Rate wage incentive plan.

5.17 Design a wage incentive system for Post Office delivery men ("mailmen"). List any assumptions concerning the jobs that are necessary to develop your system. Describe your plan in terms of the major components of a wage incentive system.

5.18 The Davis Wire-Drawing Company hopes to increase the productivity of its wire-drawing department by implementing an individual wage incentive system for drawing machine operators. Operators receive a base rate of $4.30/hr and work an 8-hr shift. The incentive plan is based on a standard of machine utilization of 6 hours out of an 8-hr shift, and guarantees the base rate below 80-percent utilization. Between 80- and 100-percent utilization, the plan gives an operator an incentive pay of 2 percent of his base rate for each 6 percent increase in machine utilization achieved during the shift. Also, the plan calls for a step increase of 10 percent for utilization above the standard, as well as a gain-sharing provision in which 25 percent of incentive earnings beyond the standard goes to the firm.

(a) Give a graphical model of this plan.
(b) Give a numerical, mathematical model of the plan.
(c) What would be the shift take-home pay for a worker who keeps his machine busy for 7.5 hours?

5.19 Compare and contrast the incentive plan given in Problem 5.18 with Taylor's plan in terms of (a) motivational effect on highly productive workers, (b) motivational effect on new employees, and (c) direct labor component of unit cost.

Chapter 6

Engineering Economy[1]

6.1 THE NATURE OF ENGINEERING ECONOMY

Engineering economy involves the analysis of the economic worth of an activity or investment, and the use of this information to make decisions regarding these activities. This analysis usually centers around the costs and financial benefits of an activity such as a proposed project to be undertaken by a firm. The analysis attempts to reduce these costs and benefits to one or more measures that reflect the overall value of an activity. This enables a decision to be made as to whether the project merits investment or should a competing activity be chosen.

Engineering economy studies or problems have a broad scope of application. For example, a list of typical problems is given below.

1. If I invest a fourth of my monthly paycheck in a savings account that earns 6 percent interest, how much money will I have at the end of two years?
2. Is the land deal my friend wants me to invest in a better use of my money than the tax-free bonds in which this money is currently tied up?
3. Should we replace the manual drill press we are now using in the metal shop with a new, high-volume, automated drill press?
4. Our company's funds are limited; which of the available 15 projects should we select in which to invest these funds?

This chapter deals with how problems such as these are analyzed and resolved. The discussion begins by highlighting several concepts and principles that are fundamental in identifying and using the basic data for the analysis—the costs and benefits that are associated with the project or activity. The use of these data in resolving typical engineering economy problems is then discussed. This includes a description of several of the key principles of engineering economy, such as the time value of money. We

[1]This material was contributed by D. L. Kimbler.

also discuss the use of graphical models of the cash flows inherent in the cost data, as well as several criteria or measures of the overall value of these cash flows. Finally, we discuss the procedures used to compare several competing activities in order to select the best one(s) in which to invest.

6.2 COST CONCEPTS

A key to a useful analysis of project costs is *relevance*. While it may seem obvious that only relevant costs need to be considered, the idea of discarding all data that will not have any bearing on the solution is one that is frequently overlooked. A principle of engineering economy is that only those data items which will affect the solution should be included in the analysis. This idea is extended to the philosophy that only differences between alternatives need to be considered. These differences usually come down to cost, making project costs of the utmost importance in economic analyses.

Before costs can be used in the analysis, they must be made available in a usable form. These data come, in part, from accounting information. The accounting function quantifies real costs into categories so that projects can be easily analyzed and compared. For example, the extent of labor involved in a project is measured by the labor costs. The other costs, such as material, equipment, and utilities, are tabulated and assigned to projects or cost centers. Accounting data are a very useful source of information for engineering economy studies. For this reason, some knowledge of accounting principles and terminology is important for a thorough understanding of engineering economy.

6.2.1 Accounting Costs

Most of the costs incurred in carrying out the business of an organization are usually recorded in the accounting books of the firm. Manufacturing costs are usually easily identified in the accounting system. Costs are associated with labor, materials, capital equipment, inventory, furniture, fringe benefits, and many other system components and are reported along these lines. In an operation with no associated tangible product, such as a service industry, the same classifications still apply. The point here is that the structure of the project or business defines the cost structure. In a manufacturing plant, the process line and its products are the focal points. Relevant costs will be defined in terms of products and manufacturing centers. Each cost category, such as labor or material, may be allocated to the products being manufactured or the services provided. Some companies will have comprehensive cost-accounting systems, where the cost of production is allocated to products, people, and machines at a highly detailed level. In the absence of a cost-accounting system, costs will still be accounted for, but will be aggregated within categories.

Costs may be classified by the kind of relation they have with production. Fixed costs are those that are incurred regardless of production rate. For example, the rent or building cost must be paid in order to remain open. It does not typically fluctuate with the production rate. The cost of basic machines and tooling, along with modifications to the space they occupy, is also a fixed cost. Variable costs are the costs that change with production volume. The cost of materials, utilities for machine operation, and production labor are variable costs. Some costs may be fixed or variable, depending on the environment. Ordinary machine maintenance is usually a fixed cost. When machines are highly utilized, however, the maintenance cost can change.

Direct costs are those attributable to the production effort, and overhead or indirect costs are further removed from production. For example, machine operator wages are direct costs. Wages of a supervisor's secretary would be overhead costs. The distinction here is between operating the production facility and the support function that must be present. In general, overhead is the cost of the support function necessary for business operation. Direct costs may be fixed or variable, but overhead costs are usually fixed.

6.2.2 Implicit Costs

While many of the necessary cost data are available from accounting records, some costs may not be recorded or may be subject to different interpretations. These implicit costs may be difficult to find and classify, but they may also be extremely important. Part of the industrial engineer's function in engineering economy is the identification and classification of the costs that may be implied by the operation of a project, but not otherwise accounted for. In fact, the idea of implicit costs is one of the things that brings economic analysis into the realm of engineering. The consideration of these costs requires the creativity and flexibility of analysis that may not be present in a rigid tabulation of accounting files.

One interpretation of implicit cost is as a source of profit. Many production functions are frequently regarded as cost centers. Quality control, for example, is frequently thought to generate costs with no direct contribution to value or profit. It is true that it is difficult to attach to each unit of production the value of its inspection. The alternative of no quality control, however, leads to the consideration of increased costs due to return of product, warranty service, and customer dissatisfaction. Maintenance cost can be considered from the same point of view, with lack of maintenance bringing about more frequent machine replacements. It is no coincidence that the design of quality control plans and maintenance policies is based on the costs involved.

Implicit cost can also be considered as a pro rata allocation of expenditure. Costs not directly associated with production, such as floor space and inventory, can be spread across units of production or time in the project analysis. In an analysis based strictly on accounting costs, these costs might be ignored. For instance, training is an area that may be considered as not contributing directly to profit and not associated with production. When the benefits of continuing training are considered in terms of productivity, it is obvious that some of the value of each unit of production depends on the skill levels of the operators. In this context, it makes sense to include training on a unit of production or unit time basis.

One additional kind of cost is important because it should *not* be included in the analysis. This is the "sunk cost." A sunk cost is a cost that has been incurred in the past that has no economic bearing on the future. Typically, sunk costs are related to the project under study and appear to affect it. For instance, consider an obsolete machine that is being replaced. It has a value based on its marketability, and this value can be important as a means of partly paying for the replacement. The machine's initial cost, however, is a sunk cost because it was incurred due to a previous decision. The new decision cannot result in the old machine never having been purchased. Many economy problems have related sunk costs, and these must be kept out of the analysis. A good rule-of-thumb in identifying sunk costs is whether money or something of value exchanges hands; if not, that item is a sunk cost and should not be included.

Example 6.1

As a continuing example of engineering economy principles, the case of the hypothetical company Amalgamated Conglomerate Enterprises (ACE Inc.) will be analyzed from different points of view. One of the ACE subsidiaries is a manufacturer of material-handling equipment. A matter of some concern is whether to replace some manual transfer operations in the plant with robotic loaders and transfer machines. The starting point in the analysis is the identification and classification of pertinent costs. Obviously, labor cost is a potential savings if the manual content of the operation is to be reduced. In this case, labor is probably the only direct cost involved.

Indirect costs associated with the robotic handler must be carefully enumerated if the analysis is to be meaningful and accurate. The addition of equipment will require additional maintenance time and maintenance spares, both of which may be classified as overhead. Depending on the sophistication of the system, additional training may be required, especially during the system startup. These costs indicate that the timing of cash flows may be important; maintenance costs will be incurred during the operational life, but training must be done in order to start the system. Another initial cost might be special handling fixtures, depending on the unit load that the robot will handle. These fixtures may require capital expenditures, increasing the first cost of the system. If the robot selected is electrically powered, utilities may be considered as a variable indirect cost; if it is powered by shop air, utilities may be insignificant.

These costs and their possible classifications are summarized in Table 6.1. This summary should be regarded as the starting point in the problem analysis. ■

Table 6.1 Summary of costs for ACE

Cost	Direct	Overhead	Fixed	Variable	Capital	Expense
Labor	X	X	X			
Training		X	X			X
Maintenance		X	X			
Utilities		X	X	X		
Fixtures			X		X	
Robot			X		X	

6.3 TIME VALUE OF MONEY

As illustrated in the ACE case (Example 6.1), the specific time at which various costs are incurred can have a bearing on the economic evaluation of a project. This idea that money has some worth other than its face value is fundamental to engineering economy and is the beginning of the general concept of utility. One can think of utility as a subjective worth, depending on one's attitudes and goals. An ocean-going yacht has little utility in Kansas, even though it might be manufactured there. Value is added through transportation, raising the utility and cost of the yacht by bringing it to its point of use. The yacht itself has little utility to someone uninterested in sailing. To this individual, the market value of the yacht in cash has more value than the yacht itself. To a business, money today has more value than the same sum tomorrow because of the profit that can be made through its investment. The difference in the value of money through the passage of time can be described by an interest rate, which is one way of expressing its time value.

Probably the most obvious example of the time value of money is the interest charged for a loan. A lender provides a sum of cash, in exchange for repayment of a larger sum over a period of time. The extent to which the payments exceed the initial sum is determined by the interest rate on the loan. Although this rate is fixed by the lender, the borrower has his own, more flexible idea about time value. If the borrower can predict or estimate the profit in proportion to investment, and express the increase in value by an interest rate, a direct comparison can be made to determine whether borrowing money to engage in business is profitable.

Another useful example of time value of money is a savings account. Suppose an amount P is deposited in an account earning money at a rate i per period. At the end of the first period, the amount available has increased to $P(1 + i)$. Simple interest means that the amount of increase depends only on the initial amount, and the amount in the account, assuming no withdrawals, is expressed as:

$$\begin{aligned} F &= P + Pi + Pi + \ldots \\ &= P(1 + ni) \end{aligned} \tag{6.1}$$

where n is the number of periods of earnings.

Simple interest, while useful for illustration, is seldom used. A more reasonable assumption would be that, since the interest of the first period is left in the account, then interest in the second period should be computed on the initial deposit plus earnings. This is called compound interest. The amount of time during which the basic earnings occur is called the compounding period. For the first period, the amount is still $P(1 + i)$. During the second period, however, the amount in the account earns interest, so after two periods the amount is $P(1 + i)(1 + i)$. The amount in the account after n compounding periods is:

$$F = P(1 + i)^n \tag{6.2}$$

where i is the rate per period.

In order to determine the amount of growth in the account, then, three quantities must be known: the initial deposit, the rate per period, and the number of compounding periods. The rate and its period go hand in hand, so it is important always to ensure that the interest rate in use is consistent with the length of the compounding period. A common example of this is bank charge cards. The stated interest rate might be 18 percent on an annual basis. Charges are made during the year, though, and interest is added monthly, at a periodic rate of 1.5 percent per month. The annual rate is sometimes referred to as the nominal rate, which is simply divided by the number of periods in the year to determine the periodic rate.

Another way to express interest when it is compounded other than annually is by the effective rate. The effective rate represents the true growth put on an annual basis. That is, if the monthly rate is 1.5 percent and this rate is used to compound an amount over a full year, the actual growth on an annual basis is 19.56 percent. These interest rates are simply related by the following identity:

$$(1 + i_e)^k = (1 + r)^{nk} \tag{6.3}$$

where i_e = effective rate
r = rate per period
n = periods per year
k = years.

When working on an annual basis with $k = 1$, the relationship is much simpler.

$$i_e = (1 + r)^n - 1 \tag{6.4}$$

In both cases, the nominal rate is literally translated as the name of the rate on an annual basis, and is simply $i = rn$. When the compounding is done annually, then i_e, i, and r are identical.

6.3.1 Cash-Flow Profiles and Interest Factors

As part of the methodical approach to problem description, cash-flow profiles are used to graphically describe transactions involved over a period of time. The cash-flow profile is simply a linear time scale with arrows representing sums of money. It is a graphical model of the costs and benefits involved in a project.

A typical convention is that arrows pointing up are positive (i.e., benefits) and arrows pointing down are negative (i.e., costs). Another convention that is often used is that all cash flows are represented as occurring at the *end* of a period rather than during the period. Figure 6.1 is a cash-flow profile for an investment of \$10,000 (negative, money going out) and revenues of \$4000, \$5000, and \$6500 at the ends of years 1, 2, and 3 (positive, money coming in). The annual time scale implies that the interest is compounded annually. The cash-flow profile is a good way to ensure that the interest rate is consistent with its compounding period. Simply draw the time scale first, using as its divisions the compounding period. Figure 6.2 shows the same cash flows as in Figure 6.1, but the implied interest rate is now semiannual.

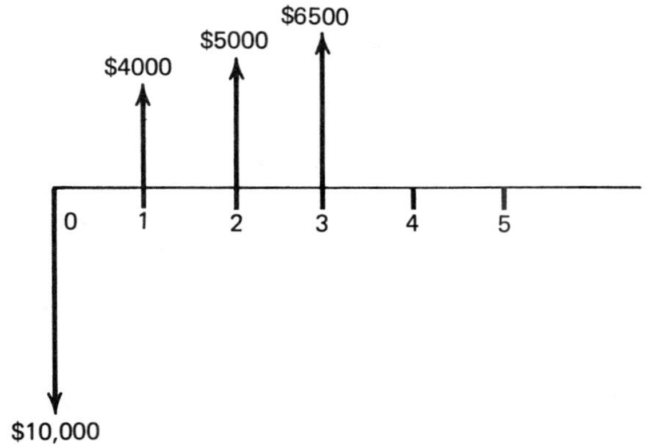

Figure 6.1 Cash-Flow Profile, Annual Interest

Many engineering projects result in specific patterns in the cash-flow profiles. One of the most common patterns is the annual series of cash flows illustrated in Figure 6.3. The initial investment P has some relationship with the annual amounts A. The relationship is not obvious in this cash flow, but if the annual series could be represented by an equivalent single sum at time 0, the comparison would be simple. Figure 6.4 is the same cash flow, with dotted arrows representing equivalent amounts. That is, the annual series over several periods has an equivalent value at some rate i when it is concentrated at a single point.

In comparing one project to another (as one cash-flow series to another), it is useful to convert a cash-flow pattern into a single sum at a prespecified point in time. If the single sums at $t = 0$ and $t = 5$ are equivalent, we already have a relationship between them. Equation (6.2) describes the growth of a single sum over n periods at a rate i, and says that the future amount F is the present amount P adjusted for interest and time. Stated another way, the term $(1 + i)^n$ makes F and P equivalent. This term

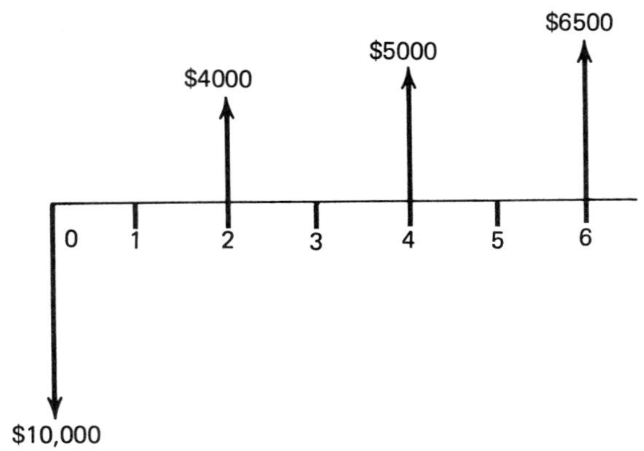

Figure 6.2 Cash-Flow Profile, Semiannual Interest

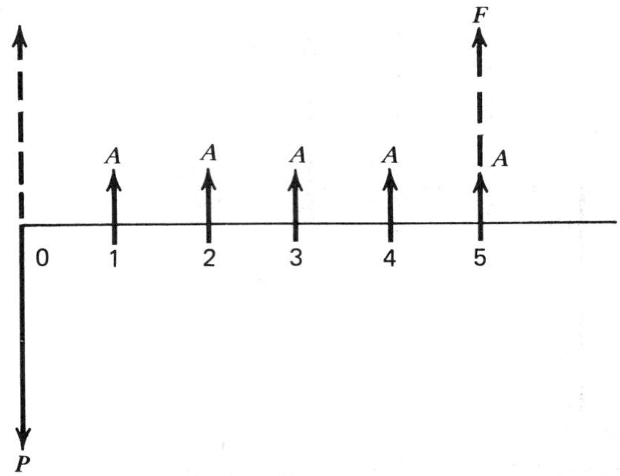

Figure 6.3 Uniform Series

is called the single-sum future worth factor, and is generally expressed as

$$(F/P, i, n) = (1 + i)^n \tag{6.5}$$

When Eq. (6.2) is solved for P, the result is the single-sum present worth factor

$$(P/F, i, n) = (1 + i)^{-n} \tag{6.6}$$

With these two factors, single sums may be moved from point to point in time and represented by their equivalent amounts, based on time and interest.

The annual series is a little more difficult. If each amount is treated individually and converted to an equivalent amount at $t = 0$, the present worth factor could be used:

$$P = A\sum_{k=1}^{n} (1 + i)^{-k} \tag{6.7}$$

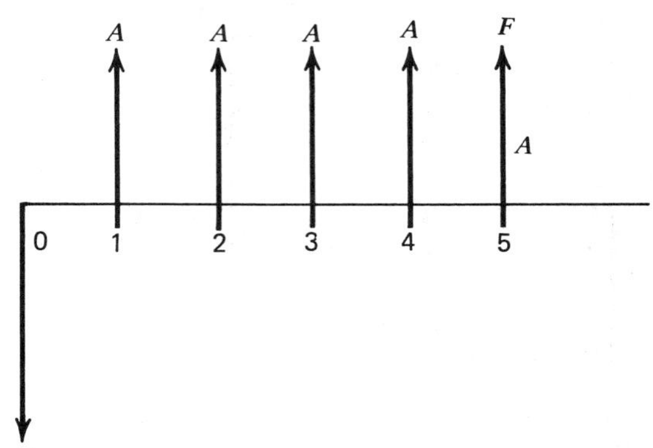

Figure 6.4 Equivalent Single Sums

This finite sum can be simplified and expressed as a single algebraic term,

$$\sum_{k=1}^{n} (1 + i)^{-k} = \frac{(1 + i)^n - 1}{i(1 + i)^n} \tag{6.8}$$

This formula is called the uniform series present worth factor, $(P/A, i, n)$, and the present equivalent of the series is computed by

$$P = A(P/A, i, n) \tag{6.9}$$

If Eq. (6.7) is solved for A, the result is the capital recovery factor,

$$(A/P, i, n) = \frac{i(1 + i)^n}{(1 + i)^n - 1} \tag{6.10}$$

By combining factors, the future equivalent F can be found for the uniform series:

$$(F/A, i, n) = (F/P, i, n)(P/A, i, n) \tag{6.11}$$

This is called the uniform-series future worth factor; its inverse, used to solve for A given F, is called the sinking fund factor:

$$(A/F, i, n) = (A/P, i, n)(P/F, i, n) \tag{6.12}$$

These factors and their formulas are summarized in Table 6.2. In practice, values for the factors are tabulated for selected interest rate and various time periods. Tabulated factors for compounding at 10 percent and 15 percent can be found in Tables 6.3 and 6.4.

Example 6.2

Cash flow A consists of amounts at $t = 1, 3, 4, 6$ of \$1000, \$3000, \$3000, and \$5000, respectively. Cash flow B consists of a four-year uniform series of \$5000 with amounts at end of year 3 through 6. Find the equivalent single sum at $t = 0$ for cash flows A, B, and $A + B$.

Table 6.2 Interest Factor Formulas

$$(P/F, i, n) = (1 + i)^{-n}$$

$$(P/A, i, n) = \frac{(1 + i)^n - 1}{i(1 + i)^n}$$

$$(A/P, i, n) = \frac{i(1 + i)^n}{(1 + i)^n - 1}$$

$$(A/F, i, n) = \frac{i}{(1 + i)^n - 1}$$

$$(F/P, i, n) = (1 + i)^n$$

$$(F/A, i, n) = i^{-1}[(1 + i)^n - 1]$$

$$(A/G, i, n) = i^{-1} - (n/i)(A/F, i, n)$$

$$(P/G, i, n) = (1 - (1 + ni)(1 + i)^{-n})i^{-2}$$

Table 6.3 Interest Factors for i = 10 percent

n	$(F/P, i, n)$	$(P/F, i, n)$	$(F/A, i, n)$	$(A/F, i, n)$	$(P/A, i, n)$	$(A/P, i, n)$	$(A/G, i, n)$	$(P/G, i, n)$
1	1.1000	0.9091	1.0000	1.0000	0.9091	1.1000	0.0000	0.0000
2	1.2100	0.8264	2.1000	0.4762	1.7355	0.5762	0.4762	0.8264
3	1.3310	0.7513	3.3100	0.3021	2.4869	0.4021	0.9366	2.3291
4	1.4641	0.6830	4.6410	0.2155	3.1699	0.3155	1.3812	4.3781
5	1.6105	0.6209	6.1051	0.1638	3.7908	0.2638	1.8101	6.8618
6	1.7716	0.5645	7.7156	0.1296	4.3553	0.2296	2.2236	9.6842
7	1.9487	0.5132	9.4872	0.1054	4.8684	0.2054	2.6216	12.7631
8	2.1436	0.4665	11.4359	0.0874	5.3349	0.1874	3.0045	16.0287
9	2.3579	0.4241	13.5795	0.0736	5.7590	0.1736	3.3724	19.4215
10	2.5937	0.3855	15.9374	0.0627	6.1446	0.1627	3.7255	22.8913
11	2.8531	0.3505	18.5312	0.0540	6.4951	0.1540	4.0641	26.3963
12	3.1384	0.3186	21.3843	0.0468	6.8137	0.1468	4.3884	29.9012
13	3.4523	0.2897	24.5227	0.0408	7.1034	0.1408	4.6988	33.3772
14	3.7975	0.2633	27.9750	0.0357	7.3667	0.1357	4.9955	36.8005
15	4.1772	0.2394	31.7725	0.0315	7.6061	0.1315	5.2789	40.1520
16	4.5950	0.2176	35.9497	0.0278	7.8237	0.1278	5.5493	43.4164
17	5.0545	0.1978	40.5447	0.0247	8.0216	0.1247	5.8071	46.5819
18	5.5599	0.1799	45.5992	0.0219	8.2014	0.1219	6.0526	49.6395
19	6.1159	0.1635	51.1591	0.0195	8.3649	0.1195	6.2861	52.5827
20	6.7275	0.1486	57.2750	0.0175	8.5136	0.1175	6.5081	55.4069
21	7.4003	0.1351	64.0025	0.0156	8.6487	0.1156	6.7189	58.1095
22	8.1403	0.1228	71.4028	0.0140	8.7715	0.1140	6.9189	60.6893
23	8.9543	0.1117	79.5430	0.0126	8.8832	0.1126	7.1085	63.1462
24	9.8497	0.1015	88.4973	0.0113	8.9847	0.1113	7.2881	65.4813
25	10.8347	0.0923	98.3471	0.0102	9.0770	0.1102	7.4580	67.6964

Table 6.4 Interest Factors for $i = 15$ percent

n	$(F/P, i, n)$	$(P/F, i, n)$	$(F/A, i, n)$	$(A/F, i, n)$	$(P/A, i, n)$	$(A/P, i, n)$	$(A/G, i, n)$	$(P/G, i, n)$
1	1.1500	0.8696	1.0000	1.0000	0.8696	1.1500	0.0000	0.0000
2	1.3225	0.7561	2.1500	0.4651	1.6257	0.6151	0.4651	0.7561
3	1.5209	0.6575	3.4725	0.2880	2.2832	0.4380	0.9071	2.0712
4	1.7490	0.5718	4.9934	0.2003	2.8550	0.3503	1.3263	3.7864
5	2.0114	0.4972	6.7424	0.1483	3.3522	0.2983	1.7228	5.7751
6	2.3131	0.4323	8.7537	0.1142	3.7845	0.2642	2.0972	7.9368
7	2.6600	0.3759	11.0668	0.0904	4.1604	0.2404	2.4498	10.1924
8	3.0590	0.3269	13.7268	0.0729	4.4873	0.2229	2.7813	12.4807
9	3.5179	0.2843	16.7858	0.0596	4.7716	0.2096	3.0922	14.7548
10	4.0456	0.2472	20.3037	0.0493	5.0188	0.1993	3.3832	16.9795
11	4.6524	0.2149	24.3493	0.0411	5.2337	0.1911	3.6549	19.1289
12	5.3503	0.1869	29.0017	0.0345	5.4206	0.1845	3.9082	21.1849
13	6.1528	0.1625	34.3519	0.0291	5.5831	0.1791	4.1438	23.1352
14	7.0757	0.1413	40.5047	0.0247	5.7245	0.1747	4.3624	24.9725
15	8.1371	0.1229	47.5804	0.0210	5.8474	0.1710	4.5650	26.6930
16	9.3576	0.1069	55.7175	0.0179	5.9542	0.1679	4.7522	28.2960
17	10.7613	0.0929	65.0751	0.0154	6.0472	0.1654	4.9251	29.7828
18	12.3755	0.0808	75.8364	0.0132	6.1280	0.1632	5.0843	31.1565
19	14.2318	0.0703	88.2118	0.0113	6.1982	0.1613	5.2307	32.4213
20	16.3665	0.0611	102.4436	0.0098	6.2593	0.1598	5.3651	33.5822
21	18.8215	0.0531	118.8101	0.0084	6.3125	0.1584	5.4883	34.6448
22	21.6447	0.0462	137.6316	0.0073	6.3587	0.1573	5.6010	35.6150
23	24.8915	0.0402	159.2764	0.0063	6.3988	0.1563	5.7040	36.4988
24	28.6252	0.0349	184.1678	0.0054	6.4338	0.1554	5.7979	37.3023
25	32.9190	0.0304	212.7930	0.0047	6.4641	0.1547	5.8834	38.0314

The present worths can be found individually and then added. For cash flow A,

$$P(A) = 1000(P/F, i, 1) + 3000(P/F, i, 3) + 3000(P/F, i, 4) + 5000(P/F, i, 6)$$

For cash flow B,

$$P(B) = 5000(P/A, i, 4)(P/F, i, 2)$$

Notice that the uniform series in cash flow B is first converted to a single sum at $t = 2$. This is then treated as a future sum with respect to $t = 0$. Using interest factors from Table 6.3 for $i = 10$ percent, the $t = 0$ sum of A and B is \$21,132.53. The reader should compute the intermediate results and confirm the single equivalent sums.

■

6.3.2 Other Interest Factors

Another common cash-flow pattern in projects is the gradient. This pattern, illustrated in Figure 6.5, is the result of each successive amount increasing by an additive constant. This is a common method of representing constant growth in a project or accounting for an aging machine, which may require maintenance expenditures that increase with use. The formulas in Table 6.2 for $(P/G, i, n)$ and $(A/G, i, n)$ are based on an important assumption about the timing of the cash flows. The time scale at $t = 1$ in Figure 6.5 has no associated cash flow. This is because of the convention for computing the gradient amount:

$$A_k = G(k - 1) \tag{6.13}$$

where A_k is the cash flow amount in period k. The amount at $t = 1$ is 0, and the series begins at $t = 0$. A present worth equivalent would therefore be placed one period before the first real amount in the gradient. This results in a simpler computational formula for the gradient series.

In some cases, the discrete flow assumption is not as descriptive as it should be. One of these cases is continuous compounding. Suppose that the number of compounding periods in a year were allowed to increase without bound. Then the length

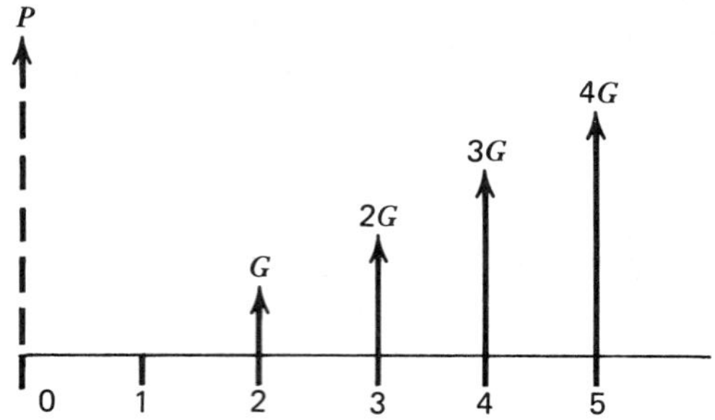

Figure 6.5 Gradient Series

of the compounding period would become infinitesimally small. These interest factors are based on the limiting relationship:

$$e^{in} = \lim_{k \to \infty} \left(1 + \frac{i}{k}\right)^{kn} \tag{6.14}$$

where i is the nominal interest rate, k is the number of periods in a year, and n is the number of years. The effective rate for continuous compounding is simply

$$i_e = e^i - 1 \tag{6.15}$$

In fact, continuous compounding factors may be derived from the discrete factors in Table 6.2 by replacing $(1 + i)$ with e^i. These factors are subscripted with an infinity symbol to denote continuous compounding.

Example 6.3

In anticipation of a steady growth in the stock room of ACE, a new forklift truck will be purchased each year for the next five years, at a cost of $8000 each. Operator salaries are expected to be $15,000 annually. At $i = 15$ percent, what is the present worth at $t = 0$ of the resulting cash flow?

The forklift purchases generate a uniform annual series of $8000 for 5 years. The annual salaries begin at $15,000 in year 1 and increase by $15,000 each year. It can be seen from the cash-flow diagram in Figure 6.6 that the annual totals can be described by an annual series of $23,000 and a gradient of $15,000. By including the first-year salary in the annual series, the remaining gradient has a present worth at $t = 0$. The equivalent present worth is

$$\begin{aligned} P &= 23000(P/A, 15, 5) + 15000(P/G, 15, 5) \\ &= 23000(3.3522) + 1500(5.7751) \\ &= \$163{,}727.10 \end{aligned}$$

Since the arrows point down, this amount represents a cost to the company. ■

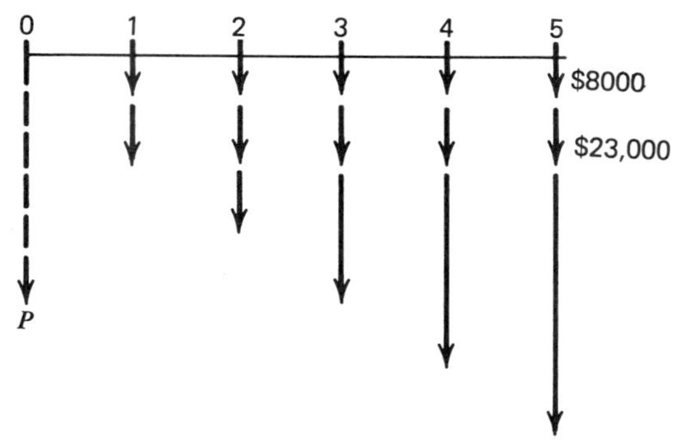

Figure 6.6 Cash Flow for Example 6.3

Example 6.4

Assume that the $8000 part of the cash flow in the previous example is subject to continuous compounding at 15 percent. What is the present worth of this series?

The uniform-series present worth factor can be modified by $e_i = 1 + i$. Then the continuous factor is

$$(P/A, r, n)_\infty = \frac{(e^{rn} - 1)}{e^{rn}\,(e^r - 1)}$$

At $r = 15$ percent, the factor is 3.2603. Then the present worth of the $8000 series is $26,082.40. Compare this with the discrete compounding amount from the previous example. ■

6.4 MEASURES OF ECONOMIC WORTH

One of the objectives of engineering economy is the development of simple descriptors of value. Ideally, the merit or worth of a project can be expressed as a single number, representing a single sum of money or an interest rate. By simplifying a cash-flow profile and expressing it as a single number, several projects can be compared, or a single project can be evaluated as to its value. The most common measures of worth are the present value of a project and the annual cost, both evaluated at some specified interest rate. An alternative is to assume that investment is equivalent to returned revenue less costs, and to find the interest rate that makes the equivalence statement true. This interest rate is the rate of return, or return on investment. Finally, other measures may be used in special cases, such as the benefit–cost ratio and the payback period. These measures of worth are described in the following section.

6.4.1 Present Value

The present value is one of the simplest economic measures. It is found by simply taking each element in the cash-flow profile and converting it, by the present worth factor, to an equivalent amount at $t = 0$. The sum of these amounts is the present worth of the entire cash flow, and is said to be equivalent to the cash flow on a present worth basis.

6.4.2 Annual Cost

A useful measure of merit is the equivalent uniform annual cost (EUAC). By convention, costs are taken to be positive, whereas returns are negative. The cash flow is converted to a uniform series that represents the equivalent annual outflow of cash.

Example 6.5

ACE is considering a project with the cash-flow profile shown in Figure 6.7. Find the present worth (P) and annual cost (EUAC) for this cash flow at $i = 10$ percent.

The cash flow begins at $t = -1$, but by convention P is placed at $t = 0$. The

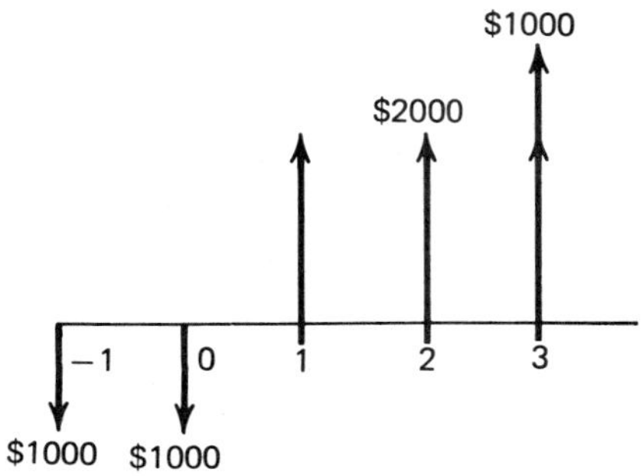

Figure 6.7 Cash Flow for Example 6.5

present worth calculation is

$$\begin{aligned} P &= -1000(F/P, 10, 1) - 1000 + 2000(P/A, 10, 3) + 1000(P/F, 10, 3) \\ &= \$4204.80 \end{aligned}$$

For the EUAC calculation, the length of the series must be known. For a series of three years, the calculation is

$$\begin{aligned} \text{EUAC} &= (1000(F/P, 10, 1) + 1000)(A/P, 10, 3) - 2000 - 1000(A/F, 10, 3) \\ &= -\$1457.69 \end{aligned}$$

Since part of the cash flow is already on an annual basis, it is added to the uniform equivalents at $t = 0$ and $t = 3$. Since the present worth is positive, by convention EUAC is negative. ■

6.4.3 Rate of Return

The rate of return of a project is the interest rate that describes the project's economic merit. For a simple cash flow in which there is an initial investment followed by a series of revenue cash flows, rate of return has a simple interpretation. It is the rate of interest that would have been earned on an initial deposit if the periodic cash flow represented withdrawals from a bank account.

As cash flows become more complicated, the definition of rate of return needs to be generalized. Rate of return (ROR) is the interest rate that makes negative cash flows equivalent to positive cash flows at a specified point on the time axis. The specific time point is selected to facilitate ROR calculation. Once two cash flows are determined to be equivalent at a value of t, they will be equivalent for all values of t. Rate of return can be found by solving the following equality:

$$\sum_{t=0}^{n} R_t(1 + r)^{n-t} = \sum_{t=0}^{n} C_t(1 + i')^{n-t} \tag{6.16}$$

This equality specifies equivalence of revenues (R_t) and costs (C_t) on a future worth basis, at $t = n$. It assumes that revenues are reinvested at the rate r as they are received. The rate of return is i'. Solving for i' involves finding the positive root of a polynomial of degree n, possibly by trial and error.

Another, older version of rate of return assumes that $i' = r$; that is, revenues are reinvested at the rate of return. This version, while useful, is not as realistic as Eq. (6.16), and has the more serious defect of not allowing a unique solution for i' when the cash flows alternate in sign. Rate of return in this textbook will refer to the interest rate found by solving Eq. (6.16) or its equivalent.

Example 6.6

A cash flow consists of an initial investment of $12,000 and uniform annual returns of $4000 for 4 years. At a reinvestment rate of 15 percent, what is the rate of return?

In a simple problem the ROR can sometimes be computed directly. First compute F, the future worth of revenue at 15 percent:

$$\begin{aligned} F &= 4000(F/A,\ 15,\ 4) \\ &= \$19{,}973.60 \end{aligned}$$

This is set equal to the future worth of investment at rate i':

$$\$19.973.60 = 12000(F/P,\ i',\ 4)$$

Now solve for the interest factor:

$$(F/P,\ i',\ 4) = 1.6645$$

and substitute the computational formula for future worth, Eq. (6.5). The rate of return can be found by simple algebra and logarithms:

$$\begin{aligned} 4 \ln (1 + i) &= \ln 1.6645 \\ \exp(\ln(1 + i)) &= \exp(0.25 \ln 1.6645) \\ 1 + i &= 1.1358 \\ i &= 13.58 \text{ percent} \end{aligned}$$

An alternative method is trial and error. In many cases the interest factors containing i' will be too complicated to allow a direct solution. The first step in this case is to assume a value for i' and compute the error E in the desired equality. In this case, assume $i' = 10$ percent. Then

$$\begin{aligned} \$19{,}973.60 + E &= \$12000(F/P,\ 10,\ 4) \\ E &= -2404.4 \end{aligned}$$

Now try a different interest rate, say 15 percent:

$$\begin{aligned} \$19{,}973.60 + E &= \$12000(F/P,\ 15,\ 4) \\ E &= 1014.4 \end{aligned}$$

The reversal of sign indicates that the true ROR is between 10 percent and 15 percent.

If the second trial did not result in a change of sign in E, the new E would be compared with the old E. If the new value is less, then you are moving in the right direction. Once i' is bracketed, it can be found by interpolation.

$$i' = 10 + \frac{5(2404.4)}{(2404.4 + 1014.4)}$$
$$= 13.52 \text{ percent}$$

This is close to the exact calculation of 13.58 percent. The difference is due to the nonlinearity of the interest factors. Error can be minimized by interpolating between interest rates that are as close as possible. If interest tables for 12 percent were available, interpolating between 12 percent and 15 percent would be preferred to using 10 percent and 15 percent. ■

6.4.4 Other Measures

Two related measures are the savings–investment ratio and the benefit–cost ratio. The savings–investment ratio is the ratio of equivalent single-sum revenues to equivalent single-sum cost, both using the same interest rate. If the ratio is greater than 1, then the rate of return would also be greater than the interest rate used in the calculation and the project would have a favorable evaluation.

The benefit–cost ratio has been used since the 1930s to evaluate public projects. The equivalent value of benefits is divided by equivalent cost, taking care that numerator and denominator are consistent as to interest rate and timing. The resulting real fraction can be used as a relative measure of worth.

The payback period is a measure of merit that ignores the time value of money. Although it is not mathematically elegant, it has a place in economic analysis, especially in short-duration projects. The payback period is simply the amount of time required to recover the initial investment completely at no interest. This measure is frequently used to evaluate small projects, especially in times of high interest or in companies with severely restricted cash reserves. In these cases the emphasis is put on short-term cash flow, not on long-term performance. While the payback period is sometimes criticized for being inexact, it has a more serious philosophical defect. The method frequently works in the short term, and its exclusive use can discourage a company from considering longer-term goals and conditions. This focus on short-term rather than long-term gains can be severely limiting on a small scale and is a major contributor to low productivity on a large scale.

6.5 COMPARISON OF ALTERNATIVES

One of the most powerful uses of engineering economy is in comparison of alternatives. Capital for expansion is typically a scarce resource, and it must be effectively allocated over alternatives for investment. Without a methodical approach free of inherent bias, the results of comparison can be misleading. This section deals with methods that, when properly used, lead to sound economic decisions for selection of alternative investments.

A common thread running through all engineering disciplines is that of consistency. For a comparison to be valid, the alternatives must be compared on an equal basis against a common set of conditions. The first step in the comparison process is the development of the conditions under which the comparison is valid. The analyst must define the environment of the problem to the extent that it can affect the solution, and then express each alternative within this environment.

An obvious first step in the analysis environment is the specification of the time value of money. The same interest rate must be used for all alternatives. Less obvious but just as important is the selection of a planning horizon. The planning horizon is essentially a snapshot of time during which the comparison is to take place. This may be a matter of policy; the company might always use a span of five years, for example. The horizon may be determined by the alternatives themselves. The longest or shortest project life may be chosen as a cutoff point. When all alternatives have the same life, the problem is simple. When lives differ from the planning horizon, some assumptions must be made to reconcile the difference.

Simplification of assumptions should be done whenever possible and valid, to avoid cluttering the problem with unnecessary numbers. One easy way to do this is to focus on differences between alternative cash flows. It is in their differences that some alternatives will be better than others. Subtracting one cash flow from another leads to finding the marginal advantage, which is really the objective of comparison.

Once the environment has been determined, the alternatives themselves are developed. A single alternative may consist of several projects combined into a single cash flow. It is important that alternatives be mutually exclusive, so that the selection of one will not imply the selection of another. A good methodical approach is first to develop all possible combinations of projects into alternatives. If there are m projects,

Table 6.5 Mutually Exclusive Alternatives

	Projects[a]		
Alternatives	I	II	III
Do nothing	0	0	0
A	0	0	1
B	0	1	0
C	0	1	1
D	1	0	0
E	1	0	1
F	1	1	0
G	1	1	1

[a]A "1" in a column denotes that the project in question is included in the associated alternative; a zero denotes that the project is not included in the alternative.

these can be arranged into 2^m alternatives, as shown in Table 6.5. An alternative that is easily overlooked is the "do nothing" alternative, which is adopted if no other alternative is sufficiently profitable.

Once all possible alternatives are known, infeasible ones can be eliminated. A limit on available capital, incompatibility between projects, and dependencies between projects can make alternatives infeasible. The remaining alternatives are then compared economically, and the best of the set is recommended. The actual methods of comparison are described in the following sections.

6.5.1 Alternatives with Equal Lives

A typical problem in engineering economy is that of selecting the best of several alternatives for investing a limited amount of capital over a fixed period of time. This problem could occur in an industrial setting in which the most profitable alternative is desired to accomplish a specific objective, such as plant expansion. If all alternatives have the same economic life, the planning horizon can be specified as the duration of the project and the cash flows compared without modification to bring them into a common time span. Care must be taken to ensure that this planning horizon is valid, and that the alternatives are consistent in all other respects.

Once the cash-flow profiles have been developed, alternatives may be compared, using one of the measures of economic worth described in Section 6.4. This comparison is illustrated in the following examples.

Example 6.7

ACE has been awarded a contract to produce mechanical hardware assemblies for the new HAL microcomputer. This contract is expected to run for five years, at which time the contract must be renegotiated. Alternative A involves a completely automated facility, with an initial cost of \$370,000, annual costs of \$150,000, and a salvage value of \$125,000 at the end of five years. Alternative B is a labor-intensive facility with an initial cost of \$125,000, annual costs of \$320,000, and salvage value of \$40,000. Alternative C is a partially automated compromise, with an initial cost of \$200,000, annual costs of \$190,000, and salvage value of \$90,000. At a minimum acceptable rate of return of 15 percent, which alternative is preferred?

Since each alternative is expected to generate the same revenue, the decision may be made by comparing costs. The ranking method may be used to compare equivalent uniform annual costs at 15 percent. The measures of merit for each alternative are:

$$\begin{aligned}
\text{EUAC(A)} &= \$370{,}000(A/P,\ 15,\ 5) + \$150{,}000 - \$125{,}000(A/F,\ 15,\ 5) \\
&= \$370{,}000(0.2983) + \$150{,}000 - \$125{,}000(0.1483) \\
&= \$241{,}833.50 \\
\text{EUAC(B)} &= \$351{,}355.50 \\
\text{EUAC(C)} &= \$236{,}313.00
\end{aligned}$$

At an interest rate of 15 percent, alternative C is preferred because it has the lowest equivalent annual cost. Ranking other amounts, such as present worth, will yield the same result. ■

Example 6.8

For the data in Example 6.7, select the preferred alternative using an incremental rate of return analysis. Assume that one of the alternatives must be selected.

The incremental method begins by ranking alternatives in ascending order of initial investment, BCA. Ordinarily, the rate of return would be found for alternatives until one is found with $i \geq$ MARR. This allows the "do nothing" alternative to be selected if none of the others are profitable. Since one of the three must be selected, we begin by using B as a basis for comparison and finding the incremental rate of return for cash flow $C - B$. Computing the future worth at $i = 15$ percent will tell whether the rate of return is greater or less than the MARR:

$$F(C - B) = -75{,}000(F/P, 15, 5) + 130{,}000(F/A, 15, 5) + 50{,}000$$
$$= \$775{,}657$$

At $i =$ MARR, this future worth would be zero, by Eq. (6.16). Since it is positive, alternative C is better than B, earning more than 15 percent on the additional investment. Repeating the procedure on cash flow $A - C$,

$$F(A - C) = -170{,}000(2.0114) + 40{,}000(6.7424) + 35{,}000$$
$$= -37{,}242$$

Since this amount would be zero at $i =$ MARR and positive if $i >$ MARR, the additional investment required to fund alternative A does not earn at least 15 percent. The preferred alternative is C. ■

6.5.2 Alternatives with Unequal Lives

In some cases the assumption of equal lives of alternatives is obviously not justified. A common example is a replacement problem where an existing machine with only a short useful life remaining may be replaced by one of several challenging machines with longer useful lives.

The classical approach to this problem is to use an annual cost comparison assuming a horizon of the least common multiple of lives. The drawback to this approach is its implicit assumption of repeated lives. Consider two alternatives, one with a three-year life and one with a five-year life. The classical solution is to compare annual costs, calculated over three years for one and five years for the other. If these costs are truly representative, then the planning horizon must have been 15 years (that is, the least common multiple of three and five). Further, the three-year cash flow must be exactly replicated four times and the five-year cash flow two times to fill out the 15-year horizon. This repetition of cash flows must occur if this approach is to be valid.

A preferred approach is the cash-flow approach, specifying every cash-flow element over a fixed horizon. The major assumption in this approach is one of estimated salvage value. Returning to the replacement lives of three and five years, one possible horizon is three years. If this horizon is used, longer-lived cash flows must have a salvage estimate at the three-year point, as if the project were terminated at the end of three years. If a five-year horizon is used, some assumption must be made about the fourth and fifth years of the equipment with a three-year life. In summary, all alternatives must be consistent in time scale and function.

Example 6.9

ACE has an injection molding machine that was bought 5 years ago and was expected to last for 10 years. The purchase price was \$26,000 and the expected salvage was \$8000. The machine has a present market value of \$14,000. Based on operating costs and production rate, this machine generates net revenue of \$35,000 annually. This machine can be replaced by a more modern machine costing \$40,000. The new machine can generate annual net revenue of \$40,000 annually. It has estimated salvage values of \$18,000 at end of year 5 and \$8000 at end of year 10. A more sophisticated machine can generate \$45,000 annually. It costs \$50,000 and can be salvaged for \$30,000 at end of year 5 or \$12,000 at end of year 10. At MARR = 15 percent, should the existing machine be replaced?

First, the relevant cash flows need to be defined. Note that the purchase price of the existing machine is not pertinent to this study, and the market value is pertinent only if the machine is sold. Cash flow E (for the existing machine) consists of annual revenue of \$35,000 for 5 years and salvage value of \$8000. The first challenger cash flow C_1 consists of an initial cost of \$24,000 (purchase price less sale of the old machine), annual revenue of \$35,000, and salvage value of \$18,000 in year 5. Challenger C_2 cash flow has an initial cost of \$36,000, revenue of \$45,000, and salvage in 5 years of \$30,000. (Note: Increased annual revenues are valid only if the company can utilize the increased capacity or take advantage of reduced costs. The five-year horizon is chosen because salvage estimates for the new machines are available.)

The incremental analysis begins by ranking the alternatives in order of investment, EC_1C_2. The next step is to find the first cash flow with a rate of return greater than 15 percent. Since cash flow E is all positive, it has an infinite rate of return. To find the future worth of $C_1 - E$ at 15 percent:

$$\begin{aligned} F(C_1 - E) &= -26{,}000(F/P, 15, 5) + 5{,}000(F/A, 15, 5) + 10{,}000 \\ &= -26{,}000(2.0114) + 5{,}000(6.7424) + 10{,}000 \\ &= -\$8{,}584.40 \end{aligned}$$

Challenger 1 is not preferred over the existing machine. For challenger 2,

$$\begin{aligned} F(C_2 - E) &= -36{,}000(F/P, 15, 5) + 10{,}000(F/A, 15, 5) + 22{,}000 \\ &= \$17{,}0103.60 \end{aligned}$$

So, the incremental investment for $C_2 - E$ yields more than 15 percent. The second challenger is the preferred alternative. ■

6.5.3 Incremental Analysis versus Ranking

Although ranking equivalent amounts for different cash flows is a logical procedure, the need for an incremental analysis of rate of return is not obvious. The key to understanding the need for incremental analysis is in the assumptions implicit in comparison of cash flows. If a fixed amount is available for investment, the entire amount must be accounted for. This assumption causes no problem in a ranking analysis, since the present worth (or annual or future) of an investment and its return is zero at i = MARR when the investment is made at MARR. This investment of the remainder is assumed whenever the initial cost of an alternative is less than the total available. In ranking analysis, the assumption is equivalent to adding a zero amount to each alternative.

When the measure of merit is rate of return, savings–investment ratio, or benefit–cost ratio, an incremental analysis must be used unless the initial investments for all alternatives are identical. The difference between alternatives' initial investments must earn at least the MARR, since it is assumed that any funds not invested in an alternative are invested to earn the MARR. For an illustration of incremental versus ranking analysis, see Problem 6.5.

6.6 TAX EFFECTS

Business enterprises are subject to many kinds of taxes. Some of these are on income (state and federal), real property, inventory, and capital gains. Taxes cannot be ignored in engineering economy studies, as they can have a substantial effect on actual cash flow. Tax payments result in cash transfers that may depend on project performance.

Actual tax liability depends on time, locale, and circumstance. Corporate federal (and some state) income taxes are based on a graduated scale up to a marginal rate of 46 percent for income over $100,000. Because the marginal tax is of prime importance in comparing alternatives, 46 percent is used here as the basic tax rate. The typical assumption in the absence of specific tax information is that the aggregate tax rate is 50 percent. This rate is assumed to include federal, state, and local taxes.

6.6.1 Taxes as a Cost of Doing Business

Taxes can best be incorporated into economic analyses by explicitly including them in cash flows. An analysis of before-tax cash flow can be useful, but the after-tax cash flow is a more realistic depiction of an alternative within the overall corporate environment.

An important concept is that of the contribution of negative taxes as a tax savings. Taxes are based on net corporate income: revenues less deductible expenses. If this taxable income is negative for the corporation, then no taxes are due. If, however, the taxable income for a project is negative but the corporation is otherwise profitable, the contribution of negative tax is to reduce other tax liabilities in profitable projects. This tax savings is attributed to the project that generates the savings, and should be included in the measure of merit of this project.

The proper handling of taxes is illustrated in Table 6.6. The first column defines the years in which cash flows occur, using an end-of-year (EOY) convention. The second column is simply the net before-tax cash flow (BTCF). The entry at year zero is an investment in capital equipment; following entries are net yearly revenues. Column three is the taxable income column. Note that the initial investment is not included. This is because there is no change in the company's assets; the investment is exchanged for equipment of equal value. In general, exchanges that do not cause a change in assets do not affect tax liability.

Taxes, in column four, are calculated at 50 percent of taxable income. The after-tax cash flow (ATCF) is then the before-tax cash flow less taxes. The assumption in this example is that the net yearly revenue is taxable. If nontaxable adjustments were made, then the after-tax amounts would be the before-tax amounts less taxes and adjustments. The after-tax cash flow can now be used in economic analyses just like the before-tax cash flow.

One of the more significant adjustments to before-tax income is for borrowed money. Interest on loans is deductible from income, but the principle payment is not. The taxable income calculations must then take into account the interest paid in each year. A common type of loan is one that is repaid in equal installments. The loan payments are just the uniform equivalent of the loan amount computed over the life of the loan at the loan rate. The interest amount in each year is given by

$$I_t = A[1 - (P/F, i, n - t + 1)] \tag{6.17}$$

where A is the annual loan payment and n is the life of the loan at rate i per period.

Example 6.10

A computer system costing \$500,000 is purchased with \$200,000 financed by a loan at 10 percent over 5 years. Expected revenues are \$100,000 annually, and operating costs are \$38,000 annually. Develop the after-tax cash flow.

Table 6.6 After-Tax Cash Flows

EOY	BTCF	Taxable	Taxes	ATCF
0	−500,000			−500,000
1	62,000	62,000	31,000	31,000
2	62,000	62,000	31,000	31,000
3	62,000	62,000	31,000	31,000
4	62,000	62,000	31,000	31,000
5	62,000	62,000	31,000	31,000

Table 6.7 Computations for Example 6.10

		Adjustments				
EOY	BTCF	Principle	Interest	Taxable	Taxes	ATCF
0	−300,000					−300,000
1	62,000	32,759	20,001	41,999	20,999	−11,759
2	62,000	36,035	16,725	45,275	22,638	−13,398
3	62,000	39,638	13,121	48,879	24,439	−15,199
4	62,000	43,601	9,159	52,841	26,420	−17,180
5	62,000	47,964	4,796	57,204	28,602	−19,362

Refer to Table 6.7. The before-tax cash flow includes all actual cash transactions other than loans. The annual loan payment at 10 percent is $52,760. Notice that the net in year 0 is $300,000, the purchase price less the loan amount. Two new columns, for loan adjustment, are included in the table. The new fourth column is used in computing taxable income in column five. Taxes are computed at 50 percent as before, based on taxable income. The loan makes a substantial difference in the after-tax cash flow, which is found by subtracting columns three, four, and six from column two. Compare this to Table 6.6, which is the same project with equity financing. ■

6.7 SUMMARY

Engineering economy is an aspect of industrial engineering that has applications in every other branch of the discipline. In both the public and private sector, the value of a project is an important basis for decision making. This chapter is intended to introduce the reader to the fundamentals of economic analysis of engineering projects. The principles presented here are a beginning in the study of engineering economy, but they form the basis for all of the extensions that can follow.

In summary, engineering economy studies begin with an objective and the need for data. These data typically come, at least in part, from accounting information. Although accounting data are static in time, it is necessary to analyze projects from both a time and a dollar point of view. The principle that money has a time value makes it possible to attach to cash flows a value based on both their amounts and their timing. The basis of engineering economy is thus the ability to develop descriptive cash-flow profiles and then, using time and interest relationships, to evaluate them.

When cash-flow profiles are analyzed on a consistent basis, they can be expressed as meaningful measures of worth, such as present worth and rate of return. The expression of a cash flow as a single value then allows direct comparison of alternatives on a consistent basis. For many engineers, comparison and ranking of alternatives is the most useful aspect of engineering economy. Many more complicated problems, such as replacement analysis and tax effects, are in reality just methods for modifying cash flows. The heart of engineering economy is the analysis of cash-flow profiles and comparison of economic alternatives.

Of necessity, some important topics must be left out of a limited discussion. For example, many problems involve cash flows that occur according to some probability distribution. Cost accounting and capital budgeting occupy a major part of economic analysis, as do decision theory and value analysis. Intricacies of tax law and depreciation merit several chapters alone. These topics, and more detailed treatment of topics covered here, may be found in the references for this chapter.

REFERENCES AND SUGGESTED READING

1. J. R. Canada, *Intermediate Economic Analysis for Management and Engineering*, Prentice-Hall, Englewood Cliffs, NJ, 1971.

2. W. J. Fabrycky and G. J. Theusen, *Economic Decision Analysis*, Prentice-Hall, Englewood Cliffs, NJ, 1974.

3. E. L. Grant, W. G. Ireson, and R. S. Leavenworth, *Principles of Engineering Economy*, John Wiley, New York, 1982.

4. J. L. Riggs, *Engineering Economics*, McGraw-Hill, New York, 1982.

5. J. A. White, M. H. Agee, and K. E. Case, *Principles of Engineering Economic Analysis*, John Wiley, New York, 1977.

REVIEW QUESTIONS AND PROBLEMS

6.1 An engineering student wishes to buy a microcomputer that costs \$850. If she saves \$50 per month for six months in an account that earns 1 percent per month, how much will she accumulate for a down payment?

6.2 Find the future worth at $t = 6$ of a 6-year uniform annual series of \$1000 at 10 percent.

6.3 Find the future worth at $t = 6$ of a gradient series of \$1000 at 10 percent.

6.4 The nominal interest rate is 21 percent compounded monthly. Find the monthly rate and the annual effective rate.

6.5 \$20,000 is available for investment in alternatives A (\$10,000), B (\$15,000), and C (\$20,000). (a) Find the rate of return for each alternative for revenue of \$1770 (A), \$2441 (B), and \$3116 (C) occurring as annual series over 10 years at $r = 10$ percent. (b) Find the annual equivalents for the *uninvested* part of the \$20,000 for each alternative at $i = 10$ percent. (c) Add the amounts in part b to the revenues in part a to find the composite revenues. Using these amounts, repeat part a. What does this tell you about ranking rate of return?

6.6 Projects A, B, C, and D have initial investments of \$50,000, \$70,000, \$75,000, and \$65,000, respectively. Only \$150,000 is available for investment. What are the mutually exclusive alternatives?

6.7 Given the following cash flows, select the preferred alternative at MARR = 15 percent. Use the present worth and rate of return methods.

EOY	A	B	C
0	−25,000	−42,000	−50,000
1–6	6,000	13,000	14,000
6	1,000	4,000	5,000

6.8 ACE has a machine that is used on a contract job that has three years to run. The machine has a present market value of $9000 and operating costs of $6000 annually. A new machine could be leased for three years for $3000 per year, with operating costs of $1000 per year. Should the old machine be sold and replaced by the lease machine? Use MARR = 15 percent.

6.9 Solve Problem 6.8 on an after-tax basis. How does the present worth of the decision change when taxes are considered?

SECTION THREE

OPERATIONS RESEARCH METHODOLOGY

Operations research involves the use of quantitative or mathematical models and procedures in the analysis and solution of decision problems. These problems are often concerned with the best way to *operate* a piece of equipment, a facility, or a system (a system being a collection of people, equipment, and facilities). However, these decision problems also include how to *design* equipment, facilities, or complete systems. For example, an excellent illustration of an OR problem is the derivation of the optimal design of a pipeline system to move coal slurry from a mining area to several power-generating plants scattered throughout a region. The problem involves determining the best pipe characteristics (diameter, thickness, etc.) as well as the best pipeline network (that is, which source/sink points to link directly, and the specific path to use in delivering each plant's needs).

The procedures used to analyze and solve OR problems involve direct and indirect application of basic mathematics—calculus, probability theory, and so on. Some OR analyses use these mathematical techniques by applying them directly to a model, such as using the first partial derivative to find the most economical order quantity in an inventory problem. Other OR analyses use algorithms or iterative, multistep procedures that involve a sequence of logic operations ("if this happens, then do this; if not, do . . ."). However, these individual logic operations normally involve use of basic mathematical procedures, such as basing the logic branching on whether the solution of an equation is positive or negative.

Note that many of the traditional IE topics discussed in Section Two also involve the use of basic mathematics. Determination of the proper sample size to use in a work sampling study involves basic probability and statistics. Derivation of the discount factor to use in converting a future cash flow into an equivalent annuity in engineering economy involves basic calculus. So, then, how do IE and OR methods differ with respect to the use of mathematics? Part of the answer is that OR techniques generally involve more complex, more sophisticated mathematical principles and procedures. Compare, for example, the theory underlying the optimization techniques discussed in Chapter 7 with the theory underlying engineering economy methods in Chapter 6.

You will find that the optimization material directly uses advanced calculus theory whereas the mathematical formulations of engineering economy methods are less rigorous (although very precise).

Another point to note in understanding the difference between the OR material in this section and the IE topics in Section Two is that *general* OR techniques and procedures, not application uses, are discussed in Section Three. The IE topics were oriented to specific analytical tasks or activities rather than to general purpose methods or techniques. For example, the discussion of work measurement in Chapter 4 concentrated on the philosophy of the task of measuring work and on how work can and is measured, not on the mathematical or nonmathematical methods underlying the "how." The method used in the derivation of the proper number of cycles to time in a stopwatch time study is not the major issue in work measurement. More central issues include how work elements can and should be defined, what allowances should be made, and how the overall time study should be put together.

The OR topics in this section are the fundamental *methods* used in the analysis of a wide spectrum of applications. For example, the classical optimization techniques discussed in Chapter 7 are the basic analytical tools used in applications such as economical control of inventory systems (Chapter 13), certain facility location problems (Chapter 11), and several approaches to production smoothing (Chapter 12). Section Four will discuss these application areas, with an emphasis on how OR techniques are used to solve the underlying decision problems, as well as on how traditional IE methods have been applied to the same problems. Section Three focuses on the general *techniques* themselves.

The methods described in the first chapter in this section are used to solve unconstrained optimization problems. That is, these techniques are used to derive the answer to a mathematical model of a decision problem in which there are no major constraints restricting the possible range of answers. In this type of problem, *optimization* often implies finding the best balance between several conflicting relationships—that is, finding the best trade-off among several variables.

Chapter 8, Mathematical Programming, is concerned with those techniques used to solve mathematical models of decision problems in which one or more constraints are included in the model. These techniques, called algorithms, involve iterating through a set of steps, each of which is designed to help shed light on the solution to the model. Normally, each iteration of the collection of steps moves closer and closer to the final answer.

The problems addressed in Chapter 8 are general in nature—their mathematical model or representation can apply to a wide variety of actual decision problems or applications. For example, linear programming (LP) problems all have the same general form. However, depending on the structure of the specific application, the basic LP model can often be reduced to a simpler format, such as the classic transportation problem. The techniques and models discussed in Chapter 8 will apply to any math programming problem. *Application* of these techniques will be pointed out in the topics in Section IV, such as in solving facility design problems and in production planning.

The OR techniques described in Chapters 7 and 8 are used to address *deterministic* models. Two basic classes of methods used in analyzing *probabilistic* models will be discussed in Chapters 9 and 10. In Chapter 9, the application of probability theory to the analysis of waiting line or queueing situations will be discussed. Unlike mathematical programming, the techniques used in queueing theory are not directly aimed at solving a normative mathematical model of a decision problem. Instead, queueing theory is used to analyze *descriptive* models of a system in an attempt to get information about how the system operates. Sometimes, this information is desired in order to indirectly solve a decision problem. Other times, the information is an end in itself, such as knowing the average time a customer spends waiting for service at a bank.

Chapter 10 is concerned with another technique used to analyze descriptive models: simulation. Whereas queueing theory derives information about the descriptive model in question through closed form, rigorous mathematical analysis, simulation derives information through a "try and see" approach. Unlike queueing theory, simulation normally does not involve the actual construction of a mathematical model of the system being analyzed. Rather, a descriptive flow chart or a computer program is developed and functions as the simulation model. In many cases, simulation is an alternative technique to queueing theory for analyzing a given situation.

As you study the material in Section Three, note that the use of models is very direct and is an integral part of the methodology in question. This is in contrast to the traditional IE topics discussed in Section Two. There, the role of a model is very subtle and not always apparent. For instance, although it was not pointed out, each of the formulas for investment measures in Chapter 6, such as those for present worth and for future worth, were actually descriptive mathematical models. You will see that the development and manipulation of descriptive and normative mathematical models is the foundation of the OR techniques presented in this section. In fact, the reader should review the discussion of modeling in Chapter 2 before proceeding.

Chapter 7

Foundations of Optimization

7.1 THE MEANING OF OPTIMIZATION

In Chapter 2 we discussed the role of models in the analysis of organizational systems; application of these models was illustrated in subsequent chapters. In this and succeeding chapters we center much of our attention on the applications of mathematical models, with the exception of Chapter 10, which will deal with simulation models. In Chapters 11 to 14 we discuss the analysis of a variety of manufacturing and operating systems through the development of mathematical and simulation models that may be used to describe the behavior of those systems. In some cases the objective of the analysis consists of estimation of the value of the measure of system performance through the mathematical model under each of a variety of conditions to which the system may be subjected. However, the objective of the analysis may also require the identification of those values of certain variables under management control, termed *decision variables*, that will optimize a specified measure of system performance. The purpose of this chapter is to explore and illustrate some of the methods available for system optimization through the manipulation of mathematical models.

To place the optimization problem in perspective, let us consider an example.

Example 7.1

A company manufactures a particular product. The number of units produced each year, x, must be scheduled at the beginning of the year. The setup cost for production is \$85,000 no matter how many units are produced. The variable cost of production is \$27 per unit. Thus the cost of manufacturing x units, $C(x)$, is

$$C(x) = 85{,}000 + 27x \tag{7.1}$$

If $P(x)$ is the selling price per unit for units manufactured in lots of size x, then

$$P(x) = 40 + \frac{100{,}000}{x} \tag{7.2}$$

and the total revenue $T_R(x)$ from the sale of x units is

$$T_R(x) = 100{,}000 + 40x \tag{7.3}$$

The manufacturer realizes that the number of units sold is dependent upon the selling price of each unit, and any units unsold at the end of the year are disposed of at half price. Let $D[P(x)]$ be the demand for the product if each unit is sold at a selling price $P(x)$, where

$$D[P(x)] = \frac{100{,}000}{1 + P(x)} \tag{7.4}$$

or

$$D[P(x)] = \frac{100{,}000}{14 + 65{,}000/x} \tag{7.5}$$

The objective of the analysis is to determine the number of units produced per year, x, such that total annual profit is maximized.

Let $T_P(x)$ be the total annual profit realized from the production of x units. If, during the year, the demand is greater than the number of units produced, x, $T_P(x)$ is given by

$$T_P(x) = \left[\frac{T_R(x) - C(x)}{x}\right]x \qquad x < D[P(x)] \tag{7.6}$$

However, if demand does not exceed production, total annual profit is given by

$$T_P(x) = \left[\frac{T_R(x) - C(x)}{x}\right] D[P(x)] - \left[\frac{C(x) - \frac{1}{2}T_R(x)}{x}\right]\{x - D[P(x)]\} \tag{7.7}$$

Thus

$$T_P(x) = 15{,}000 + 13x \qquad x < \frac{100{,}000}{14 + 65{,}000/x}$$

$$= \left(\frac{15{,}000 + 13x}{x}\right)\left(\frac{100{,}000}{14 + 65{,}000/x}\right) - \left(\frac{35{,}000 + 7x}{2x}\right)\left(x - \frac{1000{,}000}{14 + 65{,}000/x}\right)$$

$$x \geqslant \frac{100{,}000}{14 + 65{,}000/x} \tag{7.8}$$

or

$$T_P(x) = 15{,}000 + 13x \qquad x < \frac{100{,}000}{14 + 65{,}000/x}$$

$$= \frac{100{,}000(65{,}000 + 33x)}{28x + 130{,}000} - 17{,}500 - 3.5x \qquad x \geqslant \frac{100{,}000}{14 + 65{,}000/x} \tag{7.9}$$

The expression in Eq. (7.9) is the mathematical model or objective function that describes the behavior of the measure of system performance, annual profit, as a function of the decision variable x, annual production. As already indicated, the objective of the analysis is to determine the value of x that will maximize annual profit. Since we have a mathematical expression that defines the value of annual profit as a function of x, the objective of the analysis can be achieved by manipulating the expression for $T_P(x)$ such that the value of x that maximizes $T_P(x)$ is identified. One such approach would consist of a trial-and-error solution where various values of x are substituted into Eq. (7.9) until that value is found that maximizes $T_P(x)$. This approach is illustrated in Table 7.1, where the optimum value of x is any value between 4840 and 4849 and the function $T_P(x)$ is shown graphically in Figure 7.1. However,

Table 7.1 Trial-and-Error Solution for the Value of x that Maximizes $T_P(x)$ in Example 7.1

x	$D[P(x)]$	$T_P(x)$	Remarks
1000	1266	$28,000.00	Maximum thus far
2000	2151	41,000.00	Maximum thus far
3000	2804	48,635.51	Maximum thus far
4000	3306	49,904.96	Maximum thus far
5000	3704	50,185.19	Maximum thus far
6000	4027	49,755.03	
5500	3873	50,045.77	
4500	3516	50,148.44	
5250	3791	50,135.83	
4750	3612	50,190.59	Maximum thus far
4875	3659	50,193.60	Maximum thus far
4938	3681	50,190.76	
4813	3636	50,193.56	
4844	3647	50,193.93	Maximum thus far
4860	3653	50,193.84	
4852	3650	50,193.91	
4836	3644	50,193.91	
4850	3649	50,193.92	
4838	3645	50,193.92	
4839	3645	50,193.92	
4840	3646	50,193.93	Same profit as x = 4844
4841	3646	50,193.93	Same profit as x = 4844
4842	3646	50,193.93	Same profit as x = 4844
4843	3647	50,193.93	Same profit as x = 4844
4845	3648	50,193.93	Same profit as x = 4844
4846	3648	50,193.93	Same profit as x = 4844
4847	3648	50,193.93	Same profit as x = 4844
4848	3648	50,193.93	Same profit as x = 4844
4849	3649	50,193.93	Same profit as x = 4844

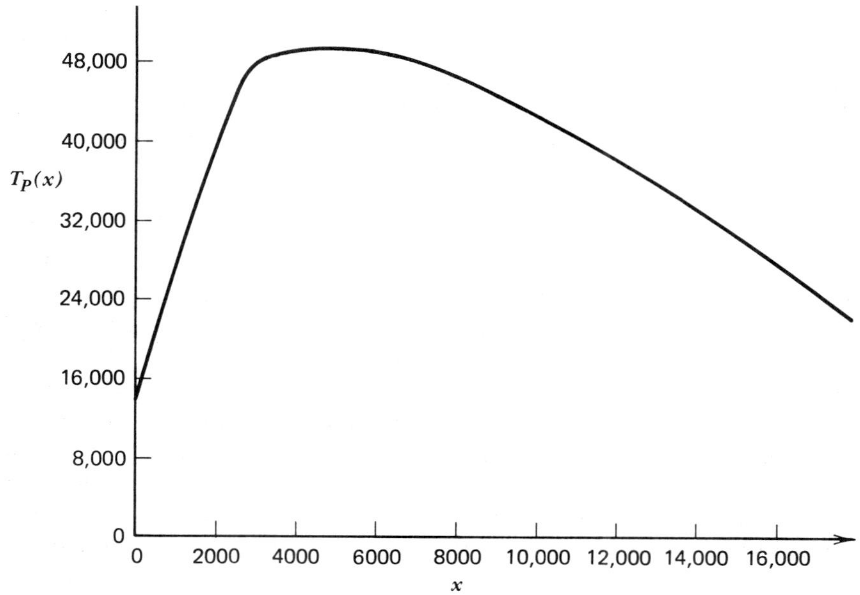

Figure 7.1 Profit Function for the System Described in Example 7.1

as we demonstrate in the remainder of this chapter, more efficient techniques are available for this undertaking. ■

The classical approach to the optimization of functions of continuous variables requires the use of the derivative. However, in some cases the function to be optimized is sufficiently complex that the classical approach becomes cumbersome to apply. An alternative to classical optimization procedures the analyst may wish to consider is a search technique. A search technique is an exploratory optimization procedure that attempts to find the optimum values of the decision variables by evaluating the measure of system performance at a systematically selected set of values of the decision variables. Based upon the value of the measure of system performance at each point in the search, the search procedure identifies the next point to be examined in the search. Proceeding iteratively in this manner, the search technique attempts to identify a set of values of the decision variables that is as close to optimal as possible.

7.2 FUNCTIONS OF A SINGLE UNCONSTRAINED CONTINUOUS VARIABLE

In this section we focus on the optimization of functions of continuous variables. The methods of analysis presented here are limited to functions of up to three variables, but these techniques can be extended to functions of an arbitrary number of variables without difficulty. These extensions do, however, require a limited knowledge of matrix algebra. For a general discussion of optimization of functions of continuous variables, the reader should see References 1, 3, 6, 10, or 12.

To put the problem we are addressing in perspective, let us consider an example. Suppose that the demand placed upon an electric power source $d(t)$ is a function of the time of year t and can be epxressed by

$$d(t) = 200 + 10000\,(0.5 - t)\sin(2\pi t) \qquad 0.0 \le t \le 1.0 \qquad (7.10)$$

and the points in time of maximum and minimum demand are to be determined. Demand $d(t)$ is shown graphically as a function of time t in Figure 7.2, where t is measured in years. For this function maxima occur at $t = 0.1771$ and $t = 0.8229$, whereas minima are found at $t = 0.0000$, $t = 0.5000$, and $t = 1.0000$.

A point at which a function attains either a maximum or a minimum is called an *extreme point*. Thus the points $t = 0.000, 0.1771, 0.5000, 0.8229$, and 1.0000 are extreme points for $d(t)$ as defined in Eq. (7.10). The problem we deal with here entails the development of a method for identifying the extreme points of a function and determining the nature (minimum or maximum) of each. The methods discussed here apply to *interior extreme points* only. If an extreme point is such that the function under consideration is defined throughout an open interval including the extreme point, then that extreme point is said to be an *interior extreme point*. Consider the extreme points at $t_1 = 0.1771$ (maximum), $t_2 = 0.5000$ (minimum), and $t_3 = 0.8229$ (maximum) for $d(t)$ in Eq. (7.10). In each case, we can define an open interval containing

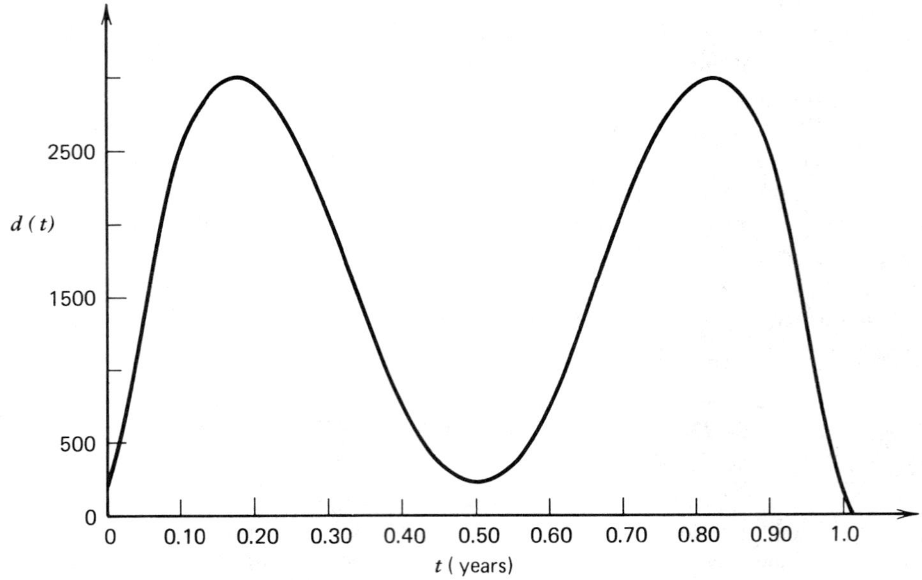

Figure 7.2 Demand for Electric Power $d(t)$ as a Function of Time

the extreme point such that $d(t)$ is defined throughout that interval. For example, if we choose the following intervals containing t_1, t_2, and t_3, respectively:

$$0.16 < t_1 < 0.18$$
$$0.49 < t_2 < 0.51$$
$$0.82 < t_3 < 0.83$$

we find that in each case $d(t)$ is defined throughout each. Since t_1, t_2, and t_3 are extreme points for $d(t)$ and since t_1, t_2, and t_3 each belongs to open intervals throughout which $d(t)$ is defined, t_1, t_2, and t_3 are interior extreme points for $d(t)$.

The points $t_0 = 0.0000$ and $t_4 = 1.0000$ were also defined as extreme points since each is a minimum for $d(t)$. However, $d(t)$ was defined only for $0.0 \leq t \leq 1.00$. Therefore, there exist no open intervals containing t_0 or t_4 such that $d(t)$ is defined throughout. Hence, t_0 and t_4 are not interior extreme points but are referred to as boundary points for $d(t)$.

To identify the interior extreme points of a function we must examine the behavior of the function at such points. To illustrate, consider the continuous function $f(x)$ in Figure 7.3. The interior extreme points for $f(x)$ are located at x_1 (maximum) and x_2 (minimum). To identify the interior extreme points of a continuous differentiable function, we may use the first derivative $f'(x)$. The derivative of a function $f(x)$ at a point x_0, $f'(x_0)$, may be interpreted as the slope of the line tangent to the function at x_0 or as the rate of change of the function at that point. To illustrate, consider the derivative of $f(x)$ in Figure 7.3 at x_1 and at points to the left and right of the interior

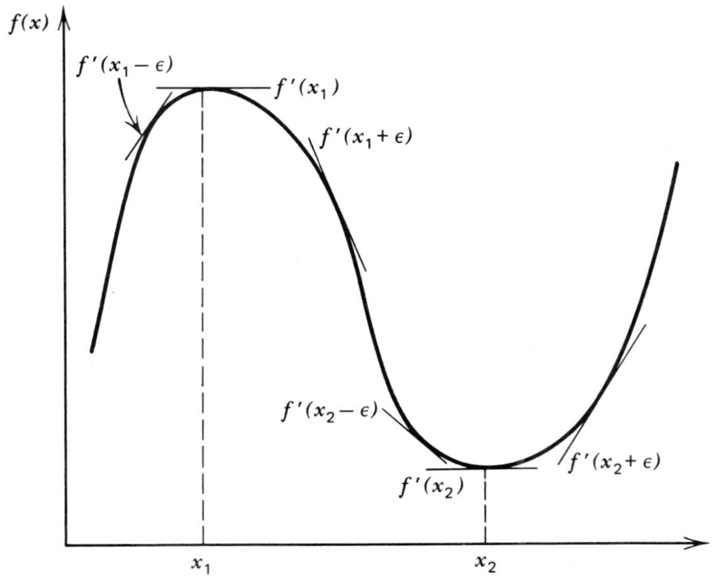

Figure 7.3 Interior Extreme Points for the Continuous Function $f(x)$

extreme point x_1 (maximum). For values of x slightly less than x_1, $f(x)$ increases; thus, the rate of change of $f(x)$ in this region must be positive. That is:

$$f'(x_1 - \epsilon) > 0 \qquad \epsilon > 0 \tag{7.11}$$

At x_1, $f(x)$ ceases to increase and starts to decrease. Hence, the rate of change in $f(x)$, measured by $f'(x)$, is negative for points sufficiently close to but to the right of x_1 and

$$f'(x_1 + \epsilon) < 0 \tag{7.12}$$

for at least some $\epsilon > 0$. Finally, at x_1 the function is neither increasing nor decreasing and

$$f'(x_1) = 0 \tag{7.13}$$

Now let us examine the behavior of $f'(x)$ at points near x_2, which is a minimum for $f(x)$. To the left of x_2 $f(x)$ decreases and

$$f'(x_2 - \epsilon) < 0 \tag{7.14}$$

at least for some $\epsilon > 0$. To the right of x_2, the function is increasing and

$$f'(x_2 + \epsilon) > 0 \tag{7.15}$$

At x_2 the function is neither increasing nor decreasing, and

$$f'(x_2) = 0 \tag{7.16}$$

From this discussion we may conclude that if $f(x)$ is a continuous, differentiable function of x and if x^* is an interior extreme point for $f(x)$, then

$$f'(x^*) = 0 \tag{7.17}$$

Since the condition in Eq. (7.17) is a requirement for the existence of an extreme point at x^*, Eq. (7.17) is called a *necessary condition* for a local maximum or minimum at x^*. Thus if the continuous, differentiable function $f(x)$ achieves a maximum and/or a minimum on the interior of the interval of definition of $f(x)$, then all such maxima and minima must satisfy Eq. (7.17).

An interior point x^* that satisfies Eq. (7.17) is called a *stationary point*. Every interior extreme point for a continuous, differentiable function is a stationary point, but every stationary point for such a function is not necessarily an interior extreme point. To illustrate, consider the function described in Figure 7.4. The point x^* satisfies Eq. (7.17) but is neither a maximum nor a minimum. Such points are called *points of inflection*. Thus, a stationary point may be an interior minimum, an interior maximum, or a point of inflection.

A function may possess an arbitrary number of stationary points. To illustrate, consider the function in Figure 7.5, where $f(x)$ possesses maxima at x_1, x_3, and x_6, minima at x_2, x_5, and x_7, and a point of inflection at x_4. When a function possesses several maxima (minima), each is referred to as a *local maximum* (*local minimum*). The point x' for which $f(x') \leq f(x)$ for all x on the interval of definition of $f(x)$ is

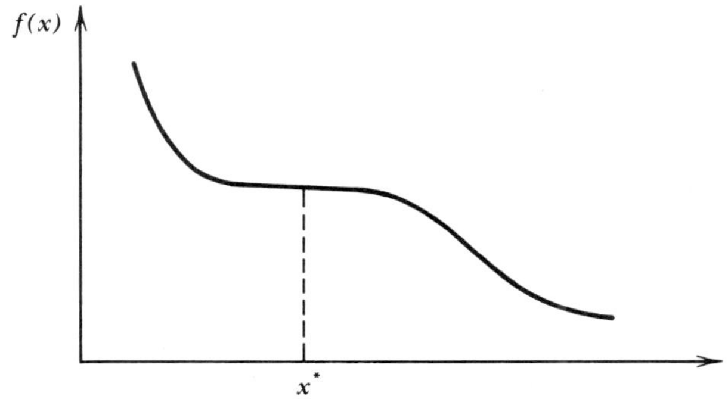

Figure 7.4 Point of Inflection for $f(x)$

called a *global minimum*. Similarly, the point x' for which $f(x') \geq f(x)$ for all x on the interval of definition of $f(x)$ is referred to as a *global maximum*. It should be noted that the global maximum (minimum) may or may not be an interior extreme point. For example, in Figure 7.5 x_0 is a global minimum for $f(x)$ on the closed interval $[x_0, x_8]$ but is not an interior extreme point. However, the global maximum for $f(x)$ on $[x_0, x_8]$ is x_3 and is an interior extreme point.

The above discussion suggests the sequence of steps necessary to identify the global maximum (global minimum). First, one must identify all stationary points for the function. From among the stationary points the analyst must identify those points that are local maxima and those that are local minima. Finally, the function is evaluated at each local minimum, each local maximum, and the boundary points. By comparing the values of the function at each local minimum, local maximum, and the boundary points, one can identify the global maximum and global minimum.

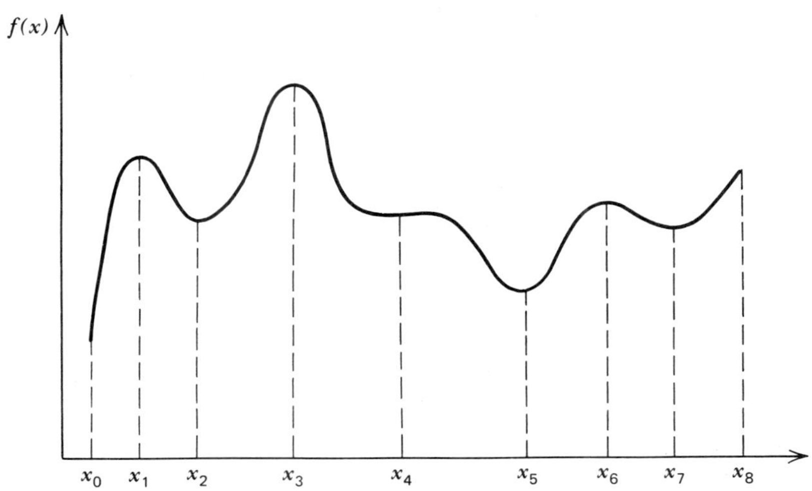

Figure 7.5 Function Possessing Several Local Maxima and Minima

Example 7.2

Find the stationary points for the following function.

$$f(x) = \frac{1}{4}x^4 - \frac{1}{3}x^3 + 5$$

Each stationary point for $f(x)$ must satisfy Eq. (7.17). That is:

$$\begin{aligned} f'(x) &= x^3 - x^2 \\ &= 0 \end{aligned}$$

Solving for the values of x for which $f'(x) = 0$ yields

$$\begin{aligned} x^3 - x^2 &= x^2(x - 1) \\ &= 0 \end{aligned}$$

Hence $x^* = 0$ and $x^* = 1$ are stationary points for $f(x)$. ■

Example 7.3

A manufacturer utilizes a particular chemical in the production of one of its product lines. However, the volume of the chemical required per month, x, is not known with certainty but has a uniform distribution between a and b gallons, where $0 < a < b$. The volume of the chemical purchased at the beginning of the month is s. If the requirement per month, x, is less than the quantity purchased, s, then the oversupply is disposed of at a cost $c_1(s - x)$. On the other hand, if x exceeds s, the cost of underestimation is $c_2(x - s)$. The expected cost of operating the inventory system, $C(s)$, is given by:

$$C(s) = \frac{1}{2(b - a)}[s^2(c_1 + c_2) - 2s(c_1 a + c_2 b) + a^2 c_1 + b^2 c_2] \tag{7.18}$$

Find the value of s that will minimize the cost of operating the inventory system.

The first step in finding the value of s that minimizes $C(s)$ is to identify the stationary points for $C(s)$. From Eq. (7.17):

$$\begin{aligned} C'(s) &= \frac{1}{2(b - a)}[2s(c_1 + c_2) - 2(c_1 a + c_2 b)] \\ &= 0 \end{aligned} \tag{7.19}$$

Solving for s yields:

$$s^* = \frac{c_1 a + c_2 b}{c_1 + c_2} \tag{7.20}$$

Since s^* is a stationary point it may be a maximum, a minimum, or a point of inflection. To identify the nature of s^* we evaluate $C(s)$ at s^*, $s^* - \epsilon$, and $s^* + \epsilon$, where ϵ is an arbitrarily small positive number. If $C(s^* + \epsilon) > C(s^*) < C(s^* + \epsilon)$, s^* is a minimum. If $C(s^* - \epsilon) < C(s^*) > C(s^* + \epsilon)$, s^* is a maximum. Finally, if $C(s^* - \epsilon) < C(s^*) < C(s^* + \epsilon)$ or $C(s^* - \epsilon) > C(s^*) > (s^* + \epsilon)$, s^* is a point of inflection.

Since s^* is given by Eq. (7.20), $C(s^*)$ is given by:

$$C(s^*) = \frac{c_1 c_2 (b - a)}{2(c_1 + c_2)} \tag{7.21}$$

Now, substituting $s^* - \epsilon$ into Eq. (7.18) we have:

$$C(s^* - \epsilon) = \frac{c_1 c_2 (b - a)}{2(c_1 + c_2)} + \epsilon^2 \frac{(c_1 + c_2)}{2(b - a)} \tag{7.22}$$

Finally, substituting $s^* + \epsilon$ into Eq. (7.18) yields:

$$C(s^* + \epsilon) = \frac{c_1 c_2 (b - a)}{2(c_1 + c_2)} + \epsilon^2 \frac{(c_1 + c_2)}{2(b - a)} \tag{7.23}$$

Since $\epsilon^2[(c_1 + c_2)/2(b - a) > 0$,

$$C(s^* - \epsilon) > C(s^*) < C(s^* + \epsilon) \tag{7.24}$$

and s^* as given in Eq. (7.20) minimizes $C(s)$. Since s^* is the only interior extreme point for $C(s)$, the boundary points for $C(s)$ must be maxima. ■

As we have already mentioned, a stationary point is not necessarily an extreme point. Thus, having determined the location of the stationary points we are faced with the problem of determining which of these points are extreme points and then determining which of the extreme points are local maxima and which are local minima. While this determination can be accomplished experimentally as illustrated in Example 7.3, more efficient methods are available.

To determine the nature of an interior extreme point, we use the second derivative of the function evaluated at the extreme point. To illustrate, consider the function

$$f(x) = 8x^3 - 30x^2 + 36x - 12 \tag{7.25}$$

The first and second derivatives of $f(x)$, $f'(x)$ and $f''(x)$, respectively, are given by:

$$f'(x) = 24x^2 - 60x + 36 \tag{7.26}$$

$$f''(x) = 48x - 60 \tag{7.27}$$

The functions $f(x)$, $f'(x)$, and $f''(x)$ are shown graphically in Figure 7.6. As we expect, $f'(x) = 0$ at the interior extreme points $x^* = 1.0$ and $x^* = 1.5$. Now let ϵ be a small positive number such that $x^* \pm \epsilon$ is a point quite close to x^*. In Figure 7.6 note that $f''(1.0) < 0$ and $f''(1.5) > 0$. This would imply that if x^* is a stationary point and $f''(x^*) < 0$, then x^* must be a local maximum. On the other hand, if x^* is a stationary point such that $f''(x^*) > 0$, then x^* must be a local minimum. It should be noted, however, that the converse is not necessarily true. The point x^* may be a stationary point such that $f''(x^*) = 0$. In this case, x^* may be a local maximum, a local minimum, or a point of inflection. Hence, if $f''(x^*) = 0$ we are unable to identify

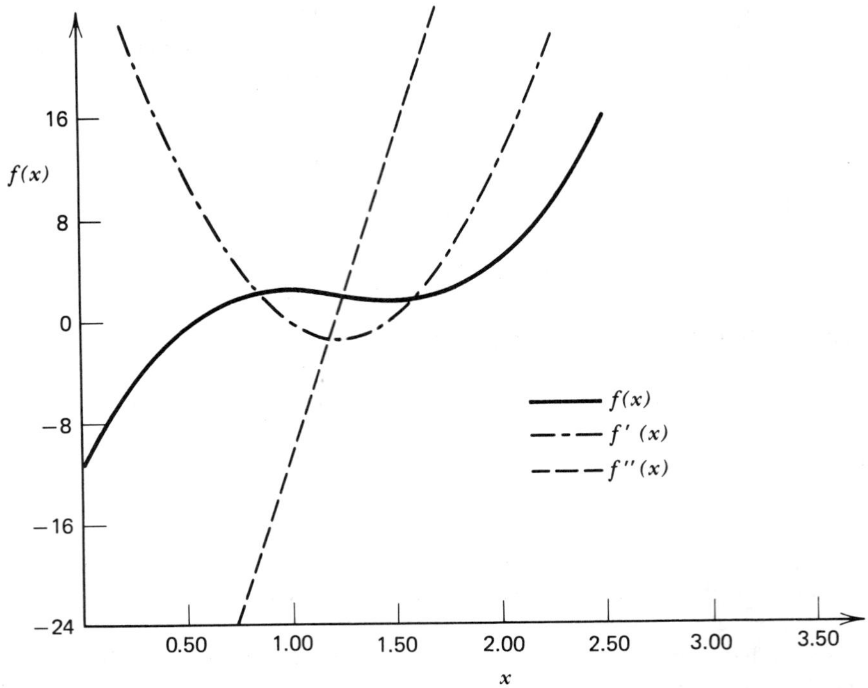

Figure 7.6 Graphs of $f(x)$, $f'(x)$, and $f''(x)$ Where $f(x)$ Is Defined by Eq. (7.25)

the nature of the stationary point x^*. In cases such as this, we examine successively higher-order derivatives of $f(x)$ evaluated at x^* until we find a value of i such that $f^i(x^*) \neq 0$, $i \geq 2$. If i is even and $f^i(x^*) < 0$, then x^* is a local maximum. If i is even and $f^i(x^*) > 0$, then x^* is a local minimum. If i is odd, x^* is a point of inflection. Since the condition $f^2(x^*) = f^3(x^*) = \cdots = f^{i-1}(x^*) = 0$ and $f^i(x^*) < 0$ for even i leads to the conclusion that x^* is a local maximum for $f(x)$ if $f'(x^*) = 0$, this provision is referred to as a *sufficient condition* for the existence of a local maximum at x^*. In a similar manner the provision that $f^2(x^*) = f^3(x^*) = \cdots = f^{i-1}(x^*) = 0$ and $f^i(x^*) > 0$ for even i if $f'(x^*) = 0$ is a sufficient condition for a local minimum at x^*. Finally $f^2(x^*) = f^3(x^*) = \cdots = f^{i-1}(x^*) = 0$ and $f^i(x^*) \neq 0$ for odd i is a sufficient condition for a point of inflection at x^*. The necessary and sufficient conditions for the existence of local maxima, local minima, and points of inflection are summarized in Table 7.2.

Example 7.4

Determine whether each of the stationary points for the function in Example 7.2 is a local minimum, local maximum, or a point of inflection.

The stationary points for the function in Example 7.2 are $x^* = 0$ and $x^* = 1$.

Table 7.2 Conditions for Identification of the Nature of a Stationary Point x^* for the Function $f(x)$

Necessary Condition	Sufficient Condition	Conclusion
$f'(x) = 0$	$f'(x^*) = f^2(x^*) = \cdots = f^{i-1}(x^*) = 0$ $f^i(x^*) < 0$ i even	Local maximum at x^*
$f'(x^*) = 0$	$f'(x^*) = f^2(x^*) = \cdots = f^{i-1}(x^*) = 0$ $f^i(x^*) > 0$ i even	Local minimum at x^*
$f'(x^*) = 0$	$f'(x^*) = f^2(x^*) = \cdots = f^{i-1}(x^*) = 0$ $f^i(x^*) \neq 0$ i odd	Point of inflection at x^*

To determine the nature of these stationary points we evaluate successively higher-order derivatives until a value of i is found such that $f^i(x^*) \neq 0$. Now

$$f''(x) = 3x^2 - x$$

and

$$f''(0) = 0 \qquad f''(1) = 2$$

Since $f^i(1) > 0$ for $i = 2$, $x^* = 1$ is a local minimum for $f(x)$. However, since $f''(0) = 0$ we cannot determine the nature of the stationary point $x^* = 0$ at this point. Evaluating $f^3(0)$ leads to

$$f^3(x) = 6x - 1$$

and

$$f^3(0) = -1$$

Since $f^i(0) \neq 0$ for odd i, $x^* = 0$ is a point of inflection for $f(x)$. ■

Example 7.5

Using the sufficient conditions presented in Table 7.2, determine whether the stationary point s^* in Example 7.3 is a local minimum, local maximum, or point of inflection.

The second derivative of $C(s)$ as defined in Example 7.3 is given by

$$C''(s) = \frac{c_1 + c_2}{b - a}$$

Since $C''(s) > 0$ for all s including s^*, s* as defined by Eq. (7.20) is a local minimum for $C(s)$. ■

Example 7.6

Identify the interior extreme points and points of inflection for the density function of the normal random variable defined by

$$f(x) = \frac{1}{\sigma\sqrt{2\pi}} e^{-(x-u)^2/2\sigma^2} \qquad \sigma > 0 \qquad -\infty < u < \infty \tag{7.28}$$

The stationary points for $f(x)$ must satisfy Eq. (7.17).

$$f'(x) = -\frac{(x - u)}{\sigma^3\sqrt{2\pi}} e^{-(x-u)^2/2\sigma^2} \tag{7.29}$$

Setting $f'(x)$ equal to zero and solving for x^* yields:

$$x^* = u$$

To determine the nature of x^* we evaluate $f''(x^*)$.

$$f''(x) = -\frac{1}{\sigma^3\sqrt{2\pi}} e^{-(x-u)^2/2\sigma^2} + \frac{(x - u)^2}{\sigma^5\sqrt{2\pi}} e^{-(x-u)^2/2\sigma^2}$$

Hence

$$f''(x^*) = -\frac{1}{\sigma^3\sqrt{2\pi}} \tag{7.30}$$

Since $\sigma > 0$, $f''(x^*) < 0$ and x^* is a local maximum for $f(x)$, x^* is the only interior extreme point for $f(x)$ and

$$f(x^*) = \frac{1}{\sigma\sqrt{2\pi}} \tag{7.31}$$

Since

$$\lim_{x \to -\infty} f(x) = 0 \tag{7.32}$$

$$\lim_{x \to \infty} f(x) = 0 \tag{7.33}$$

we have

$$\lim_{x \to -\infty} f(x) < f(x^*) > \lim_{x \to \infty} f(x)$$

Thus x^* is a global maximum for $f(x)$. ■

Example 7.7

A subcontractor has agreed to supply a particular solvent to a truck manufacturer. The following information is available:

c_1 = cost/day of holding a unit of solvent in stock.

c_2 = cost/day for failing to deliver a unit of solvent on schedule.

R_1 = number of units of the solvent to be supplied per day.

where

$$c_1 = \$5.00 \qquad c_2 = \$100.00 \qquad R_1 = 25$$

The subcontractor receives x units of the solvent at the beginning of each month. The cost of carrying x units per month is given by

$$C(x) = \frac{c_1 x^2}{2R} + \frac{c_2(30R - x)^2}{2R} \tag{7.34}$$

Determine the order quantity x that minimizes the cost of inventory.

The necessary condition for a minimum at x^* is

$$C'(x^*) = 0$$

Now

$$C'(x) = \frac{c_1 x}{R} - \frac{c_2(30R - x)}{R} \tag{7.35}$$

Solving for x^* such that $C'(x^*) = 0$ leads to

$$\frac{x(c_1 + c_2)}{R} = 30c_2$$

or

$$x^* = \frac{30c_2 R}{c_1 + c_2} \tag{7.36}$$

The nature of the stationary point given by x^* is determined as follows:

$$\begin{aligned} C''(x) &= \frac{c_1 + c_2}{R} \\ &> 0 \end{aligned} \tag{7.37}$$

Therefore x^* is a local minimum and

$$x^* = 714.29$$

Since

$$\begin{aligned} C(x^*) &= \$53{,}571.43 \\ C(0) &= \$1{,}125{,}000 \\ \lim_{x \to \infty} C(x) &= \infty \end{aligned}$$

$x^* = 714$ is a global minimum for $C(x)$. ■

It should be noted that the necessary and sufficient conditions for an extreme point assume that the function $f(x)$ is twice differentiable at that extreme point. However, an extreme point x^* may not satisfy these conditions. This occurs when $f'(x^*)$ does not exist. For example, consider the function shown in Figure 7.7. The point x^* is a maximum for $f(x)$. However, in this case $f'(x^*)$ does not exist and thus x^* does not satisfy the necessary condition in Eq. (7.17).

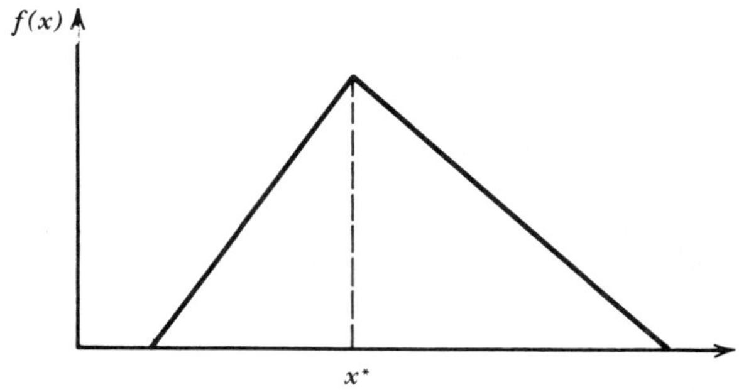

Figure 7.7 Extreme point x^* for the Function $f(x)$ where $f'(x^*)$ Does Not Exist

7.3 FUNCTIONS OF SEVERAL UNCONSTRAINED CONTINUOUS VARIABLES

The principles and methods introduced in the previous section for optimization of functions of a single continuous variable may be extended to functions of several variables without difficulty. Although our discussion here will be limited to functions of two and three variables, the extension of the methods presented to n variables is direct. However, such generalization requires a knowledge of matrix algebra and is therefore beyond the scope of this text. As in the case of functions of a single continuous variable we must identify conditions that will isolate the extreme points of the function and identify those that are maxima and those that are minima. As in the case of functions of a single variable, we present necessary conditions for the existence of a stationary point that may or may not be an extreme point. Sufficiency conditions are then applied in an attempt to determine the nature of the stationary point.

To initiate our search for necessary and sufficient conditions for a local maximum or minimum consider the function shown in Figure 7.8, where $f(x_1, x_2)$ is considered to be continuous and differentiable. The point (x_1^*, x_2^*) is a local minimum for $f(x_1, x_2)$ and therefore an extreme point. As in the case of functions of a single variable, the rate of change of $f(x_1, x_2)$ at (x_1^*, x_2^*) is zero. Therefore

$$\frac{\partial}{\partial x_1} f(x_1^*, x_2^*) = 0 \tag{7.38}$$

$$\frac{\partial}{\partial x_2} f(x_1^*, x_2^*) = 0 \tag{7.39}$$

These relationships also hold for local maxima if $f(x_1, x_2)$ is differentiable at the local maxima. To generalize, if $f(x_1, x_2, \ldots, x_n)$ is a continuous differentiable function and $(x_1^*, x_2^*, \ldots, x_n^*)$ is an interior extreme point for $f(x_1, x_2, \ldots, x_n)$, then

$$\frac{\partial}{\partial x_i} f(x_1^*, x_2^*, \ldots, x_n^*) = 0 \qquad i = 1, 2, \ldots, n \tag{7.40}$$

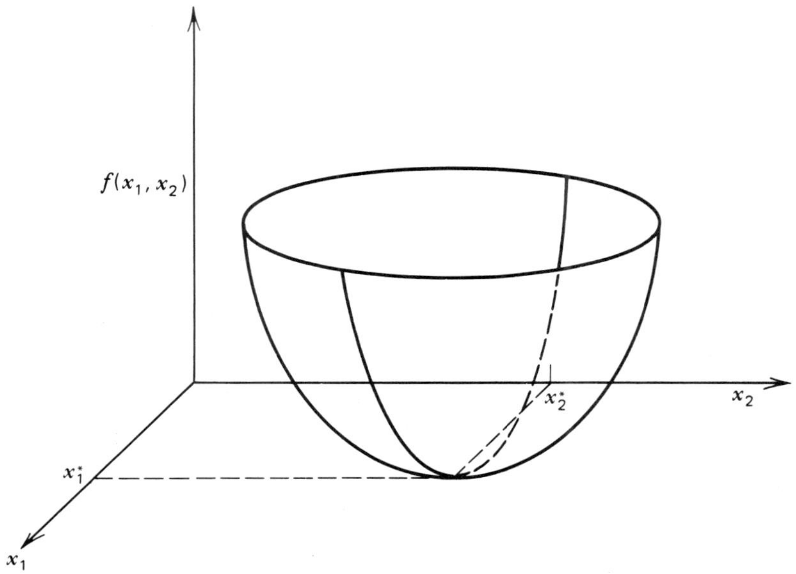

Figure 7.8 Function of Two Continuous Variables with a Minimum at (x_1^*, x_2^*)

Equation (7.40) is a *necessary condition* for the existence of an extreme point at $(x_1^*, x_2^*, \ldots, x_n^*)$. However, every point satisfying Eq. (7.40) is not necessarily an extreme point. Any point satisfying Eq. (7.40) is a *stationary point* for $f(x_1, x_2, \ldots, x_n)$. Thus an interior extreme point is a stationary point, but a stationary point is not necessarily an extreme point. For example, consider the function in Figure 7.9. The point (x_1^*, x_2^*) satisfies the conditions for a stationary point but is a *saddle point* instead of an interior extreme point. A saddle point may be considered to be the multivariate analogue of the point of inflection for a function of a single variable.

Example 7.8

Identify the stationary points of the following function:

$$f(x_1, x_2) = x_1^2 - 4x_1 + 4x_2^2$$

If (x_1^*, x_2^*) is a stationary point for $f(x_1, x_2)$, then

$$\frac{\partial}{\partial x_i} f(x_1^*, x_2^*) = 0 \qquad i = 1, 2$$

Now

$$\frac{\partial}{\partial x_1} f(x_1, x_2) = 2x_1 - 4 = 0$$

$$\frac{\partial}{\partial x_2} f(x_1, x_2) = 8x_2 = 0$$

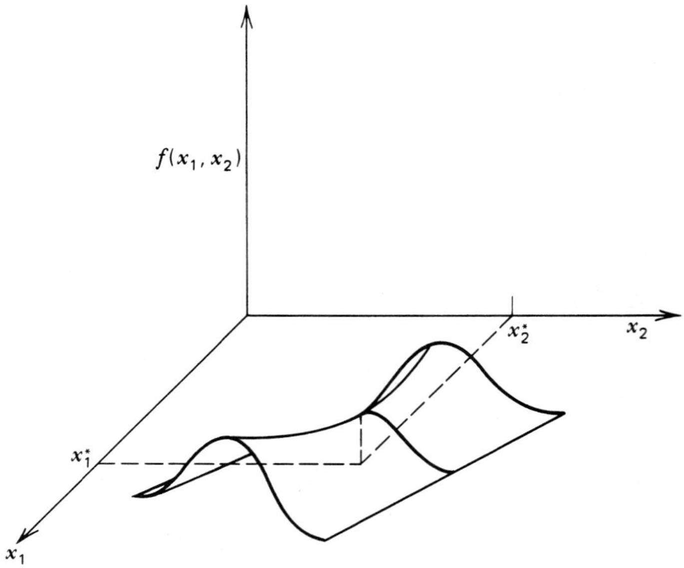

Figure 7.9 Saddle Point for $f(x_1, x_2)$ at (x_1^*, x_2^*)

and

$$(x_1^*, x_2^*) = (2, 0)$$ ■

Example 7.9

Find the stationary points of $f(x_1, x_2)$ where

$$f(x_1, x_2) = x_1^3 + 2x_1^2 - 3x_1 + 4x_2^2 - 3x_1x_2$$

The first partial derivatives of $f(x_1, x_2)$ are given by

$$\frac{\partial}{\partial x_1} f(x_1, x_2) = 3x_1^2 + 4x_1 - 3x_2 - 3$$

$$\frac{\partial}{\partial x_2} f(x_1, x_2) = 8x_2 - 3x_1$$

Setting these partial derivatives equal to zero, we have the following solution for (x_1^*, x_2^*).

$$x_2 = \frac{3}{8} x_1$$

and

$$3x_1^2 + \frac{23}{8} x_1 - 3 = 0$$

Thus

$$x_1^* = \frac{-23/8 \pm \sqrt{(23/8)^2 - 4(3)(-3)}}{6}$$
$$= -1.588, 0.630$$
$$x_2^* = -0.596, 0.236$$

The stationary points for $f(x_1, x_2)$ are then $(-1.588, -0.596)$ and $(0.630, 0.236)$.

■

Each stationary point for a function $f(x_1, x_2, \ldots, x_n)$ is potentially an extreme point for the function. Thus we require conditions sufficient for the identification of those stationary points that are maxima and minima for the function. Although the necessary conditions given in Eq. (7.40) hold for an arbitrary number of variables, we examine sufficient conditions for functions of two and three variables only.

Let

$$\begin{aligned} f_{ij} &= \frac{\partial}{\partial x_i\, \partial x_j} f(x_1, x_2) \qquad i = 1, 2 \qquad j = 1, 2 \\ &= \frac{\partial}{\partial x_i\, \partial x_j} f(x_1, x_2, x_3) \qquad i = 1, 2, 3 \qquad j = 1, 2, 3 \end{aligned} \tag{7.41}$$

and

$$\begin{aligned} f_{ij}^* &= \frac{\partial}{\partial x_i\, \partial x_j} f(x_1^*, x_2^*) \qquad i = 1, 2 \qquad j = 1, 2 \\ &= \frac{\partial}{\partial x_i\, \partial x_j} f(x_1^*, x_2^*, x_3^*) \qquad i = 1, 2, 3 \qquad j = 1, 2, 3 \end{aligned} \tag{7.42}$$

Let

$$H_1 = f_{11}^* \tag{7.43}$$

$$H_2 = f_{11}^* f_{22}^* - (f_{12}^*)^2 \tag{7.44}$$

$$H_3 = f_{11}^* f_{22}^* f_{33}^* + 2 f_{12}^* f_{23}^* f_{13}^* - f_{22}^*(f_{13}^*)^2 - f_{11}^*(f_{23}^*)^2 - f_{13}^*(f_{12}^*)^2 \tag{7.45}$$

In the case of continuous, differentiable functions of two variables, if $(\partial/\partial x_i)$ $(x_1^*, x_2^*) = 0$, $i = 1, 2$, then (x_1^*, x_2^*) is a local minimum if

$$H_1 > 0 \qquad H_2 > 0 \tag{7.46}$$

and is a local maximum if

$$H_1 < 0 \qquad H_2 > 0 \tag{7.47}$$

For continuous differentiable functions of three variables, if $(\partial/\partial x_i)$ $f(x_1^*, x_2^*, x_3^*) = 0$, $i = 1, 2, 3$, then (x_1^*, x_2^*, x_3^*) is a local minimum if

$$H_1 > 0 \qquad H_2 > 0 \qquad H_3 > 0 \tag{7.48}$$

and is a local maximum if

$$H_1 < 0 \qquad H_2 > 0 \qquad H_3 < 0 \tag{7.49}$$

The relationships given in Eqs. (7.46) through (7.49) are the *sufficient conditions* for a local minimum and maximum for continuous, differentiable functions of two and three variables. The necessary and sufficient conditions for a local minimum and maximum are summarized in Table 7.3.

Example 7.10

Determine the nature of the stationary point for the function in Example 7.8.

For the function given in Example 7.8, the second partial derivatives are given by

$$f_{11} = 2 \qquad f_{12} = 0 \qquad f_{22} = 8$$

Therefore, f_{11}, f_{12}, and f_{22} are constants for all x_1 and x_2 and

$$H_1 = 2 \qquad H_2 = 16$$

Since $H_1 > 0$ and $H_2 > 0$, (2, 0) is a local minimum for $f(x_1, x_2)$ ■

Example 7.11

Determine the nature of the stationary points for the function in Example 7.9.

From Example 7.9

$$f_{11} = 6x_1 + 4$$
$$f_{12} = -3$$
$$f_{22} = 8$$

Table 7.3 Necessary and Sufficient Conditions for a Local Maximum, Local Minimum, or a Saddle Point

Number of Variables	Necessary Conditions	Sufficient Conditions	Conclusion about (x_1^*, x_2^*) or (x_1^*, x_2^*, x_3^*)
2	$\frac{\partial}{\partial x_i} f(x_1^*, x_2^*) = 0$	$H_1 > 0, H_2 > 0$	Local minimum
		$H_1 < 0, H_2 > 0$	Local maximum
	$i = 1, 2$	$H_1 \lesseqgtr 0, H_2 < 0$	Saddle point
		$H_1 = 0$ or $H_2 = 0$	None
3	$\frac{\partial}{\partial x_i} f(x_1^*, x_2^*, x_3^*) = 0$	$H_1 > 0, H_2 > 0, H_3 > 0$	Local minimum
		$H_1 < 0, H_2 > 0, H_3 < 0$	Local maximum
	$i = 1, 2, 3$	$H_i \neq 0, i = 1, 2, 3$, but neither of the above	Saddle point
		$H_1 = 0, H_2 = 0$, or $H_3 = 0$	None

At $(-1.588, -0.596)$

$$f_{11}^* = -5.528$$
$$f_{12}^* = -3$$
$$f_{22}^* = 8$$

Therefore

$$H_1 = -5.528$$
$$H_2 = (-5.528)(8) - (3)^2$$
$$= -53.224$$

Since $H_1 < 0$, $H_2 < 0$ the point $(-1.588, -0.596)$ is a saddle point. At (0.630, 0.263) we have

$$f_{11}^* = 7.78$$
$$f_{12}^* = -3$$
$$f_{22}^* = 8$$

and

$$H_1 = 7.78$$
$$H_2 = 53.24$$

Since $H_1 > 0$ and $H_2 > 0$, (0.630, 0.263) is a local minimum. ■

Example 7.12

A company manufactures a product at a constant and continuous rate of ψ units per year. Each unit produced requires three raw materials. The number of units of raw material j required to produce one unit of the final product is m_j, $j = 1, 2, 3$. Every t_j years q_j units of raw material j are received and placed in inventory. C_{0j} is the cost of handling an order and is independent of the quantity received. C_{Ij} is the cost of carrying one unit of raw material j in inventory for one year. The total annual cost of the inventory system is given by $C_T(q_1, q_2, q_3)$, where:

$$C_T(q_1, q_2, q_3) = \sum_{j=1}^{3} \left[\frac{m_j \psi}{q_j} C_{0j} + \frac{q_j}{2} C_{Ij} \right] \tag{7.50}$$

Find the values of q_1, q_2, and q_3 that minimize the total annual cost of the inventory system.

The stationary points for $C_T(q_1, q_2, q_3)$ must satisfy:

$$\frac{\partial}{\partial q_j} C_T(q_1, q_2, q_3) = -\frac{m_j \psi C_{0j}}{q_j^2} + \frac{C_{Ij}}{2}$$
$$= 0$$

Therefore

$$q_j^* = \sqrt{\frac{2m_j\psi C_{0j}}{C_{Ij}}} \qquad j = 1, 2, 3 \tag{7.51}$$

To determine the nature of (q_1^*, q_2^*, q_3^*) we examine the second partial derivatives of $C_T(q_1, q_2, q_3)$.

$$f_{jj} = \frac{2m_j\psi C_{0j}}{q_j^3} \qquad j = 1, 2, 3$$

$$f_{ij} = 0 \qquad i \neq j$$

Therefore

$$f_{jj}^* = \frac{C_{Ij}^{3/2}}{\sqrt{2m_j\psi C_{0j}}}$$

$$f_{ij}^* = 0$$

and

$$H_1 = \frac{C_{I1}^{3/2}}{\sqrt{2m_1\psi C_{01}}}$$

$$H_2 = \frac{(C_{I1}C_{I2})^{3/2}}{2\psi\sqrt{m_1m_2C_{01}C_{02}}}$$

$$H_3 = \frac{(C_{I1}C_{I2}C_{I3})^{3/2}}{(2\psi)^{3/2}\sqrt{m_1m_2m_3C_{01}C_{02}C_{03}}}$$

If we take the positive roots in the expressions for H_1, H_2, H_3, then $H_1 > 0$, $H_2 > 0$, $H_3 > 0$ and (q_1^*, q_2^*, q_3^*) as defined in Eq. (7.51) is a local minimum for $C_T(q_1, q_2, q_3)$. If we take the negative root in the expressions for H_1, H_2, and H_3 then q_1^*, q_2^*, and q_3^* are negative and therefore infeasible. ■

The sufficient conditions presented here are adequate for identification of local minima and maxima for functions of two and three variables. However, general conditions for identification of local minima and maxima have been developed for continuous functions of an arbitrary number of variables and are discussed in References 1, 6, 10, 12, and 14. A simple approach to this determination is to evaluate $f(x_1, x_2, \ldots, x_n)$ at each stationary point. The stationary point yielding the largest value of $f(x_1, x_2, \ldots, x_n)$ is the global maximum and that yielding the smallest value of $f(x_1, x_2, \ldots, x_n)$ is the global minimum.

In our discussion of classical optimization techniques we have focused on problems where the optimum values of the decision variables may assume unrestricted values. That is, limitations were not placed upon the range of values permissible for any decision variable. Many situations arise, however, where such limitations are in order. To illustrate, suppose that warehouse space was limited to 500 units of solvent in Example 7.7. Since the optimal order quantity derived in Example 7.7 was 714.29 units of solvent, this solution would have to be rejected as infeasible since the maximum volume of solvent that could be stored in inventory at any single point in time is 500 units.

The optimization methods presented in Sections 7.2 and 7.3 have been extended to include problems where constraints are placed upon the values that one or more decision variables may assume. These methods include the Lagrange multiplier technique and procedures referred to as mathematical programming methods. For a discussion of the Lagrange multiplier technique, the reader should see References 1, 6, 10, and 12. In Chapter 8 we discuss mathematical programming, where we shall place emphasis on methods for solving a particular class of constrained optimization problems.

7.4 SEARCH TECHNIQUES

The optimization techniques discussed thus far have been analytic in nature, using the derivative to locate and identify the extreme points for the function. For functions of continuous variables we require that the functions be continuous and differentiable, requirements that may not be satisfied in practice. Even when the requirements of continuity and differentiability are satisfied the solution of the system of equations in (7.40) may prove sufficiently cumbersome that an alternative approach is desirable.

To circumvent the difficulties associated with optimization through analytic means search techniques are often used. A *search technique* is an iterative procedure for systematically exploring the solution space in an effort to find the combination of values of the decision variables that maximizes or minimizes the objective function for the problem. All search procedures are founded upon the same principle. Let $f(x_1, x_2, \ldots, x_n)$ be the objective function to be minimized (maximized), where $x_1, x_2, \ldots, x_n$ are the decision variables. An initial value is assigned to each x_i, $i = 1, 2, \ldots, n$, and this set of values represents the starting point for the search, denoted $(x_{10}, x_{20}, \ldots, x_{n0})$. The objective function is then evaluated at the starting point. Based upon the algorithm for the specific search procedure used, a new point is designated for investigation, $(x_{11}, x_{21}, \ldots, x_{n1})$. The search then compares $f(x_{10}, x_{20}, \ldots, x_{n0})$ and $f(x_{11}, x_{21}, \ldots, x_{n1})$. Depending upon whether $f(x_{10}, x_{20}, \ldots, x_{n0}) \leq f(x_{11}, x_{21}, \ldots, x_{n1})$ or $f(x_{10}, x_{20}, \ldots, x_{n0}) \geq f(x_{11}, x_{21}, \ldots, x_{n1})$ and the criterion of optimality, the search procedure designates a new point for investigation. The procedure continues iteratively in this fashion until an optimal or near optimal solution is identified.

The algorithms for most search techniques are basically heuristic and are intended to optimize functions of unconstrained variables. As such the solution indicated as optimal by a search procedure may fail to be truely optimal. However, a search procedure will usually yield a solution that is sufficiently close to the true optimum to be acceptable for practical purposes, particularly when the objective function has only one interior maximum or minimum.

7.4.1 One-at-a-Time Search

Conceptually, the simplest multivariate search technique is the one-at-a-time search. This search procedure increases and/or decreases one decision variable at a time in constant increments, holding the values of the remaining decision variables constant. When further variation of that decision variable proves unproductive, the value of that variable is fixed at its optimum value thus far and manipulation of one of the other decision variables commences. The search proceeds iteratively, varying one variable at a time in prespecified increments, until further improvement of the objective function cannot be achieved unless the increment for each variable is reduced. At this point each increment is reduced to refine the search, and the entire procedure is repeated. The search continues in this fashion until a defined number of iterations have been completed.

Let $\mathbf{x}$ be a vector of values of the decision variables $(x_1, x_2, \ldots, x_n)$ and $f(\mathbf{x})$ the objective function that defines the value of the measure of effectiveness at the point $\mathbf{x}$. As the search continues values of $\mathbf{x}$ will be identified that improve the measure of effectiveness. We will define $\mathbf{x}^0$ as the value of $\mathbf{x}$ that yields the minimum (maximum) value of $f(\mathbf{x})$ up to the current point in the search. Let $\mathbf{d}_i$ be a vector representing the magnitude of the change in $\mathbf{x}$ when the ith variable is increased. Thus

$$\mathbf{x} = (x_1, x_2, \ldots, x_i, \ldots, x_n) \tag{7.52}$$

$$\mathbf{x}^0 = (x_1^0, x_2^0, \ldots, x_i^0, \ldots, x_n^0) \tag{7.53}$$

$$\mathbf{d}_i = (0, 0, \ldots, \delta_i, \ldots, 0) \tag{7.54}$$

and

$$\mathbf{x} + \mathbf{d}_i = (x_1, x_2, \ldots, x_i + \delta_i, \ldots, x_n) \tag{7.55}$$

In the course of the search successively larger or smaller values of x_i may be investigated. The direction of the search on a given decision variable will be defined as s, where $s = 1$ implies that the variable altered is being increased in value and $s = -1$ implies a decrease in that variable.

Each step in the search is taken from the best point achieved up to the current point. Hence, if $\mathbf{x}^0$ is the best point thuse far and we wish to examine $f(\mathbf{x})$ when we increase x_i, then $\mathbf{x} = \mathbf{x}^0 + s\mathbf{d}_i$, where $s = 1$. If we wish to decrease x_i, then $\mathbf{x} = \mathbf{x}^0 + s\mathbf{d}_i$, where $s = -1$. To illustrate the fundamental operation of the one-at-a-time search consider the following example.

Example 7.13

Apply the one-at-a-time search to the minimization of the function

$$f(x_1, x_2) = x_1^2 + x_2^2$$

The starting point is to be (9, 7) and

$$\mathbf{d}_1 = (2, 0) \qquad \mathbf{d}_2 = (0, 3)$$

At the starting point $f(x_1, x_2)$ has the value 130; prior to the evaluation of further points it is the best point. Therefore $\mathbf{x}^0 = (9, 7)$ and $f(\mathbf{x}^0) = 130$. We first increment x_1. Thus $s = 1$ and

$$\begin{aligned}\mathbf{x} &= \mathbf{x}^0 + s\mathbf{d}_1 \\ &= (9, 7) + (2, 0) \\ &= (11, 7)\end{aligned}$$

and

$$f(11, 7) = 170$$

Since $f(\mathbf{x}) > f(\mathbf{x}^0)$, a further increase in x_1 is likely to be unproductive. Thus we decrease x_1. In this case $s = -1$ and

$$\begin{aligned}\mathbf{x} &= \mathbf{x}^0 + s\mathbf{d}_1 \\ &= (9, 7) - (2, 0) \\ &= (7, 7)\end{aligned}$$

and

$$f(7, 7) = 98$$

Since $f(\mathbf{x}) < f(\mathbf{x}^0)$, the best point thus far is $x^0 = (7, 7)$ and we continue to decrease x_1 until further improvement in $f(\mathbf{x})$ fails to be achieved. $\mathbf{x}$ is now given by

$$\begin{aligned}\mathbf{x} &= (7, 7) - (2, 0) \\ &= (5, 7)\end{aligned}$$

Since $f(5, 7) = 74 < f(\mathbf{x}^0)$, $\mathbf{x}^0 = (5, 7)$, and

$$\begin{aligned}\mathbf{x} &= (5, 7) - (2, 0) \\ &= (3, 7)\end{aligned}$$

$f(3, 7) = 58 < f(\mathbf{x}^0)$ and $\mathbf{x}^0 = (3, 7)$. Again decreasing x_1 we have

$$\begin{aligned}\mathbf{x} &= (3, 7) - (2, 0) \\ &= (1, 7)\end{aligned}$$

where $f(1, 7) = 50$ and $\mathbf{x}^0 = (1, 7)$. Now

$$\begin{aligned}\mathbf{x} &= (1, 7) - (2, 0) \\ &= (-1, 7)\end{aligned}$$

Since $f(-1, 7) = 50 = f(\mathbf{x}^0)$, continued reduction in $f(\mathbf{x})$ has not been achieved. Thus $\mathbf{x}^0$ is not changed; we now attempt to improve $f(\mathbf{x})$ by varying x_2. Increasing x_2 we have

$$\begin{aligned}\mathbf{x} &= \mathbf{x}^0 + \mathbf{d}_2 \\ &= (1, 7) + (0, 3) \\ &= (1, 10)\end{aligned}$$

where $f(1, 10) = 101$. Since $f(\mathbf{x}) > f(\mathbf{x}^0)$, the objective function has not been improved. Therefore $\mathbf{x}^0$ is not altered, and we try decreasing x_2. Thus

$$\begin{aligned}\mathbf{x} &= (1, 7) - (0, 3)\\ &= (1, 4)\end{aligned}$$

Since $f(1, 4) = 17 < f(\mathbf{x}^0)$, $\mathbf{x}^0$ is given by (1, 4). We reduce x_2 again:

$$\begin{aligned}\mathbf{x} &= (1, 4) - (0, 3)\\ &= (1, 1)\end{aligned}$$

and

$$f(1, 1) = 2$$

Since $f(1, 1) < f(\mathbf{x}^0)$, $\mathbf{x}^0 = (1, 1)$ and further reduction in x_2 would seem to be advisable. Thus

$$\begin{aligned}\mathbf{x} &= (1, 1) - (0, 3)\\ &= (1, -2)\end{aligned}$$

where

$$f(1, -2) = 5$$

Since increasing x_2 from 1 to 4 or decreasing x_2 from 1 to -2 yields a value of $f(\mathbf{x})$ greater than $f(\mathbf{x}^0)$, where $\mathbf{x}^0 = (1, 1)$, further variation of x_2 will not prove effective at this point. Therefore we return to alteration of x_1. Increasing x_1 from $\mathbf{x}^0$ yields:

$$\begin{aligned}\mathbf{x} &= (1, 1) + (2, 0)\\ &= (3, 1)\end{aligned}$$

where

$$f(3, 1) = 10$$

Since $f(3, 1) > f(\mathbf{x}^0)$, we will try decreasing x_1.

$$\begin{aligned}\mathbf{x} &= (1, 1) - (2, 0)\\ &= (-1, 1)\end{aligned}$$

Now since $f(-1, 1) = f(\mathbf{x}^0)$, the objective function has not been reduced in value. We have reached a point at which $f(\mathbf{x}_0 \pm \mathbf{d}_1) \geq f(\mathbf{x}^0) \leq f(\mathbf{x}_0 \pm \mathbf{d}_2)$. Thus further variation in x_1 and x_2 will not improve $f(\mathbf{x})$ unless we change the step sizes $\mathbf{d}_1$ and $\mathbf{d}_2$. ∎

In the course of the search if a point is reached such that $f(\mathbf{x}^0) \leq f(\mathbf{x}^0 \pm \mathbf{d}_i)$, $i = 1, 2, \ldots, n$, improvement of the objective function can be achieved only if the step size $\mathbf{d}_i$ is altered for at least one variable. For our purposes we will adopt the convention of halving $\mathbf{d}_i$ for all i whenever $f(\mathbf{x}^0) \leq f(\mathbf{x}^0 \pm \mathbf{d}_i)$, $i = 1, 2, \ldots, n$, although any convention that reduces at least one $\mathbf{d}_i$ may be used. After reducing the step size $\mathbf{d}_i$, $i = 1, 2, \ldots, n$, the entire search procedure is repeated again until all

$\mathbf{d}_i$ are reduced again. If M is the number of different step sizes used in the search, then the search terminates when the step size is less than $(\frac{1}{2})^M \mathbf{d}_i$, $i - 1, 2, \ldots, n$.

A flowchart for the one-at-a-time search algorithm is given in Figure 7.10. The notation for this algorithm is as follows:

$f(\mathbf{x})$ = objective function value at the point $\mathbf{x}$.

M = number of iterations of the search.

$\mathbf{d}_i$ = initial step size on variable i.

j = iteration number.

$\mathbf{x}^0$ = indicated optimum at any point in the search.

$\mathbf{x}$ = point at which $f(\mathbf{x})$ is evaluated.

i = variable number.

s = direction indicator; specifically,

$$s = \begin{cases} -1, & \text{variable value } x \text{ decreased, or} \\ 1, & \text{variable value increased.} \end{cases}$$

$$\nabla = \begin{cases} 0 & \text{if no improvement of } f(\mathbf{x}) \text{ has been achieved through variation of the current variable,} \\ 1 & \text{if improvement of } f(\mathbf{x}) \text{ has been achieved through variation of the current variable.} \end{cases}$$

N = indicator used to determine the point at which an iteration of the search should be terminated.

The procedure summarized in Figure 7.10 is intended to minimize a function of n variables. However, this algorithm may also be used for maximization by multiplying the objective function by -1. That is, if the problem posed is to maximize $f(\mathbf{x})$, then we apply the algorithm in Figure 7.10 to the minimization of $-f(\mathbf{x})$.

Example 7.14

Specification limits for a dimension of a product are to be established. Each unit manufactured is inspected, and if the dimension is outside the specification limits it is rejected and accepted otherwise. The lower and upper specification limits are denoted by L and U, respectively. The quality of an individual unit is dependent upon the deviation of the measured dimension x from the desired dimension u. Specifically, if x is the dimension of the unit then the cost of this deviation is $C_a[(u - x)/(x - a)]$ if $x < u$, where C_a and a are constants. If $x > u$, the cost is given by $C_a[(x - u)/(b - x)]$, where b is a constant. However, if such a unit is rejected this cost does not arise but a cost C_r of disposing of the unit is incurred. The cost per unit of product for specification limits L and U is given by $C_T(L, U)$, where

$$C_T(L, U) = \frac{1}{(b - a)} \left[\frac{C_a(u - L)^2 + C_r(L - a)^2}{(u - a)} + \frac{C_a(U - u)^2 + C_r(b - U)^2}{(b - u)} \right] \quad (7.56)$$

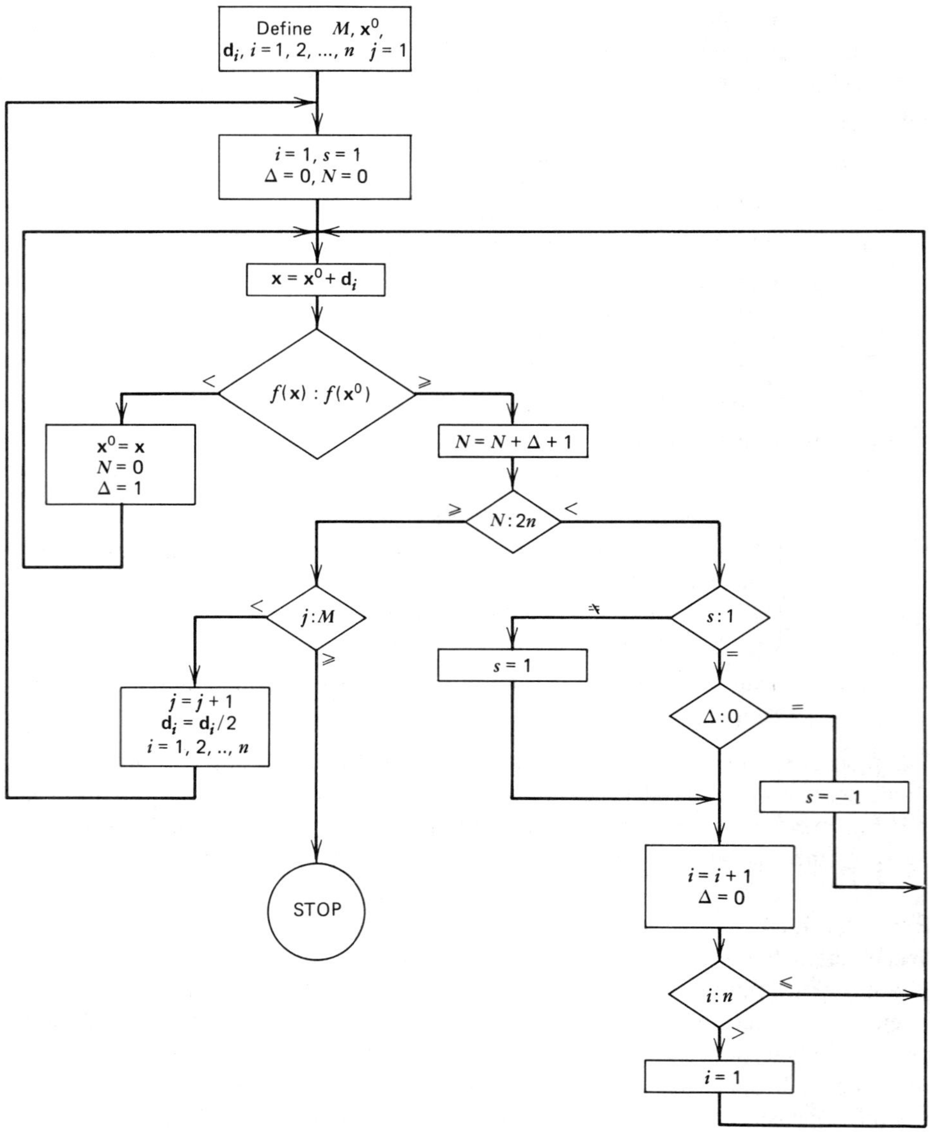

Figure 7.10 Flowchart for the One-at-a-Time Search

Table 7.4 Two Iterations of the One-at-a-time Search for the Problem in Example 7.14

M	$\mathbf{d}_1$	$\mathbf{d}_2$	i	j	N	Δ	s	$\mathbf{x}^0$	$f(\mathbf{x}^0)$	$\mathbf{x}$	$f(\mathbf{x})$
2	(0.10, 0.00)	(0.00, 0.10)	1	1	0	0	1	(8.90, 10.00)	10.46	(9.00, 10.00)	10.20
						1		(9.00, 10.00)	10.20	(9.10, 10.00)	9.98
								(9.10, 10.00)	9.98	(9.20, 10.00)	9.81
								(9.20, 10.00)	9.81	(9.30, 10.00)	9.68
								(9.30, 10.00)	9.68	(9.40, 10.00)	9.59
								(9.40, 10.00)	9.59	(9.50, 10.00)	9.550
			2		2	0		(9.50, 10.00)	9.550	(9.60, 10.00)	9.552
					0	1				(9.50, 10.10)	9.37
								(9.50, 10.10)	9.37	(9.50, 10.20)	9.24
								(9.50, 10.20)	9.24	(9.50, 10.30)	9.15
								(9.50, 10.30)	9.15	(9.50, 10.40)	9.102
								(9.50, 10.40)	9.102	(9.50, 10.50)	9.100
			1		2	0		(9.50, 10.50)	9.100	(9.50, 10.60)	9.14
					3		−1			(9.60, 10.50)	9.102
	(0.05, 0.00)	(0.00, 0.05)		2	4					(9.40, 10.50)	9.14
			1		0	0	1			(9.55, 10.50)	9.096
					0	1		(9.55, 10.50)	9.096	(9.60, 10.50)	9.102
			2		2	0				(9.55, 10.55)	9.111
					3		−1			(9.55, 10.45)	9.091
					0	1		(9.55, 10.45)	9.091	(9.55, 10.40)	9.098
			1		2	0	1			(9.60, 10.45)	9.098
					3		−1			(9.50, 10.45)	9.096
					4						

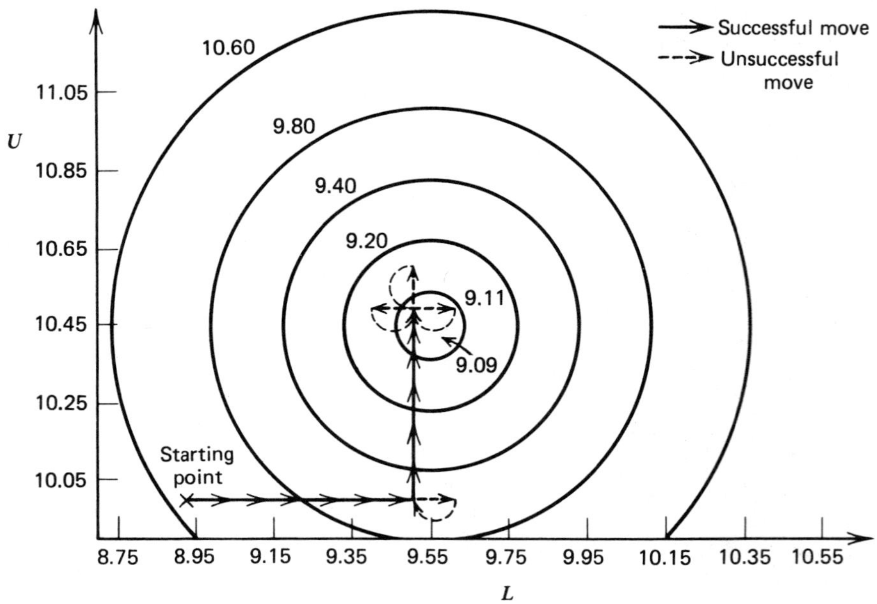

Figure 7.11 First Iteration of the One-at-a-Time Search for the Problem in Example 7.14 Where the Step Size is 0.10

The values of L and U that minimize $C_T(L, U)$ are to be determined using the one-at-a-time search procedure, where $C_a = \$100$, $C_r = \$10{,}000$, $a = 5$, $b = 15$, and $u = 10$. The starting point for the search is to be $L = 8.90$ and $U = 10.00$. The initial step size is to be 0.10.

The search for the optimum is summarized in Table 7.4, where two iterations of the search are given for step sizes of 0.10 and 0.05. The first iteration of the search is shown graphically in Figure 7.11 along with the contours of equal cost for $C_T(L, U)$. The optimum values of L and U indicated by the search are 9.55 and 10.45, respectively. ∎

7.5 SUMMARY

In this chapter we have discussed two approaches to the optimization of functions of several decision variables. The simpler of the two is the classical approach using the derivative for continuous functions. Although the classical procedures are convenient to use in many cases, the analyst may choose to use a search technique for one of the following reasons:

1. The objective function is a function of several discrete variables or several discrete and several continuous variables.
2. The objective function is not differentiable, although it is continuous.

3. The objective function is differentiable, but solution for the stationary points requires numerical methods.

The decision to optimize a given objective function through a search technique should be taken with caution when classical methods are feasible. The examples presented in this chapter deal with rather simple functions involving no more than two decision variables, and the search techniques used are performed with reasonable efficiency. However, as the number of decision variables increases, the need for a digital computer to execute the search becomes evident, and the cost of the analysis can become substantial. The analyst should exercise care in choosing between classical methods and a search procedure. Although the analysis required for solution by classical methods may be complex, the cost of such an analysis may prove to be far less than that which would result from the indiscriminate application of a search technique.

Search techniques are, however, effective optimization procedures. Undoubtedly, their application to the solution of practical optimization problems will grow, as classical methods are often infeasible or impractical for such problems.

REFERENCES AND SUGGESTED READING

1. G. S. G. Beveridge and R. S. Schechter, *Optimization Theory and Practice*, McGraw-Hill, New York, 1970.

2. L. Cooper and D. Steinberg, *Introduction to Methods of Optimization*, Saunders, Philadelphia, 1970.

3. M. M. Denn, *Optimization by Variational Methods*, McGraw-Hill, New York, 1969.

4. M. Friedman and L. J. Savage, Planning Experiments Seeking Maxima, in *Techniques of Statistical Analysis* (C. Wisenhart *et al.*, Ed.), McGraw-Hill, New York, 1947.

5. W. C. Giffin, *Introduction to Operations Engineering*, Irwin, Homewood, Ill. 1971.

6. R. L. Gue and M. E. Thomas, *Mathematical Methods in Operations Research*, Macmillan, London, 1968.

7. F. S. Hillier and G. J. Lieberman, *Introduction to Operations Research*, Holden-Day, San Francisco, 1967.

8. D. M. Himmelbau, *Applied Nonlinear Programming*, McGraw-Hill, New York, 1972.

9. R. Hooke and T. A. Jeeves, Direct Search Solution of Numerical and Statistical Problems, *J. Assn. Comp. Mach.*, **8** (2) (1961).

10. J. W. Schmidt, *Mathematical Foundations for Management Science and Systems Analysis*, Academic Press, New York, 1974.

11. H. A. Taha, *Operations Research: An Introduction*, Macmillan, New York, 1971.

12. D. Teichroew, *An Introduction to Management Science*, John Wiley, New York, 1964.

13. H. M. Wagner, *Principles of Operations Research*, Prentice-Hall, Englewood Cliffs, N.J., 1969.

14. D. J. Wilde and C. S. Beightler, *Foundations of Optimization*, Prentice-Hall, Englewood Cliffs, N.J., 1967.

15. C. R. Wylie, *Advanced Engineering Mathematics*, McGraw-Hill, New York, 1960.

REVIEW QUESTIONS AND PROBLEMS

7.1 Find the stationary points for the following function and determine which are local minima and local maxima.

$$f(x) = \frac{1}{5}x^5 - \frac{5}{2}x^4 + \frac{35}{3}x^3 - 25x^2 + 24x$$

7.2 For the function in Problem 7.1 plot the first and second derivatives and identify the local minima and maxima from the graphical display.

7.3 Show that the function

$$f(x) = x^3$$

has no interior maximum or minimum.

7.4 Find the local minima and maxima for the following function:

$$f(x) = \frac{k}{4}x^4 - \frac{k}{3}x^3 + b$$

and

a. $k < 0$
b. $k > 0$

7.5 Determine the conditions for a, b, c, and d such that the point x^* for which $f'(x^*) = 0$ is a point of inflection for the function

$$f(x) = ax^3 + bx^2 + cx + d$$

7.6 Determine whether the problem in Example 7.1 can be solved by the classical techniques of calculus.

7.7 The mode of a continuous random variable is defined as the value of the random variable at which its probability density function is a maximum. Find the mode of the random variable x if its probability density function is defined by

a. $f(x) = \dfrac{\lambda^\alpha}{(\alpha - 1)!} x^{\alpha - 1} e^{-\lambda x} \qquad 0 < x < \infty,\ \lambda > 0,\ \alpha > 1$

b. $f(x) = \dfrac{\Gamma[(n + 1)/2]}{\sqrt{n\pi}\ \Gamma(n/2)} \left(1 + \dfrac{x^2}{n}\right)^{-(n+1)/2} \qquad \infty < x < \infty,\ n \geq 1$

7.8 The mean rate of flow of vehicles $g(C)$ in vehicles per hour is related to mean traffic concentration C, jam concentration C_j, and vehicle free speed S, by

$$q(C) = CS - S\frac{C^2}{C_j}$$

Find the mean traffic concentration C that maximizes $q(C)$.

7.9 The fuel cost $C(v)$ for running a particular vehicle at a speed v is given by

$$C(v) = \frac{v}{20} + \frac{80}{v}$$

Find the speed v that will minimize fuel cost.

7.10 An air traffic controller has two aircraft on his radar scope, A and B, both flying level and at the same altitude. Aircraft A is flying along the straight line given by $x_1 + y_1 = 10$ and B along the straight line $3x_2 + y_2 = 10$. At time zero A is at the point $(-40, 50)$ and B is at $(-20, 70)$ as shown in the figure below. A is flying at 550 nautical miles per hour and B at 615 nautical miles per hour. At what point in time will the aircraft be closest to one another? If aircraft must maintain a separation distance of 5 nautical miles, will the separation standard be violated if the aircraft continue their flight paths? By how much must the velocity of B be changed at time zero so that the aircraft will not violate the 5-mile separation standard if current flight paths are maintained?

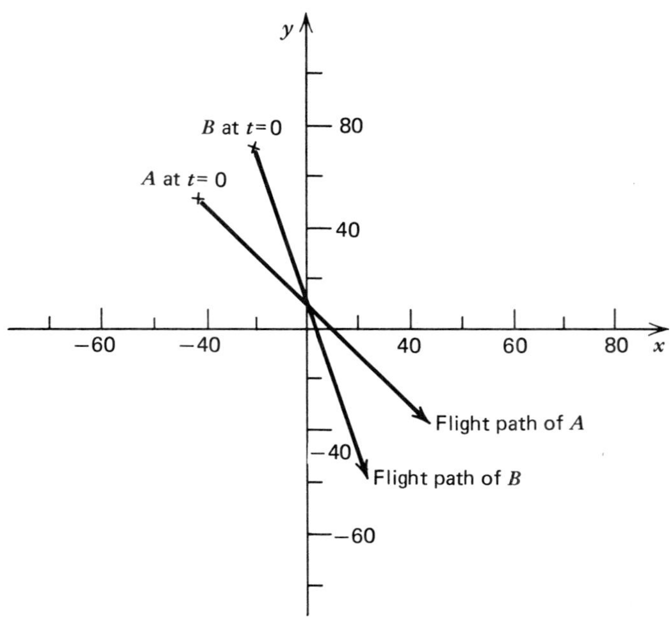

7.11 Identify the stationary points for the following functions

(a) $f(x_1,x_2) = 3x_1^2 + x_2^2 - 10$

(b) $f(x_1,x_2) = 100(x_1 - x_2^2)^2 + (1 - x_1)^2$

(c) $f(x_1,x_2) = x_1^3 + x_2^3 + 2x_1^2 + 4x_2^2$

7.12 Determine whether the stationary points for the functions in Problem 7.11 are interior minima, maxima, or saddle points.

7.13 A closed metal cylindrical container is to have a volume of 100 ft^3. Find the dimensions of the container so that the amount of metal is minimized where

$$\text{volume} = \pi r^2 h$$
$$\text{area} = 2\pi r h + \pi r^2$$

and r is the radius of the base and h is the height of the container.

7.14 Solve the problem in Example 7.13 through the classical calculus technique.

7.15 A generator supplies an electromotive force E and is connected in series with a fixed impedance Z_f and a load impedance Z_l where

$$Z_f = R_f + iX_f, \quad Z_l = R_l + iX_l$$

and R_f and R_l are the fixed and load resistances respectively, X_f and X_l are the fixed and load reactances respectively, and $i = \sqrt{-1}$. The power to the load, P, is

$$P = \frac{E^2 R_l}{(R_f + R_l)^2 + (X_f + X_l)^2}$$

Find the values of R_l and X_l which maximize the power to the load.

7.16 An equation of the form

$$y = b_0 + b_1 x$$

is to be fit to an experimentally obtained set of data. There are n observed sets of data on x and y. That is, if x_i is the ith observed value of the independent variable x, then y_i is the corresponding observed value of the dependent variable y. The values of b_0 and b_1 are to be defined such that

$$\sum_{i=1}^{n} (y_i - b_0 - b_1 x_i)^2 = \min$$

Derive expressions for calculations of b_0 and b_1.

7.17 Given the following observed values of x and y, find the equation of the form

$$y = b_0 + b_1 x$$

which best fits the data according to the criterion in Problem 7.16.

y	x
−3	−3
0	−2
1	−1
3	0
5	1
8	2
9	3

7.18 Solve the problem in Problem 7.16 if an equation of the form

$$y = b_0 + b_1 x + b_2 x^2$$

is to be fit to the data such that

$$\sum_{i=1}^{n} (y_i - b_0 - b_1 x_i - b_2 x_i^2)^2 = \min$$

7.19 The total annual cost of maintaining an inventory system is given by

$$C_T(Q,S) = \frac{d}{Q} C_O + \frac{(Q - S)^2}{2Q} C_I + \frac{Sd}{Q} C_b + \frac{S^2}{2Q} C_\beta$$

where Q is the order quantity, S is the number of backorders allowed between the receipt of orders, d is the annual demand, C_O is the cost of placing an order, C_I is the cost of carrying a unit in inventory for one year, C_b is the fixed cost associated with each unit backordered, and C_β is the time dependent cost associated with each backorder. Find the values of S and Q which minimize $C_T(Q,S)$. Treat Q and S as continuous variables.

7.20 Solve Problem 7.19 if

$d = 50{,}000 \quad C_O = \$10.00 \quad C_I = \$0.30 \quad C_b = \$0.01 \quad C_\beta = \$0.05$

7.21 Find the minimum for the function given in Example 7.8 using the one-at-a-time search.

7.22 Find the minimum for the function in Example 7.12 using the one-at-a-time search.

Chapter 8

Mathematical Programming

8.1 INTRODUCTION

In Chapter 2 we discussed the use of mathematical models in developing solutions to decision problems that confront an organization. The material discussed in this chapter, mathematical programming, is concerned with the formulation of this type of model, and the various methods commonly used to obtain a solution to the underlying decision problem. Chapter 7 was also concerned with decision problem models, but only those which involve no constraints or restrictions on the allowable values of the decision variables. The methodology described in this chapter is applicable to decision problems that *do* require the explicit consideration of constraints.

8.2 THE MEANING OF MATHEMATICAL PROGRAMMING

Mathematical programming can be defined to be the formulation and solution of normative mathematical models of decision problems involving one or more constraints. In other words, mathematical programming (or just ''math programming'') is the process of deriving recommended answers to decision problems through the development of mathematical models as representations of those problems, and the analysis and solution of these models. The model itself is referred to as the mathematical programming *problem*. The recommended answer to the underlying decision is the *solution* to the math programming problem, which is obtained through a specific sequence of calculations called the math programming *algorithm*. There are many different math programming algorithms, each one designed to solve a particular type of model or math programming problem.

The term ''mathematical programming'' means ''planning or programming the use of scarce resources through the aid of mathematical models.'' As was described in Chapter 2, a decision problem model (a math programming problem) has an objective function and a set of constraints. Each of these constraints limits the values that can be assigned to the decision variables based on the available amount of a particular resource. In solving the math programming problem, we are attempting to mathematically allocate or program the use of these available resources in such a way as to enhance our objective function. If there are no restrictions or constraints on the availability of resources in a decision problem, then there is nothing to allocate. Such a decision problem is an unconstrained optimization problem (Chapter 7), not a math programming problem.

Example 8.1

In order to illustrate the meaning of mathematical programming, consider the following capital-budgeting decision facing the ABC Corporation. Each of the three divisions of ABC has submitted its annual list of upcoming investments for which it would like funds. As is usually the case, the total capital available to ABC for funding investments is less than the combined requests from the three divisions. Therefore, ABC is faced with the following decision problem: which of the divisions' investments should be funded?

In using math programming to provide a recommended answer to ABC's decision problem, the first step would be to formulate a mathematical model to represent the objective function and constraints involved in the problem. Based on the concepts and principles discussed in Chapter 2, the following model could be formulated and used:

$$\begin{aligned} \underset{\mathbf{x}}{\text{Max}}\ x_0 &= \sum_{j=1}^{n} PV_j x_j \\ \text{S.T.:}\ \sum_{j=1}^{n} a_j x_j &\leq b \qquad x_j = 0 \text{ or } 1 \text{ for all } j \end{aligned} \tag{8.1}$$

where $x_j = 1$ if investment number j ($j = 1, 2, 3, \ldots, n$) is selected and $x_j = 0$ otherwise; PV_j is the present value of the cash flows associated with investment j; a_j is the amount of money needed to fund investment j; and b is the total amount of money available to ABC for all its investments.

This model is the symbolic form of the math programming problem to be solved. The next step in the overall analysis is to develop numerical estimates of the parameters in model (8.1)—PV_j and a_j for each j, and b. Then, a solution procedure or math programming algorithm must be found or developed that will solve the numerical version of model (8.1). One such algorithm is as follows:

Step 1. Temporarily set the value of each x_j at 0.

Step 2. For each $j = 1, 2, \ldots, n$ calculate the ratio PV_j/a_j.

Step 3. Rank the j's in decreasing order of the PV_j/a_j ratios.

Step 4. Select the first j in the ranked list. Let this j be designated as j'. If $a_{j'}$ does not exceed b, then set the final solution value of $x_{j'}$ at 1; go to Step 5. If $a_{j'}$ exceeds b, go to Step 7.

Step 5. Replace the value of b with the amount of investment funds that have not been used up yet. Remove j' from the ranked list (therefore, the j with the next highest ratio is now at the first of the list).

Step 6. Go back to Step 4.

Step 7. Stop. Let the final solution to model (8.1) be $x_j = 1$ for all those j's designated as j' in Step 4, and $x_j = 0$ for all other j's. The last j' that caused the allocation to exceed b cannot be accepted.

After applying the above algorithm, a recommended answer will be obtained for ABC's decision problem. Specifically, ABC should fund those division investments with associated $x_j = 1$. It should be noted that this may not necessarily be the best solution to ABC's problem. The recommended answer is a solution to a *representation* of the true decision problem, a representation that may not be very accurate. In addition, the solution algorithm itself may be faulty in that it may not produce the best solution to the formulated model. (The algorithm given above has this weakness, as will be evident after reading Section 8.7.) ■

8.3 APPLICATIONS OF MATH PROGRAMMING

Since the 1950s, math programming has been applied to a variety of decision problems in military organizations, businesses, and governmental agencies. For instance, a mathematical programming approach has been used to balance the work stations along an assembly line, to schedule jobs to machines in a job shop, to determine the mix of end products to produce at an oil refinery, to schedule cargo ships in delivering military supplies around the world, and to determine the best menu to serve patients in a hospital. Because of the large number of decision problems faced daily by organizations, the list of potential applications for math programming is almost endless. In Section Four of this book, we cover several specific application areas in detail.

Mathematical programming is not always a readily useable tool in resolving a given decision problem. This is because of a number of factors, such as requirements for large amounts of input data, difficulty in developing adequate descriptive mathematical models, and lack of cooperation or understanding on the part of the decision makers themselves. However, there have been many successful cases of the use of math programming reported in the literature. For example, the following reviews of specific math programming applications were reported in one issue of *Interfaces*:

1. A (nonlinear, mixed integer programming) model which helps electric utilities plan investments for power generation . . . is currently being used by the New England Electric System for planning generating facilities. For each year over a planning horizon the model determines what type and size of generating plants should be constructed so as to minimize total discounted cost while meeting (required) forecasted demand loads.
2. A (math programming) model was developed for determining the number and location of sites for the decentralization of the Los Angeles regional headquarters of a national insurance company. Based on the model's projected benefits the company has embarked upon a three-year decentralization and relocation program.

3. A mathematical programming model was developed to aid in locating fire companies in Denver. The model has led to major changes in the planned number and location of Denver fire companies.
4. A normative (math programming) methodology for modeling consumer response to innovation has led to the adoption of several practical, nonobvious recommendations as to the question of whether or not to convert the MIT health department to an HMO (health maintenance organization).

Detail descriptions of math programming applications such as these can be found in regular issues of professional journals such as *Management Science, AIIE Transactions, Operations Research*, and *Naval Logistics Quarterly*.

8.4 MATH PROGRAMMING MODELS

As was indicated in Chapter 2, a decision problem model has four basic components: decision variables (the unknowns), an objective function, a set of constraints that reflect limiting factors such as scarce resources, and parameters (the "constants"). Also as mentioned in Chapter 2, the general version of a decision problem model can be stated as:

$$\begin{aligned} \max_{\mathbf{x}} \quad & x_0 = f(\mathbf{x}) \\ \text{S.T.:} \quad & g_1(\mathbf{x}) \le b_1 \\ & g_2(\mathbf{x}) \le b_2 \\ & \vdots \qquad\quad \vdots \\ & g_m(\mathbf{x}) \le b_m \end{aligned} \tag{8.2}$$

This is also a general statement of a mathematical programming problem or model. A specific math programming problem would be obtained from model (8.2) by using particular equations for the general functions $f(\mathbf{x})$, $g_1(\mathbf{x})$, $g_2(\mathbf{x})$, and so on. For example, model (8.1) in Example 8.1 is a specific math programming problem in which $f(\mathbf{x}) = \sum_{j=1}^{n} PV_j x_j$, $g_1(\mathbf{x}) = \sum_{j=1}^{n} a_j x_j$, and $m = 1$.

Math programming problems or models tend to fall into one of four general categories depending on the basic form of the functional relationships involved in the specific problem at hand. These four classes are linear programming problems, integer programming problems, nonlinear programming problems, and dynamic programming problems. Each of these categories is briefly described below.

8.4.1 Linear Programming Models

When a math programming model has the following characteristics it is classified as a linear programming problem: (1) the decision variables are allowed to take on fractional values (as opposed to only integer values), (2) these values must not be negative, (3) all the functions in the model are linear in the decision variables, and (4) all the model's parameters are deterministic in nature. With these properties in mind, the general version of a linear programming problem can be written as

$$\begin{aligned}
\max_{\mathbf{x}} \quad & x_0 = \sum_{j=1}^{n} c_j x_j \\
\text{S.T.:} \quad & \sum_{j=1}^{n} a_{1j} x_j \le b_1 \\
& \sum_{j=1}^{n} a_{2j} x_j \le b_2 \\
& \quad \vdots \qquad\qquad \vdots \\
& \sum_{j=1}^{n} a_{mj} x_j \le b_m \\
& x_j \ge 0 \qquad \text{for } j = 1, 2, \ldots, n
\end{aligned} \tag{8.3}$$

For example, the following numerical version of a math programming problem would be classified as a linear programming problem:

$$\begin{aligned}
\max_{x_1 x_2} \quad & x_0 = 10x_1 + 2x_2 \\
\text{S.T.:} \quad & x_1 + x_2 \le 20 \\
& 2x_1 - x_2 \le 10 \\
& 3x_1 + x_2 \le 18 \\
& x_1, x_2 \ge 0
\end{aligned} \tag{8.4}$$

A short-cut version of model (8.3) is

$$\begin{aligned}
\max_{\mathbf{x}} \quad & x_0 = \sum_{j=1}^{n} c_j x_j \\
\text{S.T.:} \quad & \sum_{j=1}^{n} a_{ij} x_j \le b_i \qquad \text{for } i = 1, 2, \ldots, m \\
& x_j \ge 0 \qquad \text{for } j = 1, 2, \ldots, n
\end{aligned} \tag{8.5}$$

A third way to write the general linear programming model is with vectors and matrices. Letting $\mathbf{x}$ be the column vector $(x_1, x_2, \ldots, x_n)$, $\mathbf{c}$ be the row vector $(c_1, c_2, \ldots, c_n)$, A be an $m \times n$ matrix of a_{ij} values, and $\mathbf{b}$ be the column vector $(b_1, b_2, \ldots, b_m)$, then model (8.3) can be rewritten as

$$\begin{aligned}
\max_{\mathbf{x}} \; x_0 \quad & = \mathbf{cx} \\
\text{S.T.:} \quad & A\mathbf{x} \le \mathbf{b} \\
& \mathbf{x} \ge 0
\end{aligned} \tag{8.6}$$

As an illustration, the vectors and matricies of the numerical example given in model (8.4) would be

$$\mathbf{x} = \begin{pmatrix} x_1 \\ x_2 \end{pmatrix} \qquad \mathbf{c} = (10,\ 2)$$

$$A = \begin{bmatrix} 1 & 1 \\ 2 & -1 \\ 3 & 1 \end{bmatrix} \qquad \mathbf{b} = \begin{pmatrix} 20 \\ 10 \\ 18 \end{pmatrix}$$

Example 8.2

As an illustration of the formulation of a linear programming problem, consider the following situation. A plant that manufactures specialty metal products has recently discontinued sales of a particular product line that was no longer profitable. This created excess production capacity that can be used to manufacture more of the remaining lines. Management of the firm has identified three existing profitable products and wishes to know how it should allocate the new capacity among these three so that they enhance their profits as much as possible.

The production routing sheets for these three products all specify processing on a numerical controlled milling machine, a lathe, and a grinder. The time now available on these machines is as follows:

	Available Time (Machine Hours per Week)
NC machine	520
Lathe	210
Grinder	125

The routing sheets also provide the following standard time information (in machine hours per unit):

Machine	Product 1	Product 2	Product 3
NC machine	16	5	8
Lathe	10	5	1
Grinder	2	1	1

The firm's marketing department forecasts that the sales potential for products 1 and 2 exceeds the maximum production rate, but that the potential for product 3 is limited to 50 units per week. At the same time, the industrial engineering department has estimated that, at the volume likely to be achieved and the productivity levels being projected by management, the unit profit for the products would be \$45, \$20, and \$25, respectively.

Formulation. Our objective is to represent in mathematical terms as accurately as possible the decision problem specified above. To do so, we define and model each of the four basic components of a normative model. First, decision variables: here, the decision is how many units of each of the three products should be produced each week. Second, an objective function: maximize the weekly profit, which is a linear function of the decision variables. Third, constraints: the available time of each machine type is limited, as is the sales potential for product 3. And fourth, parameters: the standard times, machine hours available, and unit profits. Letting x_0 represent the amount of weekly profit generated, and x_i equal the number of units of product i ($i = 1, 2, 3$) produced per week, then the decision problem can be represented mathematically as follows:

$$\begin{aligned}
\underset{x_1,x_2,x_3}{\text{Maximize}}\ x_0 &= 45x_1 + 20x_2 + 25x_3 \\
\text{S.T.:}\quad & 16x_1 + 5x_2 + 8x_3 \le 520 \\
& 10x_1 + 5x_2 + x_3 \le 210 \\
& 2x_1 + x_2 + x_3 \le 125 \\
& x_3 \le 50 \\
& x_1, x_2, x_3 \ge 0
\end{aligned}$$

or, in vector and matrix notation:

$$\underset{\mathbf{x}}{\text{Max}}\ x_0 = (45, 20, 25)\begin{pmatrix} x_1 \\ x_2 \\ x_3 \end{pmatrix}$$

$$\text{S.T.:}\quad \begin{bmatrix} 16 & 5 & 8 \\ 10 & 5 & 1 \\ 2 & 1 & 1 \end{bmatrix}\begin{pmatrix} x_1 \\ x_2 \\ x_3 \end{pmatrix} \le \begin{pmatrix} 520 \\ 210 \\ 125 \end{pmatrix}$$

$$\mathbf{x} \ge 0$$

Note that the above formulation has the four characteristics cited earlier that qualify it as a linear programming problem (as opposed to some other type of mathematical programming problem). First, the decision variables—x_1, x_2, and x_3—are *not* restricted to take only integer values. Second, each of the x's is restricted to be assigned only nonnegative values. Third, all of the functions embedded in the model are linear with respect to the x's. And finally, the parameters are modeled as fixed constants rather than as random variables. ■

Linear programming models are usually not totally accurate representations of the underlying decision problems being addressed in the math programming analysis. This is because relationships between decision variables are usually not linear in nature. In addition, most problems involve decision variables for which fractional values are unrealistic (for example, x might be assigned a solution value of 2.3 television sets). But at the same time, linear programming problems are by far the most common types of math programming models used to address real-world problems. The reason stems from the trade-off discussed in Chapter 2 between a model's accuracy and its utility. For math programming models, utility means the ability to solve the model. And it is generally true that linear programming problems are much easier to solve than are integer, nonlinear, or dynamic programming problems.

8.4.2 Integer Programming Models

The primary characteristic of an integer programming problem is the restriction that the decision variables must be assigned only integer values. The functions $f(\mathbf{x})$, $g_1(\mathbf{x})$, . . . , can have any form. When they are all linear in the x_j's, the model is called an "integer linear programming" problem. A general version of such problems would look like model (8.3) except that the last set of constraints, the nonnegativity constraints $x_j \geq 0$ for all j values, would be amended to include the simple statement "and x_j integer."

Integer programming problems can be subdivided into pure integer, zero–one, or mixed integer programming problems. Pure integer models require all the decision variables in the problem to be assigned an integer value, such as 0, 1, 2, 3, ... Zero–one models require that the decision variables be assigned only a value of one or zero. An illustration of this class of integer programming problems is model (8.1) as given in Example 8.1. Mixed integer models have a subset of the decision variables that are allowed to be assigned fractional values, whereas the remaining variables must take on only integer values.

Example 8.2

As an illustration of an integer programming problem consider the following decision problem. The Floyd Wood Products Co. (FWP) is in the business of dressing out raw lumber into finished board for use in manufacturing furniture. FWP is trying to decide which of three types of lumber it should devote its facilities to in the next week: pine, oak, or maple. To dress each type of lumber requires a significant amount of labor to set up the required processing equipment, as well as a direct material and labor cost for each board-foot of lumber dressed. Specifically, FWP estimates that it cost them \$1000 to set up to dress pine lumber, \$1500 to set up for oak, and \$1100 for maple. They also estimate that the before-tax net profit from one board-foot of dressed pine is \$0.04, from oak \$0.07, and from maple \$0.06. Based on time studies conducted several years ago, the standard time to dress each product is 0.8 hr per board-foot for pine, 0.8 hr for oak, and 1.1 hr for maple, and setup times are 35 hr for pine, 50 for oak, and 50 for maple. There is a total of 400 hr for dressing available in a week.

FWP feels as though they can sell any amount of each of the types of dressed lumber. The question is how should they allocate their processing time so as to maximize their expected profit for next week.

A mathematical programming model for this problem is as follows:

$$\max_{x,y} x_0 = 0.04x_1 + 0.07x_2 + 0.06x_3 - 1000y_1 - 1500y_2 - 1100y_3 \quad (8.7)$$

$$\text{S.T.:}\quad 0.8x_1 + 0.8x_2 + 1.1x_3 + 35y_1 + 50y_2 + 50y_3 \leq 400$$
$$x_1, x_2, x_3 \geq 0$$
$$y_1, y_2, y_3 = 0 \text{ or } 1$$

where x_j is the number of board-feet of type j lumber dressed out during the week ($j = 1$ represents pine, $j = 2$ represents oak, and $j = 3$ represents maple); and y_j is 1 if any amount of type-j lumber is dressed, and 0 otherwise (i.e., $y_j = 1$ if and only if $x_j \neq 0$). Note that the definition of y_j could be included in the model as a formal constraint as follows:

$$x_j - My_j \leq 0 \qquad \text{for } j = 1, 2, 3$$

where M is a very large number, such as 10^8. (Why does this constraint, coupled with the sign of the coefficient of y_j in the objective function, force the definition of y_j to hold? For a discussion, see Reference 5.)

The above model is a mixed integer programming problem. More precisely, it is a mixed integer, *linear* programming problem. The subset of decision variables required to be integers is (y_1, y_2, y_3). The remaining variables, that is, x_1, x_2, and x_3, can be assigned any nonnegative value including a fractional amount. ■

8.4.3 Nonlinear Programming Models

If one or more of the functions in a given mathematical programming model is nonlinear in the decision variables, then that model is classified as a nonlinear programming problem. For instance, in Example 8.2, if the net unit profit of dressed lumber were not linear but rather exhibited economies of scale that caused profit to increase with the square of the amount of lumber dressed, then the following nonlinear programming problem could result:

$$\max_{x,y} x_0 = 0.04x_1^2 + 0.07x_2^2 + 0.06x_3^2 - 1000y_1 - 1500y_2 - 1100y_3 \quad (8.8)$$

$$\text{S.T.:}\quad 0.8x_1 + 0.8x_2 + 1.1x_3 + 35y_1 + 50y_2 + 50y_3 \leq 400$$
$$x_1, x_2, x_3 \geq 0$$
$$y_1, y_2, y_3 = 0 \text{ or } 1$$

The nonlinearity does not have to exist in the objective function in order that the model constitute a nonlinear programming problem. For example, if the standard times in Example 8.2 were modeled to reflect learning-curve effects, the primary constraint of model (8.7) might have appeared as

$$0.8x_1^{1/2} + 0.8x_2^{1/2} + 1.1x_3^{1/2} + 35y_1 + 50y_2 + 50y_3 \leq 400$$

so that model (8.7) would then be classified as a mixed integer, nonlinear programming problem. Notice that regardless of whether the decision variables are continuous or integer in nature, a model involving a function that is nonlinear is still a nonlinear programming problem.

There are several special types of nonlinear programming problems that have special structure. For instance, "nonconvex programming problems" involve nonlinear functions (usually only the objective function) that are not convex functions of the decision variables. "Separable convex programming problems" involve a nonlinear, convex objective function that has no interaction terms among the decision variables (that is, no terms like $x_1 x_2$), so that a separate convex function can be identified for the contribution that each decision variable makes to the overall objective. The constraints are usually linear.

A third special nonlinear programming problem is the "quadratic programming problem." This type of problem involves maximizing (or minimizing) a *quadratic* objective function subject to linear constraints. Recalling the definition of a quadratic function, notice that the quadratic programming problem differs from the linear programming problem only by the inclusion of x_j^2 and $x_j x_k$ interaction terms in the objective function. These special nonlinear programming problems are more fully discussed in References 1 and 4.

8.4.4 Dynamic Programming Models

This category of math programming problems involves a series of separate stages of decision making, all linked together by some coupling relationship(s) such as the use of a common scarce resource. For example, suppose that a plant wishes to establish a schedule of the amount of each of its product to manufacture in each week for the upcoming year. In this case, the overall annual production plan is to be made through a series of 52 separate decision problems—determining the product mix for a given week. Because of the possible carryover of inventory of each product from one week to the next, the decision made in one stage or week is dependent on the decision made in the previous week. Hence, each of the 52 decision problems is linked together by an inventory production and storage relationship.

Dynamic programming is an approach to analyzing such situations. It involves the decomposition of the overall problem into separate decision problems, but in such a way that the answer to one stage's problem feeds into or provides input for the next stage problem. In other words, the decision made at any given stage is made in view of its pyramiding effect on the remaining decision stages. (For a complete discussion of dynamic programming, the reader is referred to Hillier and Lieberman [4].)

8.4.5 Deterministic versus Probabilistic Models

Another way to classify math programming problems is in terms of the probabilistic nature of the parameters in the problem. As discussed in Chapter 2, any parameter in a model can either have a fixed value (i.e., be deterministic) or take on any one of a number of values according to some probability distribution (i.e., be a random variable). Any math programming model that has one or more parameters that are random variables is said to be a probabilistic model. Otherwise, it is a deterministic model.

An example of a class of probabilistic models is the chance-constrained programming problem. Such models usually arise when it is highly desirable but not absolutely necessary for a constraint to hold. This relaxation is handled by expressing the constraint as a probability statement rather than as a strict inequality. Otherwise, the model is unaffected. For example, suppose that in Example 8.2, the available production time were no longer exactly 400 hours, but, because of possible absenteeism, it were treated as a random variable. If management were willing to have a solution that had at least a 95-percent chance of not using more than 400 hours, then the chance-constrained formulation of the decision problem in the example would be

$$\max_{\mathbf{x},\mathbf{y}} x_0 = 0.04x_1 + 0.07x_2 + 0.06x_3 - 1000y_1 - 1500y_2 - 1100y_3$$

$$\text{S.T.:}\quad P(0.8x_1 + 0.8x_2 + 1.1x_3 + 35y_1 + 50y_2 + 50y_3 \leq 400) \geq 0.95$$

$$x_1, x_2, x_3 \geq 0$$

$$y_1, y_2, y_3 = 0 \text{ or } 1$$

There are other formats for probabilistic math programming models. Several are described in Reference 4.

8.5 SOLUTION OF MATH PROGRAMMING PROBLEMS

The recommended answer to a decision problem being analyzed through mathematical programming is the final solution to the associated math programming model. This final solution is the collection of numerical values assigned to the model's decision variables as of the last step or iteration of the solution process, or "algorithm." That is, the recommended answer is that the first decision variable be given the value that the algorithm finally assigned to x_1, the second decision variable be given the final value assigned to x_2, and so forth. At the same time, it is important to recognize that we are solving only a mathematical representation of the true decision problem rather than the actual problem itself.

8.5.1 Algorithms

Just as there are alternative processes that can be used to manufacture a certain part or alternative ways to assess or measure the worth of a particular cash-flow series, there are also alternative algorithms that can be used to solve a given math programming problem. All algorithms are essentially search procedures, like the one-at-a-time method described in Chapter 7. The search is designed to move from one trail solution to another in such a way as to eventually converge to or arrive at a final solution.

The goal of the algorithm is to produce a final solution that is feasible with respect to the constraints of the model and at the same time provides a better objective function value than does any other feasible solution. Such a solution is called an *optimal* solution. For example, for the problem represented earlier in model (8.4) the solution $x_1 = 5.6$, $x_2 = 1.2$ is the optimal solution. It is feasible since, upon substituting the point (5.6, 1.2) into the three primary inequalities, we find that each constraint holds or is met at this solution value. In addition, (5.6, 1.2) is a nonnegative point and therefore meets the $x_1 \geq 0$ and $x_2 \geq 0$ restrictions. As will be seen later, this point also provides a better—that is, higher in this case, as the objective of model (8.4) is to maximize x_0—value of the objective function, $x_0 = 10x_1 + 2x_2$, than does any other *feasible* solution. For instance, the point $x_1 = 1$, $x_2 = 1$ is feasible but produces a value of x_0 of 12, which is less than the value associated with the point (5.6, 1.2) of $x_0 = 10 \times (5.6) + 2 \times (1.2)$, or $x_0 = 58.4$. Notice that other (x_1, x_2) points can be found that yield larger values of x_0, but none of these points are feasible. This is the case with the point $x_1 = 10$, $x_2 = 10$, which satisfies the first constraint in model (8.4) but does not meet the last two primary constraints.

As was the case with the unconstrained optimization models discussed in Chapter 7, a solution point can be a *local* optimum for a math programming model rather than a *global* optimum. In the case of math programming, a local optimum can be defined as a feasible solution to the model that provides a better objective function value than any other feasible solution that is within a "local neighborhood" of the local optimum point. In other words, if the solution algorithm were at a feasible point and moved a very small distance in *any feasible* direction away from the point (so that the new point moved to is still feasible), then, if no such new point were as good as the solution point, the solution point would be a local optimum. This point would also be a *global* optimum if it were better than any feasible point, regardless of how far away it was located.

The movement of an algorithm through a series of feasible points toward a final solution is called *convergence*. Depending on how it is designed, an algorithm may converge only to a local optimum rather than to a global optimum. In fact, some algorithms may terminate with a final solution point that is perhaps better than some other points but is not even a local optimum. When we cannot rigorously prove that an algorithm will always converge to a global optimum, we classify that algorithm as an inexact procedure, or a *heuristic* algorithm. The termination point of a heuristic procedure *may* be a global optimum, but this cannot be guaranteed ahead of time.

Besides the nature of the solution found at the termination of an algorithm, another important characteristic of the convergence of a solution procedure is its speed or efficiency. Two different algorithms may both be capable of converging to the global optimum for a particular problem; but one may do so twice as fast as the other. The speed of convergence is important when large problems are solved since the solution process consumes valuable manual or computer time. Even on the fastest computers some problems require hours and hours of continuous computer time before a solution is reached.

Normally, there is an inverse relationship between the speed of an algorithm and the exactness of its final result. That is, an algorithm that is guaranteed to produce the global optimum for a particular problem may not have as rapid a convergence as would a heuristic algorithm that stops with merely a good solution to that problem. In designing an algorithm to solve a particular model this trade-off must be kept in mind. The best balance between speed and end results depends on cost of the computing resource and the need for having the very best solution as opposed to perhaps only a good solution.

8.5.2 Designing Algorithms

It was pointed out above that a consideration that must be made in constructing an algorithm to solve a particular model is the trade-off between speed of convergence and the nature of its end results. Another, more fundamental consideration is that the algorithm should take advantage of the special structure of the model.

There are two general levels of structure in a math programming model. The first is the macro, overall structure that is used to classify the model as being a particular type of math programming problem, such as a linear programming problem or a dynamic programming problem. An algorithm designed to solve one class of problems is generally not suitable for solving other classes of problems.

The second level of a model's structure is its microstructure, determined by such factors as the number of constraints involved or the range of values that the coefficients in the constraints can take on. As in the case of a macrolevel model's structure, by identifying and taking advantage of the microstructure of a model, we can design a more efficient algorithm to solve that model. For example, an algorithm developed to solve a *general* linear programming problem can be used to solve a particular linear programming problem having only one constraint. However, a more efficient algorithm takes into account the fact that there is only one constraint, and the effect that has on how the optimal solution can be identified and selected. In the next section we see, for the case of a linear model, exactly how an algorithm takes advantage of the structure of the model it is intended to solve.

8.6 LINEAR PROGRAMMING

As indicated earlier, linear programming is the special case of math programming in which the decision problem being addressed is modeled so as to involve only linear functions, deterministic parameters, and continuous nonnegative decision variables. These characteristics define the macrolevel model structure discussed in the previous section. In this section we describe how these model characteristics are used in designing linear programming algorithms, particularly the Simplex algorithm. We use a graphical analysis of a two-variable problem to motivate the ideas involved. In the final part of this section we describe several special linear programming models that have peculiar substructures that allow efficient, special-purpose algorithms to be used to solve these models.

8.6.1 Graphical Analysis and Solution Properties

Consider the linear programming problem presented in model (8.4) and repeated below:

$$\begin{aligned} \operatorname*{Max}_{x_1,x_2} x_0 &= 10x_1 + 2x_2 \\ \text{S.T.:} \quad x_1 + x_2 &\leq 20 \\ 2x_1 - x_2 &\leq 10 \\ 3x_1 + x_2 &\leq 18 \\ x_1, x_2 &\leq 0 \end{aligned} \tag{8.9}$$

In order to identify the properties of the optimal solution for this problem, it is helpful to represent the functions involved on a graph of (x_1, x_2) points. From this graph we shall be able to readily find the optimal solution point and to see how a solution algorithm can be designed to mathematically find this point.

Notice that each of the inequality constraints in model (8.9) define a separate region of the (x_1, x_2) space for which all points in that region are feasible with respect to *that particular* constraint. For example, the shaded portion of the (x_1, x_2) space shown in Figure 8.1 is the set of (x_1, x_2) points that are feasible with respect to the first constraint in model (8.9.). The *intersection* of the regions for all constraints forms the subset of (x_1, x_2) points that are feasible for the overall problem. This area is called the *feasible region*. The feasible region for model (8.9) is shown in Figure 8.2.

The optimal solution for a math programming problem obviously must be one of

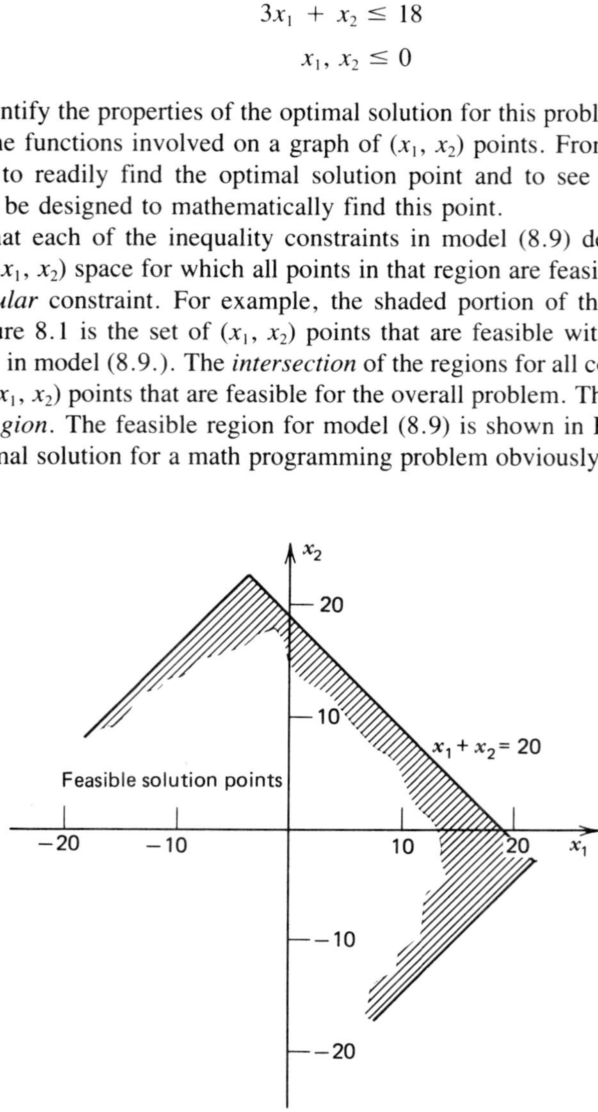

Figure 8.1 Feasible Space for Constraint One in Model (8.9)

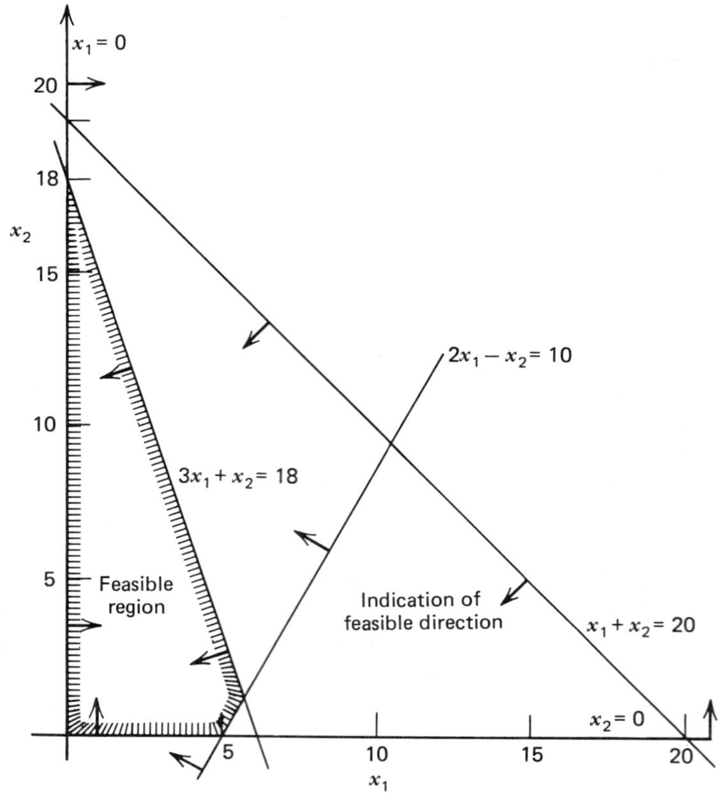

Figure 8.2 Feasible Region for Model (8.9)

the points in the problem's feasible region. But which one? For a linear programming problem, one of the points constituting an extremity of the feasible region will always be the optimal solution. To see why, it is convenient to include a representation of the objective function on the graph of the constraints.

Unlike the constraints in model (8.9), the objective function is not a function of only the two unknowns x_1 and x_2. The value of the objective function x_0 is also an unknown. Hence, the objective function for model (8.9) is a plane in the *three-dimensional* (x_1, x_2, x_0) space. As was the case with unconstrained optimization functions in Chapter 7, we can construct two-dimensional *contour lines* to show the (x_1, x_2) points providing a given function value x_0. The three-dimensional representation of the objective function for model (8.9) is given in Figure 8.3 along with an illustration of how two specific contour lines would be obtained. Such contour lines can be shown on the same graph as the feasible region in order to provide an indication of the relative value of various feasible (x_1, x_2) points. For example, the $x_0 = 40$ and $x_0 = 20$ contour lines for the objective function of model (8.9) are shown together with the feasible region in Figure 8.4.

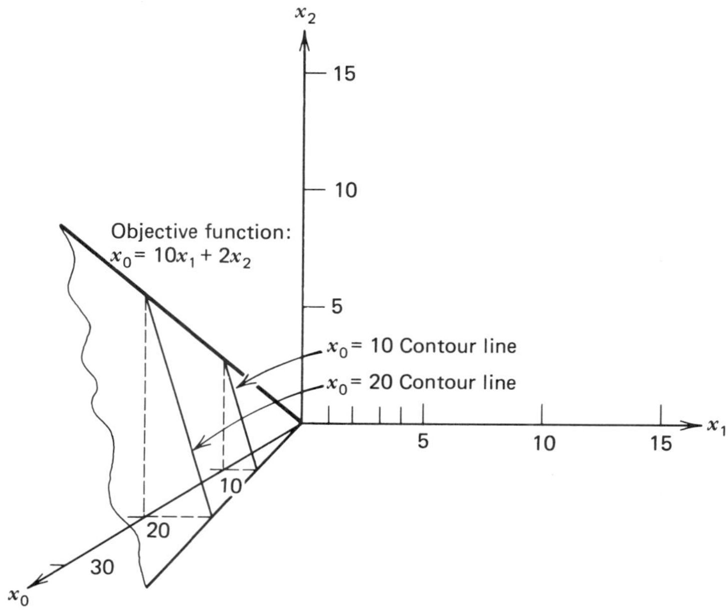

Figure 8.3 Objective Function for Model (8.9)

Notice that there are an infinite number of contour lines that could be drawn on Figure 8.4, each one representing a different value of the objective function. These lines are parallel and increase in value as they are constructed further away from the origin. Also, note that all (x_1, x_2) points lying on a given contour line yield the objective function value associated with that contour line. Therefore, the points (2, 0) and (0, 10) both provide a solution value of model (8.9) of $x_0 = 20$, which is better than points such as (1, 0) and (0, 5) along the $x_0 = 10$ contour line. It obviously follows from this observation that the optimal solution is one of the points on the boundary of the feasible region that lies on the contour line that is farthest away from the origin (if the objective were to minimize x_0 the contour line of interest would be the closest one to the origin). Sweeping the contour lines out from the origin in Figure 8.4 implies that the last contour line lies on top of the point (5.6, 1.2). Hence, the optimal solution to model (8.9) is $x_1 = 5.6$ and $x_2 = 1.2$.

The point (5.6, 1.2) is called an *extreme point* or corner point of the feasible region of model (8.9). An extreme point in a math programming problem is an intersection of the equality portion of the constraints, provided that intersection is also feasible. From Figure 8.2, note that there are 10 constraint intersections shown that lead to the four extreme points: (0, 0), (5, 0), (5.6, 1.2), and (0, 18). As is the case with model (8.9), it is true that the optimal solution to any linear programming problem, regardless of the number of variables, will always be an extreme point. As evident from the graphical analysis of model (8.9), this fact stems from the special structure

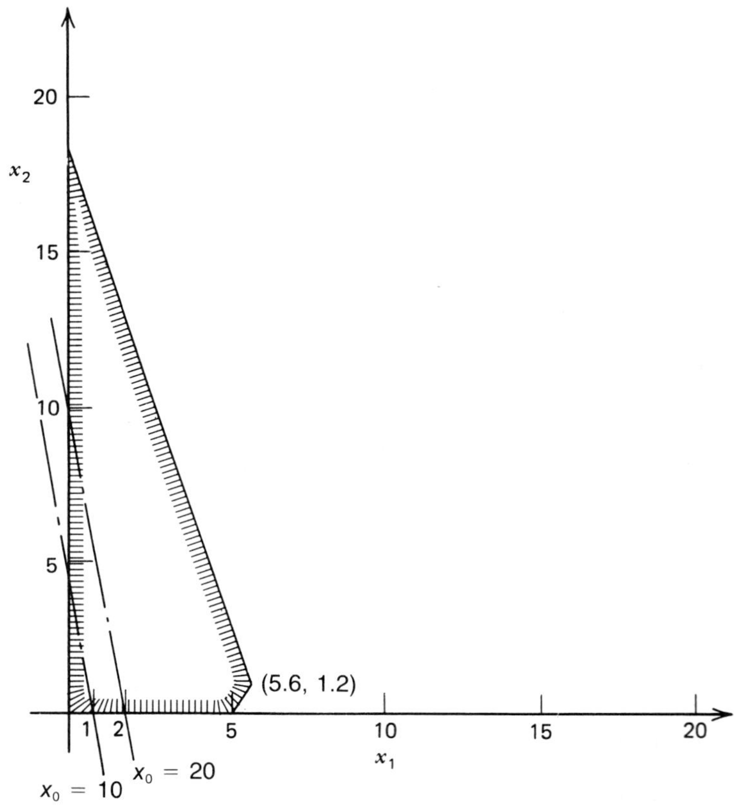

Figure 8.4 Feasible Region and Objective Function Contour Lines for Model (8.9)

of a linear programming problem—the fact that all functions are linear, and that the variables can take on fractional values.

There are several important implications of the structure of a linear programming problem that we have touched on in the previous discussion. These characteristics are given in the following properties.

Property 1. There are a finite number of extreme points of the feasible region for a given linear programming model.

Property 2. A local optimum in a linear programming problem is also the global optimum.

Property 3. One of the extreme points of the feasible region corresponds to a local optimum.

These important properties of the solution of a linear programming problem can be used in designing a solution algorithm. Specifically, an efficient solution procedure should involve searching over only those feasible points which are also extreme points. This is a basic characteristic of the most commonly used linear programming algorithm, the Simplex procedure. The next two sections briefly describe this procedure and its philosophy.

8.6.2 The Simplex Procedure

The Simplex procedure is an exact algorithm used to find the optimal solution to any linear programming problem. The approach employed in the Simplex procedure is to move from one extreme point to a neighbor or adjacent extreme point and to continue this search until a local optimal solution is found. At each extreme point in the search, a test is made to see if any improvement in the value of the objective function can be obtained by moving to one of the finite number of adjacent extreme points. If some of the adjacent points offer the chance for improvement, the one with the largest marginal gain is chosen as the next point to move to and to continue the search from. If no adjacent extreme points offer improvement, the point the search is currently at is a local optimum and the procedure terminates. It can be proven that the Simplex procedure will always terminate in a finite number of moves (see, for example, Reference 2).

As an illustration of this adjacent extreme point search, note that in Figure 8.4, if the Simplex procedure started with the origin as the initial feasible solution, it would first move to the point (5, 0). This is because there are only two extreme points that are adjacent to the origin: (5, 0) and (0, 18). The other extreme point is not adjacent to the origin. Of the two adjacent points, (5, 0) offers the larger marginal improvement in x_0 as we move away from (0, 0). Using (5, 0) as a base, the procedure would next examine the marginal change if we were to move to one of (5, 0)'s adjacent neighbors, (0, 0) and (5.6, 1.2). The change in x_0 by moving back toward (0, 0) would be a decrease and therefore offers no improvement. However, the move toward (5.6, 1.2) does offer an increase. Therefore, (5.6, 1.2) is selected as the next base point from which to continue the search. On the third iteration, we would find that moving toward neither of (5.6, 1.2)'s adjacent neighbors, (5, 0) and (0, 18), offers improvement, so the search would stop. This last base point is a local optimum since movement in any feasible direction will produce no better solution [examine a contour line of the objective fuunction through this (5.6, 1.2) point].

8.6.3 Steps In the Simplex Procedure

The specific steps of the basic Simplex algorithm are presented below. It is assumed that the problem being addressed here has a maximization objective function and only inequality constraints in which the left-hand side is constrained to be less than or equal to the right-hand side. Slight modifications to the following steps are required to solve other versions of the linear programming problem (for a complete discussion see Reference 1).

1. *Transform the model into standard form.* Convert each primary constraint into an equation by adding a new secondary decision variable, called a slack variable, to the left-hand side of the constraint. Let each slack have a zero cost coefficient in the objective function. For instance, the standard form of model (8.9) would be:

$$\begin{aligned}
\max_{x_1, x_2} x_0 &= 10x_1 + 2x_2 + 0x_3 + 0x_4 + 0x_5 \\
\text{S.T.:}\quad & x_1 + x_2 + x_3 = 20 \\
& 2x_1 - x_2 + x_4 = 10 \\
& 3x_1 + x_3 + x_5 = 18 \\
& x_1, x_2, x_3, x_4, x_5 \geq 0
\end{aligned} \tag{8.10}$$

where x_3, x_4, and x_5 are the slack variables for constraints one, two, and three, respectively. A slack variable in, say, the ith constraint, represents the excess or unused amount of the ith scarce resource after the original decision variables (e.g., x_1 and x_2) have been assigned values. For example, if the solution algorithm had moved to the extreme point (5, 0) in Figure 8.4, the value of the primary variables would be $x_1 = 5$, and $x_2 = 0$, while the slack variables would have the following values: $x_3 = 15$, $x_4 = 0$, and $x_5 = 3$. Notice that a positive slack variable indicates that the solution point [in this case, (5.0)] is away from the constraint associated with that slack. A zero-valued slack variable indicates that the solution point is *on* the associated constraint; in other words, that constraint is "binding" at the solution point. Although we are usually interested in the slack variable values only as a consequence of the values of the primary decision variables, once slacks have been added to the model they are treated by the algorithm exactly like any other decision variable.

2. *Select an extreme point as a starting trial solution.* This starting point is usually the origin. Hence, ordinarily only the slack variables would be assigned positive values in the initial feasible solution.

For bookkeeping purposes, a tableau is usually used to display a solution. The tableau provides the current value of the decision variables and is updated at each iteration of the algorithm. In terms of the parameters of the general linear programming model, the tableau representing the initial feasible solution (that is, the origin) for model (8.9) is as follows:

	x_0	x_1	x_2	x_3	x_4	x_5	RHS
x_0	1	$-c_1$	$-c_2$	$-c_3$	$-c_4$	$-c_5$	0
x_3	0	a_{11}	a_{12}	a_{13}	a_{14}	a_{15}	b_1
x_4	0	a_{21}	a_{22}	a_{23}	a_{24}	a_{25}	b_2
x_5	0	a_{31}	a_{32}	a_{33}	a_{34}	a_{35}	b_3

(8.11)

or, in terms of the numerical version of the model:

	x_0	x_1	x_2	x_3	x_4	x_5	RHS
x_0	1	−10	−2	0	0	0	0
x_3	0	1	1	1	1	1	20
x_4	0	2	−1	0	1	0	10
x_5	0	3	1	0	0	1	18

(8.12)

In constructing a tableau for any given linear programming problem, there would always be one row for each primary constraint in the problem, and a column for each variable including x_0. In addition, each row or constraint will always have a variable associated with it as indicated in the first column of the tableau. If there are m constraints in the problem, then there will be m of these variables, called *basic variables*. All other decision variables are called *nonbasic variables*.

It is through the use of these basic and nonbasic variables that we portray the particular solution at hand. Specifically, each nonbasic variable is automatically assigned a zero value, while each basic variable takes on the *updated* value of the right-hand side parameter (i.e., b_i) for the constraint with which that variable is associated. These updated b_i values are given in the last column of the tableau. For example, for tableau (8.12) we can see that the solution represented has the slack variables as the basic variables with assigned values of $x_3 = 20$, $x_4 = 10$, and $x_5 = 18$, and has the original decision variables as nonbasic variables (since they are not shown in the first column as being the basic variable for any of the constraints) with values of $x_1 = 0$ and $x_2 = 0$. Also, note from the tableau that the value of this solution is $x_0 = 0$.

3. *Select a nonbasic variable to replace a basic variable.* By definition, a particular solution or extreme point requires that one and only one decision variable be associated with each constraint or row as that contraint's basic variable. A different extreme point would still have exactly one basic variable for each constraint, but the specific decision variables that make up the set of basic variables for the new extreme point may be different. (The objective function row will always be x_0 as its "basic variable.") For instance, from Figure 8.2 we can see that at the origin, the basic variables are the slacks (x_3, x_4, and x_5), as they have positive values at that particular extreme point. In contrast, x_1 and x_2 have zero values (and, hence, are the nonbasic variables). Then, at the solution point (5, 0), the basic variables are x_3, x_1, and x_5 since at this extreme point $x_1 = 5$, $x_2 = 0$; and, by substituting $x_1 = 5$ and $x_2 = 0$ into the constraints in model (8.10), $x_3 = 15$, $x_4 = 0$, and $x_5 = 3$.

Notice that in going from the point (0, 0) to the *adjacent* point (5, 0), there was a one-for-one swap between a nonbasic variable (i.e., x_1) and a basic variable (i.e., x_4). Because the Simplex procedure is an adjacent extreme point search, it is always true that in moving from one solution and its tableau to the next solution and its tableau, a nonbasic variable and a basic variable from the old tableau will swap positions in the new tableau. As an illustration, if the current or old solution is $x_1 = 0$, $x_2 = 0$, $x_3 = 20$, $x_4 = 10$, $x_5 = 18$ as represented by tableau (8.12), and *if* x_1 were chosen to "enter the basis" (that is, to become a basic variable in the next solution) and simultaneously to replace x_4, then the new tableau would appear as

	x_0	x_1	x_2	x_3	x_4	x_5	RHS
x_0	1	0	-7	0	5	0	50
x_3	0	0	$\frac{3}{2}$	1	$-\frac{1}{2}$	0	15
x_1	0	1	$-\frac{1}{2}$	0	$\frac{1}{2}$	0	5
x_5	0	0	$\frac{5}{2}$	0	$-\frac{3}{2}$	1	3

(8.13)

We will explain shortly how and why the various numbers inside the tableau are revised. However, notice from the tableau that the new solution is $x_1 = 5$, $x_2 = 0$, $x_3 = 15$, $x_4 = 0$, and $x_5 = 3$, which is an adjacent extreme point to the origin. Also, notice in any given tableau that the basic variables can also be identified by the existence of a unit vector under them. The row position of the one in the unit vector will correspond to the constraint with which that basic variable is associated. For instance, in tableau (8.13), the basic variable x_1 is associated with the second constraint.

The choice of which currently nonbasic vector should "enter the basis" is based on our desire to move to the optimum extreme point as fast as possible. As a nonbasic variable becomes basic it changes from a zero value (thereby having a zero effect on the value of the objective function) to a positive value (thereby making some plus or minus contribution to x_0). The *relative* desirability of letting a nonbasic variable become basic is indicated by the current value of its coefficient in the x_0 row of the tableau. In the case of a maximization problem, the more negative that coefficient is, the larger will be the relative increase in x_0 as that nonbasic variable becomes basic. Therefore, our choice as to the entering nonbasic variable x_e can be given by the following rule:

Let x_e be the nonbasic variable such that

$$c'_e = \min\{c'_j \text{ for all variables } j \text{ that are currently nonbasic and are } <0 \text{ in value}\} \tag{8.14}$$

where e is the index of the entry or selected nonbasic variable, and c'_j is the current value of the coefficient of the jth variable in the x_0 row of the tableau. For example, in tableau (8.12) x_1 has the smallest or most negative c'_j value ($c'_1 = -10$) of the nonbasic variables at that solution, so $x_e = x_1$ for the next tableau. Likewise, $x_e = x_2$ in tableau (8.13).

4. *Select the basic variable to be replaced.* As indicated above, in moving from the current solution to the next extreme point, a variable that is presently "in the basis" must leave the basis and become nonbasic so that the incoming nonbasic variable can be associated with some constraint. The choice of the leaving basic variable is based on which constraint most inhibits the increase in the entering variable. The rule for selecting the leaving variable is:

Let

$$\frac{b'_r}{a'_{re}} = \min\left\{\frac{b'_i}{a'_{ie}} \text{ for each row } i \text{ such that } a'_{ie} > 0\right\} \tag{8.15}$$

where r is the row index of the basic variable selected to leave the basis, e is the column index of the entering variable, b'_i is the current tableau value of the right-hand parameter for row or constraint i, and a'_{ie} is the current tableau value of the coefficient of variable x_e in the ith constraint. For instance, applying the rule to tableau (8.13) with x_2 as the entering variable would lead to the selection of x_5 as the leaving basic variable according to the following results:

$$\frac{b'_r}{a'_{re}} = \frac{b'_3}{a'_{32}} = \frac{3}{5/2} = \frac{6}{5} = \min\left\{\frac{15}{3/2}, \frac{3}{5/2}\right\}$$

5. *Determine the new solution and update the tableau.* By choosing a specific variable to enter the basis and one to leave, we are essentially determining which adjacent extreme point to move to next. In order to find out what value each variable takes on at this new extreme point, we have to solve m simultaneous linear equations (that is, the m constraints in the model) for m specific unknowns (that is, the new basic variables, one for each constraint). Using the tableau format, this can be accomplished by using the following "Gauss–Jordan elimination" procedure.

(a) Let the index of the entering nonbasic variable be designated as e and the index of the row of the leaving basic variable be r. Also, let the elements in the current tableau be c_j, a_{ij}, and b_i.

(b) In row r of the leaving variable, let each current a_{rj} element be replaced by

$$a'_{rj} = \frac{a_{rj}}{a_{re}} \tag{8.16}$$

as well as the right-hand-side parameter be replaced by

$$b'_r = \frac{b_r}{a_{re}} \tag{8.17}$$

(c) In every other row of the tableau, let the current a_{ij} value be replaced by

$$a'_{ij} = a_{ij} - a_{ie}\left(\frac{a_{rj}}{a_{re}}\right) \tag{8.18}$$

as well as the right-hand-side parameter be replaced by

$$b'_i = b_i - a_{ie}\left(\frac{b_r}{a_{re}}\right) \tag{8.19}$$

and the x_0 row elements be replaced by

$$c_j' = c_j - c_e\left(\frac{a_{rj}}{a_{re}}\right) \tag{8.20}$$

Notice that tableau (8.13) was obtained from tableau (8.12) using this Gauss–Jordan elimination procedure. Also, note that this procedure is simply an alternative to the standard substitution method of algebraically solving simultaneous equations.

6. *Check to see if any improvement beyond the new solution is possible.* If there is, go back to step 3. If there is not, stop; the solution at hand is the optional solution. For a maximization problem, improvement in the value of the objective function x_0 can be achieved if some nonbasic variable has a current coefficient in the x_0 row that is negative. If all the elements are greater than or equal to zero, then the current solution cannot be improved by moving to an adjacent extreme point.

Example 8.3

In order to illustrate the steps in the Simplex procedure, model (8.4) will be solved.

Iteration 1

Steps 1 and 2. Transform the model into standard form and represent the starting solution in tableau format.

	x_0	x_1	x_2	x_3	x_4	x_5	RHS
x_0	1	−10	−2	0	0	0	0
x_3	0	1	1	1	0	0	20
x_4	0	2	−1	0	1	0	10
x_5	0	3	1	0	0	1	18

Step 3. Select the entering nonbasic variable. Since $-10 = \min(-10, -2)$, then let x_1 be the entering variable.

Step 4. Select the leaving basic variable. Since $10/2 = 5 = \min(20/1, 10/2, 18/3)$, then let x_4 leave the basis.

Step 5. Update the tableau.

	x_0	x_1	x_2	x_3	x_4	x_5	RHS
x_0	1	0	−7	0	5	0	50
x_3	0	0	$\frac{3}{2}$	1	$-\frac{1}{2}$	0	15
x_1	0	1	$-\frac{1}{2}$	0	$\frac{1}{2}$	0	5
x_5	0	0	$\frac{5}{2}$	0	$-\frac{3}{2}$	1	3

Notice that the new solution is $x_1 = 5$, $x_2 = 0$, $x_3 = 15$, $x_4 = 0$, and $x_5 = 3$ with an objective function value of $x_0 = 50$.

Step 6. Check for optimality. Since at least one nonbasic variable (x_2) in the solution at hand has a negative c_j' value, a better solution can be found. Therefore repeat steps 3 through 6 again.

Iteration 2

Step 3. Select the entering variable. Since x_2 is the only candidate (all other nonbasic variables in the current tableau have c_j' coefficients that are ≥ 0), let it enter the basis.

Step 4. Select the leaving variable. Since $6/5 = \min(30/3, 6/5)$, then x_5 should leave the basis.

Step 5. Update the tableau.

	x_0	x_1	x_2	x_3	x_4	x_5	RHS
x_0	1	0	0	0	$\frac{4}{5}$	$2\frac{4}{5}$	$58\frac{2}{5}$
x_3	0	0	0	1	$\frac{2}{5}$	$-\frac{3}{5}$	$13\frac{1}{5}$
x_1	0	1	0	0	$\frac{1}{5}$	$\frac{1}{5}$	$5\frac{3}{5}$
x_2	0	0	1	0	$-\frac{3}{5}$	$\frac{2}{5}$	$1\frac{1}{5}$

Step 6. Check for optimality. Since $c_j' \geq 0$ for all nonbasic variables, the current solution is optimal. Therefore, the final solution is $x_1 = 5.6$, $x_2 = 1.2$, $x_3 = 13.2$, $x_4 = 0$, and $x_5 = 0$ with an objective function value of $x_0 = 58.4$. The reader should compare the three tableaus above with the graph in Figure 8.4. ■

Example 8.4

Solve the following problem:

$$\max_{x,y,z}\ x_0 = 2x + 4y - z$$

$$\begin{aligned} \text{S.T.:}\quad x + 2y - 3z &\leq 6 \\ 3x \qquad\quad + 3z &\leq 9 \\ x,\ y,\ z &\geq 0 \end{aligned}$$

The standard form for this model is

$$\max_{x,y,z,s_1,s_2}\ x_0 = 2x + 4y - z$$

$$\begin{aligned} \text{S.T.:}\quad x + 2y - 3z + s_1 &= 6 \\ 3x \qquad\quad + 3z + s_2 &= 9 \\ x,\ y,\ z,\ s_1,\ s_2 &\geq 0 \end{aligned}$$

Iteration 1

The arrows indicate the entering and leaving variables.

	x_0	x	y	z	s_1	s_2	RHS	
x_0	1	−2	−4	1	0	0	0	
s_1	0	1	2	−3	1	0	6	←
x_2	0	3	0	3	0	1	9	
			↑					

Iteration 2

	x_0	x	y	z	s_1	s_2	RHS	
x_0	1	0	0	−5	2	0	12	
y	0	$\frac{1}{2}$	1	$-\frac{3}{2}$	$\frac{1}{2}$	0	3	
s_2	0	3	0	3	0	1	9	←
				↑				

Iteration 3

	x_0	x	y	z	s_1	s_2	RHS
x_0	1	5	0	0	2	$\frac{5}{3}$	27
y	0	2	1	0	$\frac{1}{2}$	$\frac{1}{2}$	$7\frac{1}{2}$
z	0	1	0	0	0	$\frac{1}{3}$	3

The optimal solution to the original problem is: $x = 0$, $y = 7\frac{1}{2}$, and $z = 3$. ■

8.6.4 Alternative Linear Programming Algorithms

The Simplex procedure is just one of several existing algorithms useful in solving a general linear programming problem. Other methods are the Revised Simplex, the Primal–Dual, and the Dual Simplex algorithm [2].

As does the Simplex procedure, each of these algorithms is designed to take advantage of the special solution properties of any linear programming problem. However, they use slightly different logic in an effort to be more efficient. For example, the Dual Simplex procedure first selects the leaving variable, *then* selects the entering variable, which is the opposite of the Simplex procedure. The Revised Simplex uses the same basic steps as does the Simplex procedure, but involves a streamlined bookkeeping system so that calculations to update tableau elements that are not needed at a given iteration are not made. In other words, it eliminates wasted effort hidden in the Simplex calculations. The Revised Simplex is reputed to be one of the most efficient general-purpose linear programming algorithm available.

8.6.5 Special Linear Programming Models

In the previous section, we saw how the macrolevel structure of a linear programming problem is utilized by solution algorithms. There are many subcategories of models that have special microlevel structure that lead to refined algorithms that are more efficient in solving those models than are general-purpose procedures (like the Simplex method). One such model is the transportation problem.

The decision problem involved in the classical transportation problem is the allocation of the production capacity at a set of plants (that is, ''sources'') to meet the demand of each of a set of customers (that is, ''sinks''). However, *any* decision problem that can be modeled with the same format or structure can be solved as if it were a classical plant–customer transportation problem. Using this classical context, the transportation problem model can be expressed by defining the following terms:

x_{ij} = the amount of product to transport from source i to sink j.

c_{ij} = the cost to transport one unit of product from i to j.

b_i = the amount of product on hand at source i to be shipped to customers.

d_j = the amount of product demanded by sink or customer j.

The model is as follows:

$$
\begin{aligned}
\min_{\mathbf{x}} x_0 &= \sum_i \sum_j c_{ij} x_{ij} \\
\text{S.T.:} \quad \sum_j x_{ij} &= b_i \qquad \text{for } i = 1, 2, \ldots, n \\
\sum_i x_{ij} &= d_j \qquad \text{for } j = 1, 2, \ldots, m \\
x_{ij} &\geq 0 \qquad \text{for all } i, j
\end{aligned}
\tag{8.21}
$$

Notice the implied interpretation of the two types of primary constraints in model (8.21). The first says that the total shipment of product out of plant or source i must be equal to plant i's capacity. The second restriction says that each customer or sink's demand must be satisfied. Again, the model need not be defined strictly in terms of the shipment of goods. For example, one model of the decision of how to define territories or districts for a city's fire prevention inspectors includes model (8.21) as its main component [7].

While the Simplex procedure could be used to solve model (8.21), special-purpose algorithms have been developed to take advantage of the special structure of (8.21). The features that distinguish the transportation problem model from the general linear program problem include (1) the use of only two types of constraints, (2) the fact that all coefficients in the constraints are either ones or zeros, and (3) the positioning of the ones and zeros in the constraint set. [The reader should write out model (8.21) for an example problem for illustration.]

Hillier and Lieberman [4] provide an excellent description of a commonly used algorithm for solving model (8.21) called the *transportation simplex method.* This is basically a streamlined version of the simplex method that offers "tremendous computational shortcuts . . . by exploiting the (model's) special structure."

Another class of linear programming models with a special structure that can be exploited in solution algorithms is that of *decomposition problems.* Such a model is essentially a collection of sub linear programming problems linked together by one or more overall constraints. Typical of this type problem is a capital budgeting decision in a decentralized firm. The funds available to the firm as a whole are to be allocated to each of m separate divisions. Each division has a set of activities (decision variables) that it can engage in. The use of these divisional activities is restricted among themselves due to limited production resources within a given division. Therefore, without consideration of the overall budget limitation, each division has its own linear programming problem: maximize its division-level profit, subject to the constraints on its activities necessitated by the limitations on the availability of the division's production resources.

The separate division-level linear programming problems are not independent. They are linked together by the constraint that the extent to which the separate divisional activities can be undertaken is limited by the money or capital available to the firm as a whole. This budget restriction is called the *master constraint* in the model. The constraint set in the overall decomposition model consists of the master constraint and the subset of constraints on each division's activities. The objective function is simultaneously to find the level of *all* divisional activities so as to maximize the sum of all divisional profits.

If we let x_{jk} be the level of use of the jth activity that belongs to division k (i.e., division k has its own set of activities or decision variables: $x_{1k}, x_{2k}, \ldots, x_{J_k,k}$), then the overall decomposition model for this type of capital budgeting problem can be expressed as:

$$\begin{aligned}
\operatorname*{Min}_{\mathbf{x}}\ x_0 &= \sum_{k=1}^{K}\sum_{j=1}^{J_k} p_{jk}x_{jk} \\
\text{S.T.:}\quad & \sum_{k=1}^{K}\sum_{j=1}^{J_k} d_{jk}x_{jk} \le B \\
& \sum_{j=1}^{J_k} a_{ijk}x_{jk} \le b_i \qquad \text{for each } k, \text{ and for all } i \in \{I\}_k \\
& x_{jk} \geqslant 0 \qquad \text{for all } j, k
\end{aligned} \tag{8.22}$$

where p_{jk} is the unit profit associated with acitvity j within division k, d_{jk} is the unit investment required by activity j in division k, B is the total funds available to the firm for investment, a_{ijk} is the amount of the ith resource consumed by activity j belonging to division k, b_i is the amount of resource i that is available, and $\{I\}_k$ is the set constraints i that are limitations on the available production resources for division k (for example, $\{I\}_3$ might be set $i = 6, 7, 8$, and 9).

The special structure of model (8.22) is obvious when the constraints are completely written out. The decision variables tend to cluster, as indicated by the shaded portions of Figure 8.5. The important feature to note is that if it were not for the master constraint, the overall problem could be completely *decomposed* and solved as $k = 1, 2, \ldots, K$ independent linear programming problems.

Decomposition problems arise in a number of decision situations and are not restricted to capital budgeting, or to look just like model (8.22). In fact, most large linear programming problems can be manipulated to have a block structure similar to that shown in Figure 8.5. Regardless of whether the underlying decision involves allocating capital, scheduling manpower in a hospital, or determining what stocks to include in an investment portfolio, any linear programming model that has a block appearance like Figure 8.5 can be treated as a decomposition problem.

Models like (8.22) are solved by utilizing a strategy called the *decomposition principle*. Essentially, this principle involves the iterative solution of two separate types of liinear programming problems: one using only the master constraint(s), and the other using the subproblem constraints for a given division. Information is passed back and forth between these separate problems until a solution to each subproblem is reached that is the optimal solution to the original, overall decomposition problem. Use of this decomposition principle is a more efficient solution approach for models involving the block structure of Figure 8.5 than is a general-purpose algorithm like the Simplex procedure. In fact, use of a decomposition procedure is about the only way to solve in a reasonable amount of time very large linear programming problems, those involving thousands of constraints and hundreds of thousands of rows (and such *do* occur frequently in practice).

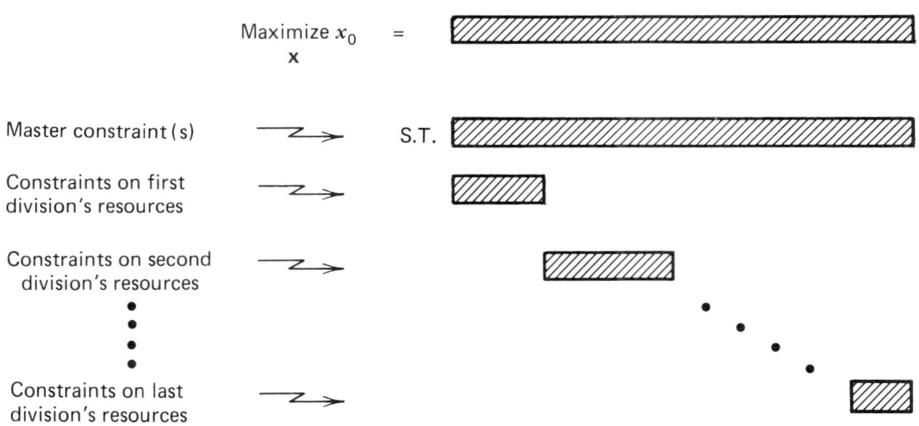

Note: The shaded areas represents the portions of the model for which decision variables have nonzero coefficients.

Figure 8.5 Graphic Portrayal of Special Structure of a Decomposition Problem

8.7 ANALYSIS INVOLVING OTHER MATH PROGRAMMING MODELS

Although linear programming is the most commonly used math programming approach, other models are often used in order to base decisions on a more accurate representation of the underlying problem. The use of several alternatives to linear programming is briefly described below.

8.7.1 Integer Programming

When the analyst formulates a model having integer restrictions on some or all of the decision variables, he is utilizing an integer programming approach to solve the underlying decision problem. Requiring a decision variable to be integer is an accurate representation of many real-world resource allocation decisions. After all, we cannot build half of a television set or assemble a third of an automobile. The units in which most decision variables are expressed necessitates that only whole quantities of those variables be used.

Accurate modeling of the nature of interrelationships present in a decision problem also often necessitate the use of integer variables. For example, consider how one might go about modeling the situation in which either one constraint or another constraint must hold. That is, the decision variables are required to satisfy *either* one of two specific constraints, but not necessarily both. For instance, in deciding how many units of each of two products to produce, it may be that two alternative processes are available to manufacture the products. However, because of space restrictions, only one of the processes can be used, and each one has a limited production capacity. In terms of a numerical illustration, the amount of product 1 made, x_1, and the amount of product 2 made, x_2, must satisfy either

$$6x_1 + x_2 \leq 24$$

or (8.23)

$$2x_1 + 3x_2 \leq 10$$

This situation can be converted into the standard format of separate constraints by defining a new, integer-valued decision variable, y. In particular, if we let y be a variable taking on only a 0 or 1 value and let M be a very large number, then constraint set (8.23) can be expressed as:

$$\begin{aligned} 6x_1 + x_2 &\leq 24 + yM \\ 2x_1 + 3x_2 &\leq 10 + (1 - y)M \end{aligned} \qquad (8.24)$$

Note that if we assign, in the math programming solution procedure, a value of 1 to y, then the second constraint in (8.24) becomes "active" while the first one holds regardless of the values of x_1 and x_2. Rewriting (8.24), we have the following familiar format that would be added to any other model constraints to produce a mixed integer programming model:

$$6x_1 + x_2 - My \leq 24$$
$$2x_1 + 3x_2 + My \leq 10 + M \qquad (8.25)$$
$$y = 0 \text{ or } 1$$

A common approach used in solving integer programming problems is to solve the analogous continuous problem (that is, the integer programming problem without the integer requirements), and round the solution to the nearest integer point. For example, if a two-variable problem were being solved and the optimal continuous solution were $x_1 = 151.2$ and $x_2 = 855.7$, then the integer solution used would be $x_1 = 151$, $x_2 = 856$. Unfortunately, this approach can result in a final integer solution that is infeasible with respect to one or more of the primary constraints in the model. And, even if the rounded solution is feasible, it may be far from true optimal solution to the original integer problem.

As an illustration, consider the decision problem as represented by model (8.4) but modified by replacing the nonnegativity constraints by the integer restriction "x_1, $x_2 \geqslant 0$ and integer." The modified problem is represented graphically in Figure 8.6. Unlike its analogous continuous problem, model (8.4), the feasible region for this

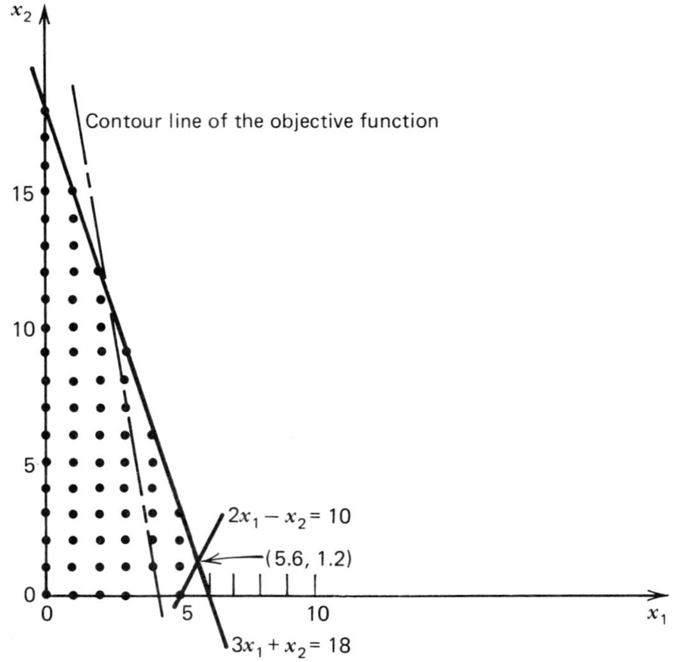

Figure 8.6 Graphic Representation of an Integer Programming Problem

integer programming problem is finite, consisting of the 63 (x_1, x_2) points (0, 0), (0, 1), (0, 2), ... The optimal solution to the continuous problem is (5.6, 1.2). Rounding this to the nearest integer yields (6, 1), which violates the constraint that $3x_1 + x_2 = 18$. If the rounding rule is modified so that the nearest feasible integer point is used, the result would be $x_1 = 5$, $x_2 = 1$. The objective function value of this point is $x_0 = 10(5) + 2(1) = 52$. However, as can easily be seen by moving the contour line shown in Figure 8.6 out away from the origin, the optimal integer solution is really $x_1 = 5$, $x_2 = 3$, with an objective function value of 56.

There are times when the rounding approach is very suitable, especially when the values of the decision variables are large numbers. In such cases, the errors introduced by rounding are not likely to have significant consequences. However, if, for example, we are trying to determine the mix of different types of ships to have in our crude oil tanker fleet, rounding from 1.4 supertankers to 1 supertanker can have a large impact on both the feasibility and the optimality of the decision.

In situations in which rounding is not appropriate, an exact solution algorithm must be used. Just as in the case of linear programming, there are general-purpose integer programming algorithms as well as special-purpose algorithms to take advantage of the particular structure of the problem at hand. Most integer programming algorithms can be classified as either implicit enumeration procedures (or "branch and bound" methods), or cutting plane methods [3]. It is interesting to note that the structure of some problems is such that the optimal solution to its continuous version will always be integer and never involve fractional answers. Such is the case with an integer version of the transportation problem described in the previous section.

8.7.2 Nonlinear Programming

Just as most real-world decisions involve integer rather than continuous variables, it is also the case that accurate normative models often involve nonlinear rather than linear functions. In fact, the degree of accuracy of most *descriptive* models (and, therefore, of the associated normative models) can usually be enhanced by increasing the nonlinearity of the terms or variables involved. However, it is also true that nonlinear programming problems are much harder to solve than linear programming problems.

Solving a nonlinear programming problem is somewhat analogous to climbing a hill in an attempt to reach its highest point. The climb is to be made in an efficient manner, and the climber must stay within a fenced-in area. The top of the hill is equivalent to the value of the problem's objective function at the optimal solution, and the fences are analogous to constraints. Most solution procedures, such as "feasible direction methods" [1], work by starting at some point inside the fenced-in area and having the climber move in the direction of greatest increase. When he runs into a fence, he merely moves back a step and proceeds in a new direction. Some methods, such as Rosen's Gradient Projection technique [1], have the climber move toward the top by walking along the fence lines once a fence is encountered.

The climbing that is done *inside* the fences is the same type of searching that is performed when solving an unconstrianed optimization problem. Recall from Chapter 7 that this might involve the use of derivatives, or the use of direct search techniques such as the "one-at-a-time" method. In fact, many nonlinear programming algorithms are nothing more than simple extensions of a particular method for solving an unconstrained optimization problem. The extension is usually a provision for redirecting the search when a constraint (or "fence") is encountered.

As in the case of linear programming, there are general-purpose nonlinear programming algorithms designed to solve any nonlinear problem and special-purpose algorithms designed to solve problems with a particular substructure. Those methods that are extensions of unconstrained technqiues are usually general-purpose algorithms. Methods for solving special classes of nonlinear models, such as nonconvex problems, separable convex problems, and quadratic problems as described in Section 8.4.3, take advantage of the nature of the nonlinearities involved in the model such as the shape of the objective function (that is, the shape of the "hill" to be climbed). For instance, because of the model structure involved, commonly used algorithms for solving quadratic programming problems and separable convex programming problems are simply applications of the Simplex method to a reformulated version of the original nonlinear model.

8.8 SUMMARY

In this chapter, we have explored the meaning of mathematical programming and its use as a tool in decision making. Math programming can be thought of as the process of mathematically planning the use of one or more scarce resources. This process involves formulating a normative model of the underlying decision problem and then developing a solution algorithm for that model. The type of model being used dictates the class or subcategory of math programming involved. These classes include linear programming, integer programming, nonlinear programming, and dynamic programming.

For any of the primary classes of math programming, general-purpose solution algorithms are available that can be used to solve any problem falling in that category. Special-purpose algorithms have also been developed to solve problems within a primary class that have a distinct substructure. Transportation problems and decomposition problems are examples of such special models within the linear programming category.

The Simplex method is one of the most commonly used linear programming algorithms. It is an adjacent-extreme-point search that is guaranteed to find the optimal solution to any linear programming problem. Integer and nonlinear programming models are more difficult to solve than linear programming models, but are usually more accurate representations of the underlying decision problem. Integer problems can be solved by ignoring the integer requirements and rounding the continuous solution. However, this can lead to an infeasible and (or) nonoptimal integer solution. Solving a nonlinear problem is somewhat like climbing a hill that is bounded by fences. Many nonlinear programming algorithms are simply extensions of solution methods for unconstrained optimization problems.

REFERENCES AND SUGGESTED READING

1. M. Avriel, *Nonlinear Programming: Analysis and Methods,* Prentice-Hall, Englewood Cliffs, NJ, 1976.
2. M. S. Bazaraa and J. J. Jarvis, *Linear Programming and Network Flows,* John Wiley, New York, 1977.
3. R. S. Garfinkel and G. L. Newhauser, *Integer Programming,* John Wiley, New York, 1972.
4. F. S. Hillier and G. J. Lieberman, *Operations Research,* Holden-Day, San Francisco, 1967.
5. Benjamin Lev, *Introduction to Mathematical Programming: Quantitative Tools for Decision Making,* North Holland, New York, 1982.
6. C. McMillan, *Mathematical Programming,* John Wiley, New York, 1970.
7. D. M. Miller and D. E. Fyffe, ''Allocating Building Inspectors for Fire Prevention,'' *Management Science,* **22,** No. 12 (August, 1976).

REVIEW QUESTIONS AND PROBLEMS

8.1 Define the following terms.
(a) Mathematical programming
(b) Linear programming
(c) Algorithm
(d) Heuristic algorithm
(e) Special-purpose algorithm
(f) Feasible region
(g) Extreme point
(h) Basic variable
(i) Adjacent extreme point
(j) Master constraint
(k) Rosen's Gradient Projection technique

8.2 Critique the algorithm given in Example 8.1. For example, is it an exact procedure? What are the properties of the final solution—feasible, local optimum, and so on?

8.3 Give an example of a decision situation that is *not* an appropriate application of math programming. State why it is not.

8.4 Give an example of a decision situation that *is* an appropriate application of math programming.

8.5 Give the specific parallels (e.g., function correspondence) between the general version of a math programming model and the model used in Example 8.1.

8.6 Find a report of a ''real-world'' application of math programming in a current journal, and prepare a summary of this report.

8.7 Give a numerical example of each of the following models.
(a) A linear programming problem involving three variables; use vector and matrix notation
(b) A mixed integer, linear programming problem
(c) A nonlinear programming problem
(d) A transportation problem

(e) A decomposition problem
(f) A chance-constrained model

8.8 Give a graphical illustration of an integer programming problem that has the same optimal integer solution as the optimal continuous solution.

8.9 Give concise, step-by-step arguments (that is, ''prove'') that the Simplex procedure converges to a global optimum.

8.10 Derive an equation (a descriptive model) that gives an upper limit on the total possible number of iterations of the Simplex procedure for any given linear programming problem. Express this limit as a function of the total number of variables (say, n) and the total number of constraints (say, m) in the problem.

8.11 Consider the following problem:

$$\begin{aligned} \underset{\mathbf{x}}{\text{Max}}\ x_0 &= 2x_1 + 3x_2 \\ \text{S.T.:}\quad 2x_1 + 4x_2 &\le 20 \\ x_1 + x_2 &\le 6 \\ x_1, x_2 &\geqslant 0 \end{aligned}$$

a. Solve this problem graphically. Identify the following:
   1. The feasible region.
   2. All extreme points.
   3. The basic feasible solution having only slack variables as basic variables.
b. Solve this problem using the Simplex method.
c. List all six possible combinations of basic and nonbasic variables. Indicate which ones correspond to basic feasible solutions. Identify which combination was found at each iteration of the Simplex procedure applied in part (b) above.

8.12 Consider the same model as given in Question 8.11, except with the following objective function: Max $x_0 = 3x_1 + 3x_2$. Solve this revised problem using the Simplex method.

8.13 Consider the same model as given in Question 8.11, except with the following objective function: Min $x_0 = 3x_1 - 3x_2$. Solve this revised problem using the Simplex method.

8.14 Solve the following problems:

a. $$\begin{aligned} \underset{\mathbf{x}}{\text{Max}}\ x_0 &= 10x_1 + 7x_2 + 5x_3 \\ \text{S.T.:}\quad 5x_1 + 3x_2 + x_3 &\le 1200 \\ 4x_1 + 3x_2 + x_4 &\le 1000 \\ x_1 + 2x_2 + 2x_3 &\le 400 \\ \mathbf{x} &\geqslant 0 \end{aligned}$$

b. $$\begin{aligned} \underset{\mathbf{x}}{\text{Max}}\ x_0 &= x_1 - 2x_2 + 3x_3 \\ \text{S.T.:}\quad 4x_1 + 2x_3 &\le 12 \\ 3x_2 + 6x_3 &\le 18 \\ x_1 + x_2 + x_3 &= 8 \\ \mathbf{x} &\geqslant 0 \end{aligned}$$

8.15 Exxo Oil Refinery manufactures and markets four main types of petroleum products: regular gasoline (RG), unleaded gasoline (UG), fuel oil (FO), and diesel fuel (DF). Because the relative sales of each product line changes from month to month, Exxo would like to have a way to mathematically determine their least-cost production plan for any given month. Unfortunately, limitations on refining capacity as well as crude oil supply restrict the product mix they can produce in any one month. Specifically, the two predominant restrictions on their monthly production are: (1) the quantity of crude oil (in barrels) available in the month must not be exceeded by the refining crude requirement of each barrel of a product times the number of barrels of that product to be manufactured, summed over all four products; and (2) the total number of production hours available in the month must not be exceeded by the number of hours required to refine a barrel of a given product times the number of barrels of that product to be produced, summed over all products.

(a) Formulate a linear programming model to represent the decision problem as implied above. Use letters and symbols to represent the parameters involved. What advantages and disadvantages could have been achieved if you could have used any math programming format for the model, rather than being forced to use a linear programming format?

(b) What real-world features of this type of decision are not included in your model?

(c) Solve your model in part (a) using the following data: (1) the unit profits of products RH, UG, and DF are all the same and are twice as much as that for FO; (2) for the month in question, 10,000 barrels of crude oil and 130 production hours are available; and (3) it takes 0.2 barrels of crude and 0.6 hours to refine one barrel of RG, 0.8 barrels of crude and 0.6 hours to refine one barrel of UG, 0.1 barrels of crude and 0.4 hours for one barrel of FO, and 0.2 barrels and 0.3 hours for one barrel of DF.

Chapter 9

Queueing Analysis

9.1 INTRODUCTION

Waiting lines or queues are a common experience throughout our personal and professional lives. They are commonplace at supermarket checkout counters, freeway toll booths, doctors' offices, hospital emergency rooms, banks, airline ticket counters, theatres, and traffic lights, to mention but a few. At one extreme, a waiting line may result in nothing more than a personal inconvenience, as in the case of standing in line at a restaurant. At the other extreme, a waiting line may lead to tragic consequences as may occur at a hospital faced with a natural disaster such as an earthquake, flood, or tornado.

In every waiting-line system, units or customers arrive and demand a service. In each system there is a mechanism for selecting the next customer to be served. Finally, once a customer is selected for service, a procedure is carried out in attempting to meet his or her requirements. A waiting-line system may be completely described by defining the arrival or input process, the queue discipline or procedure for selecting customers for service, and the service mechanism.

In this chapter we discuss the analysis of waiting-line problems with a view toward estimating the effects of waiting lines and improving the performance of the associated system. Our treatment is limited to the most elementary systems, and the models applied to their analysis are presented without derivation. (For a more intensive treatment of elementary queueing systems, see References 2, 7, and 8. For a discussion of models of more complicated queueing systems, see 1, 5, and 9.

9.2 THE WAITING-LINE SYSTEM

In general, a queueing or waiting-line system may be depicted as in Figure 9.1. Arrivals to the system may come from one or many sources. The queue may consist of one or several waiting lines. The service mechanism may be a single server or channel or

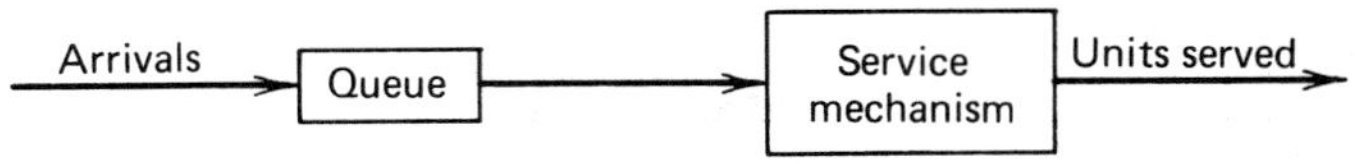

Figure 9.1 Queuing Systems

several channels in parallel or series or both. In many practical situations the service mechanism is actually a network of individual queueing systems such as that shown in Figure 9.2.

Before attempting to model and analyze a waiting-line system, the analyst should define the characteristics of the system with respect to the input process, the waiting-line or queue discipline and characteristics, and the service mechanism. Without such an analysis, model development can proceed only on the basis of untested assumptions. A model with such a questionable foundation may be of little use in describing the behavior of the system modeled.

9.2.1 Input Process

The input process may be characterized by defining:

1. The number of populations from which arrivals eminate.
2. The size of each input population.
3. The pattern of arrivals from each input population.
4. The number of customers entering the system at an arrival.
5. The conditions under which an arriving unit joins the system.

In many cases the characteristics of an arrival are determined by the input source or population. In analyzing operation of a restaurant, arrivals might be classified as those coming from the population of patrons who make advance reservations and those who do not. Arrivals from the first population may be considered essentially predictable, whereas the time of arrival of customers who do not make reservations is unpredictable and therefore random.

The size of the arrival population or populations can be an important factor in developing a model of the waiting-line system. When an arrival population is sufficiently large, the analyst may assume that arrivals will occur without interruption at discrete points in time throughout the period of analysis of the system. However, if the input population is sufficiently small, there may be periods of time where no arrival can occur.

To illustrate, consider the case of a repairman who is assigned the task of maintaining two machines. As long as one of the machines is operating, a failure and therefore an arrival requiring repair is possible at any time. However, if both machines are down at the same time, no arrival is possible until one of the machines is repaired.

The arrival pattern is usually described by the probability distribution of time between arrivals or the probability distribution of the number of arrivals per unit time.

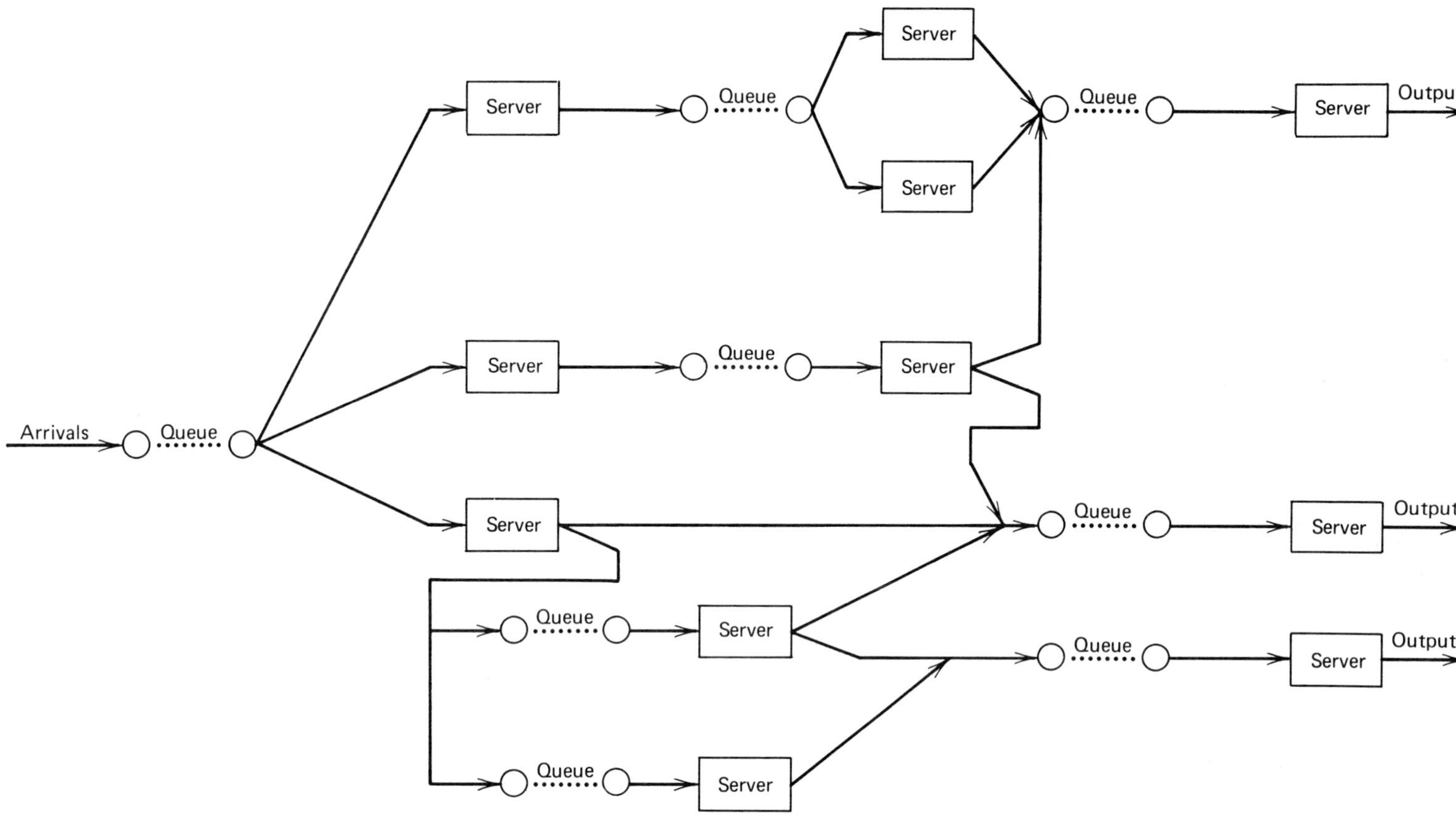

Figure 9.2 Waiting-Line System Made up of a Network of Waiting Lines

In rare cases arrivals may be deterministic, in the sense that they can be predicted with certainty. More often, the time of occurrence of arrivals cannot be predicted with certainty and is therefore considered to be random. In attempting to model a waiting-line system the variability of the time between arrivals may be sufficiently small to be considered predictable for practical purposes. For example, consider the arrival of units of product at an inspection station. If the delivery mechanism is a constant-speed conveyor and units of product are equally spaced on the conveyor, the time between the arrival of successive units of product may be considered to be deterministic and the arrival pattern must be described in probabilistic terms.

Arrivals to a waiting-line system may occur individually or in groups (bulk arrivals). Consider the arrival of customers at a service station. If we discount the case of customers traveling together but in separate cars, the waiting-line system can be treated as one where arrivals occur individually. In contrast, arrivals to a movie theater may occur individually or in groups of two or more.

Not every arrival to a waiting-line system joins the system. An arrival who does not join the system is said to balk. For example, if the waiting line is sufficiently long, the arrival may choose to return at a later date for service or go elsewhere for the service required.

Example 9.1

A maintenance department is composed of eight repairmen. The department has the responsibility for maintaining the equipment in three different production departments. The first production department includes 25 milling machines, the second department includes 32 drill presses, and the third department includes 16 punch presses. When a piece of equipment in any department fails, it is shut down and waits until a repairman is available for service. Define the characteristics of the input process.

In this case there are three populations generating arrivals. The sizes of the populations are 25, 32, and 16, respectively. As equipment failures are not likely to be predictable, the pattern of arrivals from each population could be expected to be random. The arrival pattern in the case of random arrivals is usually described by the probability distribution of time between arrivals, or the probability distribution of the number of arrivals per unit time. In this case an arrival to the maintenance system corresponds to failure of a piece of equipment. Thus the time until the next arrival occurs is the time until the next machine fails. If the various machines operate independently, the probability of two machines failing at the same time would be virtually zero, and it would be appropriate to assume that arrivals occur one at a time. Finally, each failing machine would always join the waiting line for service; otherwise, it would remain premanently inoperable. ■

Example 9.2

A major-league baseball team operates an advance-sale ticket window from 9:00 A.M. until 5:00 P.M. each day of the week. Characterize the input process.

In this case the number of populations considered might be classified as tickets sold on an individual basis and group sales. Both populations would probably be considered to be sufficiently large to be treated as infinite from a practical point of view. Since the time at which a customer buys tickets cannot be predicted with certainty, the pattern of customer arrivals from both populations would be random. The nature of the waiting-line system would imply a bulk arrival process, as customers may arrive individually or in groups of two or more. In this case the probability distribution of the number of customers requiring service per arrival would have to be determined. Finally, arriving customers may or may not join the system. For example, if the waiting line for tickets is sufficiently long, an arrival may not join the line, perhaps choosing to purchase tickets at a later date. ■

Example 9.3

A car dealer in a large urban area operates a service department. Customers are scheduled for service when they arrive. The department can service up to 75 cars per day. Describe the arrival process.

In this example customers might be considered to be one population or divided into two: those who purchased a car from the dealer and those who did not. In either case the population(s) would probably be considered to be infinite and the pattern of arrivals random. Most arrivals would include one car requiring service, although a customer might bring in two cars for service at the same time. Two conditions may arise where an arriving customer cannot obtain service quickly enough. He may decide to return at another time or go elsewhere for service. If a customer arrives and 75 cars have already been scheduled for service, he is refused entry to the system. ■

9.2.2 Waiting-Line Discipline

The waiting-line or queue discipline refers to the manner in which units awaiting service are selected for service. Many systems operate on a *first-come, first-served* discipline. Other disciplines include *priority* selection, *random* selection, and *last-come, first-served*. Priority selection occurs when units entering the system are assigned a priority for service. As an example, consider a computer facility where jobs to be run on the computer are assigned one of the following priorities: (1) urgent, (2) standard, (3) overnight, or (4) idle. When a job (customer) is submitted for processing, it is assigned one of these priorities. All jobs assigned the priority "urgent" are processed first and on a first-come, first-served basis within this priority class. All jobs with a priority standard are served after all urgent priority jobs have been processed, again on a first-come, first-served basis within the standard priority. Jobs with an overnight priority are served next; and finally, jobs with an idle priority are served last. Thus, where the queue discipline is based upon four priority classes, we have one waiting line composed of four subwaiting lines as shown in Figure 9.3.

Random selection of customers for service usually occurs in systems where it is difficult to keep track of when customers joined the waiting line. To illustrate, consider a waiting-line system where units of product are manufactured and placed in cartons for inspection in lots of size 100. The order of selection of units for inspection may be purely random. A common variant of random selection is that in which an attempt

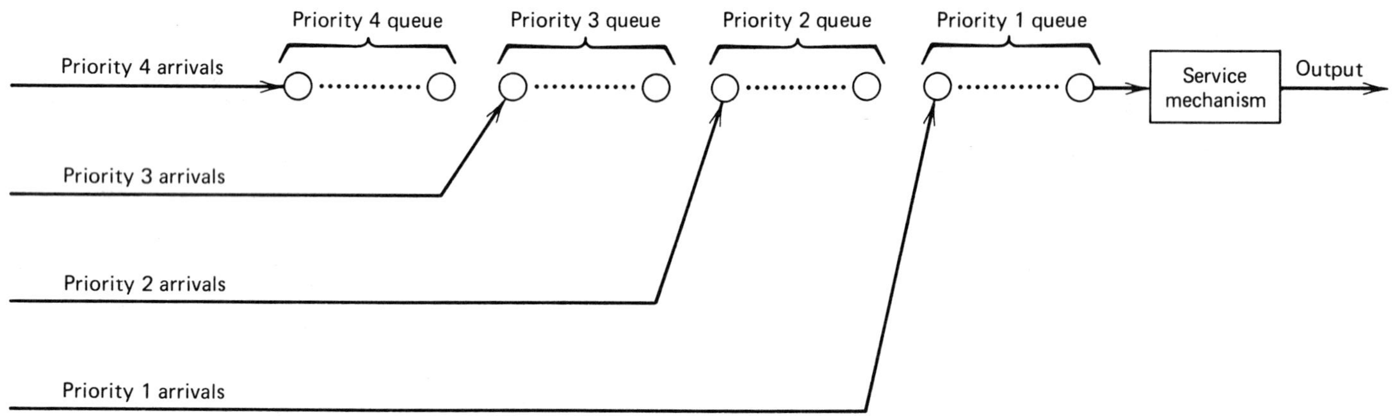

Figure 9.3 Waiting-Line System with a Priority Waiting-Line Discipline

is made to select customers on a first-come, first-served basis but does not always meet with success. This queueing discipline is common in retail stores where a clerk tries to keep track of the customer who has waited for the longest period of time. Here there is perhaps a high probability that the customer who has waited the longest will be served next, but proper selection is not a certainty. In such cases the probability of correctly selecting the next customer for service will tend to decrease as the number of customers awaiting service increases.

The last-come, first-served discipline is probably adopted by accident more often than not. Consider again manufacture of units for inspection in lots of size 100. If the units are stacked one on top of the other as they are produced, rather than packed in cartons, the selection process for inspection is likely to be last-come, first served.

Of course, there are many variations of the queue disciplines discussed so far. Some systems call for a preemptive priority system where an arriving unit has a priority that calls for interruption of the unit being served until the unit that has just arrived is served. Preemptive priority disciplines are appropriate for hospital emergency rooms. Waiting-line disciplines may call for a change in the priority of a unit after it has waited a specified period of time. Finally, units may be selected for service at random within a given priority or according to the estimated time required for service.

9.2.3 Waiting-Line Characteristics

The characteristics of a waiting line or queue refer to the number, capacity, and arrangement of waiting lines and the behavior of customers in the waiting lines. When arrivals are served by a single service channel, that channel is usually fed by a single waiting line. However, where several channels are available to provide the same service simultaneously, the service channels may be fed by a single waiting line or by individual waiting lines. To illustrate both possibilities, one need only consider ticket counters maintained by most airlines at major air terminals. Most airlines will offer ticketing and baggage-checking service at more than one counter where each or at least some of the counters offer the same service (some counters may offer both ticketing and baggage checking while others offer only baggage checking). For some airlines one waiting line will feed all counters offering the same service. Other airlines will allow a separate waiting line to form at each counter even though the counters provide the same service. Similar alternatives arise in the case of waiting lines at tellers' windows in banks.

In our discussion of the input process we mentioned that an arriving unit may choose not to enter the system because of the length of the waiting line. The system itself may prevent an arrival from joining the waiting line. For example, if a parking lot is full, an arriving customer may be forced to leave the lot, as there is no place for him to wait for a parking space to be vacated. In this case there is a limit of zero on the capacity of the waiting line. Situations also arise where there is a nonzero but finite limit on the permitted length of the waiting line.

To realistically capture the behavior of a queueing system it may be necessary to represent the behavior of customers in the waiting line. In many instances a customer will join the queue and subsequently decide to leave before being served. This phenomenon is called reneging. Where facilities providing the same service operate in parallel and are fed by independent waiting lines, customers may jockey back and forth among the lines in an effort to reduce their waiting time. Jockeying is common in banks, at airline ticket counters, and supermarket checkout counters.

9.2.4 Service Mechanism

To define the service mechanism, the analyst must determine the number and arrangement of service channels, the pattern of variation of service time, and the number of customers that can be served simultaneously by a channel. We have already mentioned that banks, airlines, and supermarkets provide *parallel service* in the sense that several channels offer the same service. Movie theaters ticket windows and vending machines are often cases where a single service facility is provided. Where parallel service is provided, the number of servers available may be constant or may vary depending upon the time of day or the number and type of customer waiting. In the case of a hospital emergency room, additional physicians may be called in during periods of high activity, or a specialist may be called upon to handle certain types of emergencies.

The service required by an arrival may actually include a sequence of services. In some cases the same individual provides each of the services. Here all phases of service in the sequence may have to be completed before the next unit can be served. An alternative form of sequential service arises when different individuals or facilities handle each of the services in the sequence. In this case a queue may or may not be permitted between successive facilities.

The time required to provide customer service is nearly always a random variable even when an identical service is provided to all customers. The range of variation in service may be sufficiently small, however, that service time can be considered constant for practical purposes. For example, let the time for a vending machine to provide its service be the elapsed time between depositing a coin and receiving the required merchandise. If we ignore machine malfunctions it would probably be appropriate to consider service time to be a constant.

Where service time must be considered a random variable, it is usually described by the probability distribution of time to provide service. The distribution of service time may depend upon the particular channel providing the service, the customer receiving the service, or both. In a supermarket, service time will depend upon the experience of the clerk and the number of purchases by the customer. Consider, however, several independent but identical minicomputers provided for student use. In this case service time would be dependent upon the student's ability to use the minicomputer and the type of job he is running, but would be independent of the facility used because all are identical.

More often than not, service at a channel is provided to one customer at a time. However, situations arise where more than one unit may be served simultaneously by a single channel. Such service is known as *bulk* service. Examples of bulk service include transportation of passengers on commercial aircraft, guided tours, rides at amusement parks, and classroom lectures. Of course, in each case the service cited may be preceded by other services that are provided to customers one at a time. Where service is provided in bulk the number of customers served may be a constant or a random variable. Where the number served simultaneously is a random variable, its probability distribution must be identified. In each of the cases just cited, the number served would probably be a random variable.

Example 9.4

A commercial airline provides service from Dallas to Chicago. Service time is considered to be the elapsed time between the start of service for ticketing, baggage checking, and seat selection until the passenger has claimed his or her baggage in Chicago. The airline operates four ticket counters: two for ticketing, baggage checking, and seat selection, and two for baggage checking and seat selection only. Describe the characteristics of the service mechanism that would have to be identified to model this waiting-line system.

Here the service mechanism is dependent upon the input process. Arrivals can be classified as follows:

A. Arrivals with a ticket and no baggage.

B. Arrivals with a ticket and baggage.

C. Arrivals without a ticket and no baggage.

D. Arrivals without a ticket but with baggage.

We assume that a separate waiting line exists before each service counter. A flow diagram for passengers taking the flight considered here is shown in Figure 9.4. Total service time in the system includes the time the passenger spends at a counter, the time he or she spends in the area awaiting boarding, the time spent on the aircraft, and the time spent claiming baggage, if any. The probability distributions with respect to time to be identified in this example include the following.

1. Time for ticketing, checking baggage, and seat assignment for type-D arrivals at counters 1 and 2 in Figure 9.4. (A different distribution may arise at each counter.)
2. Time for ticketing and seat assignment for type-C arrivals at counters 1 and 2 in Figure 9.4. (A different distribution may arise at each counter.)
3. Time for checking baggage and seat assignment for type-B arrivals at counters 3 and 4 in Figure 9.4. (A different distribution may arise at each counter.)
4. Time for seat assignment for type-A arrivals at counters 3 and 4 in Figure 9.4. (A different distribution may arise at each counter.)
5. Time between leaving the ticket counter and boarding the aircraft. This random variable and therefore its distribution is dependent upon when the passenger arrived at the terminal in Dallas and when the flight boarding occurred. The time of flight boarding is a random variable, as the flight departure may be delayed.

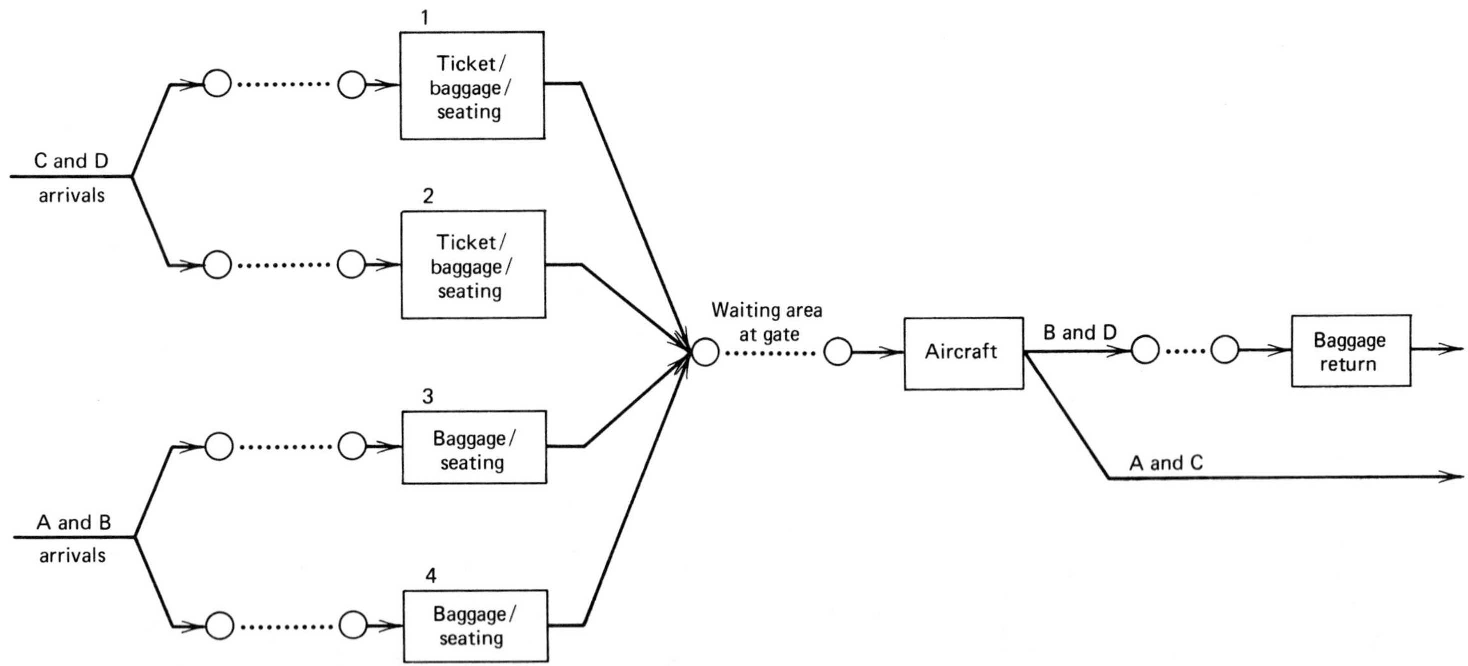

Figure 9.4 Flow Diagram for Flight Passengers in Example 9.4

6. Flight time from the gate at Dallas to the gate in Chicago. This random variable is composed of several components, each of which is random, and includes the time from the gate to the runway, the time on the runway, flight time from Dallas to terminal control in Chicago, holding time in the air at Chicago, approach time from the holding stack to the runway in Chicago, time on the runway, time from the runway to the gate, and the time between arrival at the gate and deplaning.
7. Elapsed time between deplaning and claiming baggage for type B and D passengers.

Service at the counters in Dallas may be in bulk because a customer at the counter may be holding tickets for others as well as for himself or herself. Thus the distribution of the number of passengers served simultaneously at each counter would have to be determined. Service on the aircraft is provided in bulk, but the number of passengers served is a random variable (as "no shows" can be anticipated), and the distribution of the number of passengers boarding would have to be identified. Baggage claim service at Chicago is also provided in bulk. The number served in bulk for baggage claim would be a random variable; in this case, the passengers selected for each group served simultaneously (queue discipline) may well be random. ■

9.3 MODELING QUEUEING SYSTEMS

In the discussion of waiting lines above we have given a few examples illustrating the variety of waiting-line systems commonly encountered. The number of variations is virtually limitless, leading to the need for a wide variety of models for their analysis. Our discussion of waiting-line models is limited to situations that can be treated with relative simplicity. Derivations of the models presented here can be found in References 2, 7, and 8. For a discussion of models of more complicated systems the reader should see References 1, 3, and 7.

9.3.1 Measures of Waiting-Line Performance

As with all system models, the purpose of a waiting-line model is to describe the behavior of a measure or measures of system performance as a function of those parameters and variables that influence system performance. In a business context the cost of and profit from system operation are obvious measures of performance. The total time a unit spends in the system may also be a pertinent measure of system performance in that a customer who anticipates a long stay in the system may choose to take his business elsewhere. In a similar manner, customer waiting time or length of the waiting line may be useful measures of system performance.

When a prospective customer fails to enter the system or leaves the system after he or she has joined it, income to the system is lost. Thus the likelihood of losing customers may also be a measure of system performance. Often the probability that a customer will join the system is dependent upon the length of the waiting line, and the probability that the customer will leave the system after entering it is dependent upon the time spent in the waiting line. A customer may be prevented from joining the system by a limitation on the number of customers permitted in the system at the same time. A parking lot is an example of this phenomenon.

9.3.2 System Parameters and Decision Variables

Once the measure of system performance has been defined, the task of the analyst is to develop a relationship between that measure and the parameters and variables of the system. Of particular concern are the decision variables or the variables that can be controlled from within the system itself. Those quantities which may be treated as decision variables depend upon the system itself. For example, the mean rate at which service is provided may be a decision variable for one system and a parameter for another. To illustrate, suppose that the service provided by the system is automated inspection of units of product and that there are many such processes available, each providing inspection at a different mean rate. If a decision has been made that one particular process, perhaps that which is already in use, will be used for the inspection function, then mean service rate may have to be treated as a parameter. The attitude of management, however, may permit consideration of other alternatives, and the mean service rate would be considered to be a decision variable. System capacity is another example of a quantity that may be a parameter or a decision variable.

The number of channels providing an identical service can significantly affect the length of the waiting line, the waiting time a customer spends in the system, and the cost of system operation and is often a decision variable. Furthermore, the length of time a channel is available for service may be a decision variable. For example, a service channel may be brought into service when the waiting line reaches a specified length N and removed from service when the line is reduced to a length $M(M < N)$. On the other hand, the number of service channels available may depend upon time of day where the mean rate of arrival of customers fluctuates during the day.

The mean rate of arrival of customers is treated as a system parameter if it cannot be directly controlled or influenced by the system itself. The mean rate of vehicles arriving at an interstate toll booth probably would not be affected by decisions regarding the operation of that toll booth. Cases do exist, however, where the arrival rate of customers can be directly controlled by the system itself. Consider a system where parts arrive for processing on a conveyor. If the speed of the conveyor can be controlled, then the arrival rate can be treated as a decision variable. Finally, many situations arise where the arrival rate can be influenced but not directly controlled by operation of the waiting-line system. This is particularly true in a competitive market. By increasing the rate at which a service is provided, a business may be able to lure customers away from competitors. A similar result may occur if the capacity of the system is increased. Where the arrival rate is influenced by system operation but cannot be directly controlled, it is treated as a system parameter but is expressed as a function of the decision variables of the system.

9.3.3 Assumptions

For each of the waiting line systems modeled in this chapter, the following will be assumed:

Input Process

1. Arrivals occur from a single population.
2. The size of the input population can be treated as infinite.
3. The time between successive arrivals is a random variable that is exponentially distributed.
4. Customers arrive one at a time.
5. All arriving customers join the waiting-line system unless the waiting-line length has reached its capacity.

Waiting-Line Discipline

1. Customers are served on a first-come, first-served basis.

Waiting-Line Characteristics

1. Waiting-line capacity is either infinite or a finite constant.
2. Customers do not reneg.
3. A single channel or multiple channels in parallel are fed by a single waiting line.

Service Mechandise

1. The time required to provide service is either constant or is a random variable that is exponentially distributed.
2. Customers are served one at a time in each service channel.

The results presented for the models discussed are ''steady-state'' conditions. After a waiting-line system begins operation, the status of the system will depend upon the conditions that prevailed when operation commenced. For example, suppose we have two identical waiting-line systems, A and B, each composed of a single service channel. System A starts operation empty, whereas system B starts operation with one unit in the service channel and 100 units in the waiting line. Now an arrival occurs to each system. We would expect the arrival to system B to wait longer for service than the arrival to system A. As long as system characteristics such as the mean time a unit spends waiting for service, mean time a unit spends in the system, mean waiting line length, and so on, depend upon starting conditions, the system is said to be in a ''transient state.'' When these characteristics no longer depend upon initial conditions, the system is said to be in the ''steady state.'' Theoretically, the steady state is achieved only after an infinite period of time; however, for practical purposes many systems may be treated as though they achieved steady-state conditions in a reasonably short period of time. They are then analyzed under steady-state assumptions.

We treat the time between successive arrivals and the time for customer service as exponential random variables. Although interarrival time and service time may have distributions other than the exponential, the exponential is particularly easy to work with in the analysis of basic waiting-line systems. The probability density function for exponential random variable is given by

$$f(x) = be^{-bx} \qquad 0 < x < \infty \tag{9.1}$$

where b is a positive constant. The numerical value of b will depend upon the system studied and whether the random variable described is interarrival or service time. Assume that x represents the time to serve a customer. Then

$$\begin{aligned} P(\text{service time is between } a_1 \text{ and } a_2) &= P(a_1 < x < a_2) \\ &= \int_{a_1}^{a_2} f(x)\,dx \\ &= e^{-a_1} - e^{-a_2} \end{aligned} \tag{9.2}$$

It should be noted that $f(x)$ is not a measure of probability in itself and that

$$0 < \int_{a_1}^{a_2} f(x)\,dx < 1 \tag{9.3}$$

where $0 < a_1 < a_2 < \infty$ since x cannot assume negative values.

The probability density functions of several exponential random variables are

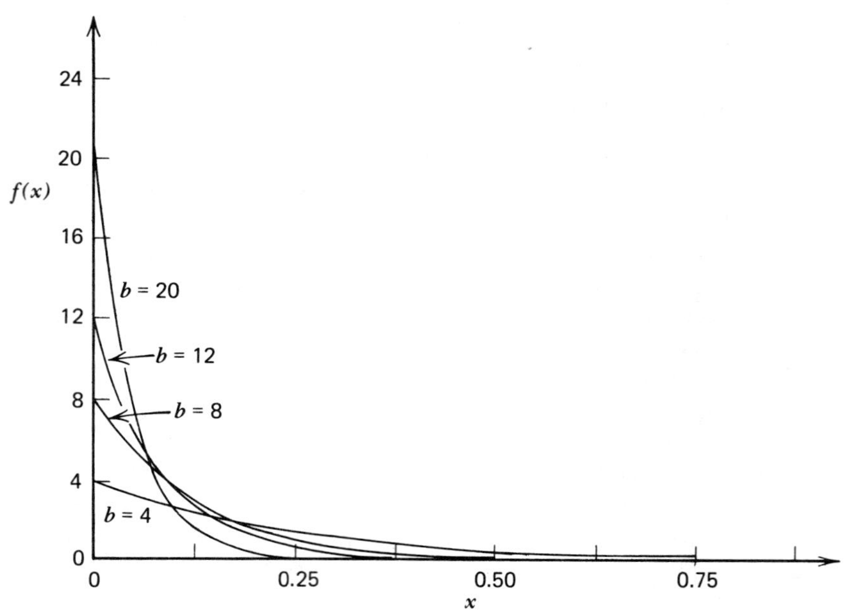

Figure 9.5 Probability Density Functions of the Exponential Random Variable

shown in Figure 9.5. As b increases the probability that x will assume a small value also increases. To illustrate:

$$P(x \text{ less than or equal to } 0.25) = P(0.0 < x < 0.25) = 1 - e^{-0.25b} \quad (9.4)$$

Now as b increases, $P(0.0 < x < 0.25)$ also increases.

If x is exponentially distributed then the mean value of x, $E(x)$, is given by

$$E(x) = \frac{1}{b} \quad (9.5)$$

Thus if x is service time in hours and $b = 5$:

$$E(x) = 0.20$$

and in the long run we would expect the average time to serve a customer to be 0.20 hr or 12 min. As Eq. (9.5) indicates, as b increases, the mean value of x decreases. The inverse of $E(x)$ is the mean rate of occurrence of events (arrivals or services). Thus if $1/b$ is the mean service time, b is the mean service rate. For $b = 5$ units are served at a mean rate of 5 per hour during periods when units are actually being served.

9.3.4 Notation

Throughout the remainder of this chapter certain parameters and variables are referred to repeatedly. The notation used to denote these quantities is as follows:

P_0 = probability that the system is empty (zero units in the waiting line and zero units in service).

λ = mean rate of arrivals to the system.

μ = mean rate of service in a channel.

ρ = utilization factor, λ/μ.

L = mean number of units in the system [number in the waiting line and service channel(s)].

L_q = mean number of units in the waiting line.

W = mean time a unit spends in the system (time in the waiting line and in the service channel).

W_q = mean time a unit spends in the waiting line.

Additional notation will be defined as necessary. In general we have:

$$L = \lambda W \quad (9.6)$$

$$L_q = \lambda W_q \quad (9.7)$$

$$W = W_q + \frac{1}{\mu} \quad (9.8)$$

These relationships are particularly important. Note that once the numerical value of either L, L_q, W, or W_q is known, the values of the remaining three quantities can be computed for given values of λ and μ.

9.4 QUEUEING MODELS

9.4.1 Single-Channel Models

In this section we consider models for situations where arriving units are served by a single service mechanism. Specifically the following three cases will be considered.

1. Constant service time and unlimited waiting-line length.
2. Exponential service time and unlimited waiting-line length.
3. Exponential service time and restricted waiting-line length.

In each case, relationships for L, L_q, W, and W_q are presented along with examples illustrating their use.

Constant Service Time and Unlimited Waiting-Line Length

In this case the time required to provide service is a known constant $1/\mu$ for all customers or units, and the waiting line is permitted to grow without limit. Given the assumptions in section 9.3.3, we have

$$P_0 = 1 - \rho \tag{9.9}$$

$$L = \frac{\rho(2 - \rho)}{2(1 - \rho)} \tag{9.10}$$

$$L_q = \frac{\rho^2}{2(1 - \rho)} \tag{9.11}$$

$$W = \frac{\rho(2 - \rho)}{2\lambda(1 - \rho)} \tag{9.12}$$

$$W_q = \frac{\rho^2}{2\lambda(1 - \rho)} \tag{9.13}$$

where $\rho = \lambda/\mu$. The results in Eq. (9.9) through (9.13) apply if and only if $0 < \rho < 1$. If $\rho > 0$ the queue length can be expected to increase without limit because $\lambda/\mu > 0$ implies that $\lambda > \mu$, which means that units tend to arrive to the system more rapidly than they can be served.

Example 9.5

A conveyor feeds units of product to a work station for packaging. Each unit is packaged separately and in the order of its arrival at the work station. Units are packaged at a constant rate of 500 per hour. The feed rate of the conveyor is a constant of 400 per hour. Units are not equally spaced on the conveyor, however. The random nature of spacing between units on the conveyor has been found to be such that the time between successive arrivals may be treated as exponentially distributed with a mean arrival rate of 400 per hour. Determine P_0, L, L_q, W, and W_q.

For this problem $\lambda = 400$, $\mu = 500$, and

$$\rho = \frac{400}{500} = 0.80$$

Hence

$$P_0 = 1 - 0.80 = 0.20$$

$$L = \frac{0.80(2.00 - 0.80)}{2(1.00 - 0.80)} = 2.400 \text{ units}$$

$$L_q = \frac{(0.80)^2}{2(1.00 - 0.80)} = 1.600 \text{ units}$$

$$W = \frac{0.80(2.00 - 0.80)}{2(400)(1.00 - 0.80)} = 0.006 \text{ hr}$$

$$W_q = \frac{(0.80)^2}{2(400)(1.00 - 0.80)} = 0.004 \text{ hr}$$

■

Example 9.6

In Example 9.5, suppose that the conveyor feed rate λ can be controlled and that the cost of operating the system per unit packaged can be expressed as $C_T(\lambda)$, where

$$C_T(\lambda) = C_1 W_q + \frac{C_2}{\lambda} \tag{9.14}$$

In Eq. (9.13) C_1 is the cost of waiting time per unit and C_2 is the cost of operating the conveyor in dollars per hour per unit. That is, the cost of operating the conveyor decreases as the feed rate increases due to increased efficiency. Let

$$C_1 = \$10.00/\text{hr}$$

$$C_2 = \$4.00/\text{hr/unit}$$

Find the conveyor feed rate λ that will minimize the total cost per hour of operating the system.

From Eq. (9.13):

$$W_q = \frac{\rho^2}{2\lambda(1 - \rho)} = \frac{\lambda}{2\mu(\mu - \lambda)} \tag{9.15}$$

Hence:

$$C_T(\lambda) = \frac{C_1\lambda}{2\mu(\mu - \lambda)} + \frac{C_2}{\lambda} \tag{9.16}$$

Taking the first derivative of $C_T(\lambda)$ with respect to λ we have:

$$\frac{d}{d\lambda} C_T(\lambda) = \frac{C_1}{2\mu(\mu - \lambda)} + \frac{C_1\lambda}{2\mu(\mu - \lambda)^2} - \frac{C_2}{\lambda^2} \tag{9.17}$$

To find the stationary points for $C_T(\lambda)$ we set Eq. (9.17) equal to zero and solve for λ, yielding

$$\lambda^* = \frac{-2\mu C_2 + \mu\sqrt{2C_1C_2}}{C_1 - 2C_2} \tag{9.18}$$

Because the conveyor feed rate must always be positive and $0 < \rho < 1$, we have

$$\lambda^* = \frac{-2\mu C_2 + \mu\sqrt{2C_1C_2}}{C_1 - 2C_2} \qquad C_1 > 2C_2 \tag{9.19}$$

and

$$\lambda^* = \frac{2\mu C_2 - \mu\sqrt{2C_1C_2}}{2C_2 - C_1} \qquad C_1 < 2C_2 \tag{9.20}$$

For the data given and noting that $\mu = 500$, from Example 9.5 we have for the stationary point:

$$\lambda^* = 236.07 \text{ units/hr}$$

To determine whether or not this solution yields the desired minimum we evaluate $d^2C_T(\lambda)/d\lambda^2$ at λ^*:

$$\frac{d^2}{d\lambda^2} C_T(\lambda) = \frac{C_1}{\mu(\mu - \lambda)^2} + \frac{C_1\lambda}{\mu(\mu - \lambda)^3} + \frac{2C_2}{\lambda^3}$$

and

$$\frac{d^2}{d\lambda^2} C_T(\lambda^*) = 1.152 \times 10^{-6}$$

Because $(d^2/d\lambda^2)\ C_T(\lambda^*) > 0$, $\lambda^* = 236.07$ is the desired minimum, and the cost per unit packaged is

$$C_T(\lambda^*) = \$0.026$$ ■

Example 9.7

In Example 9.5, suppose that the work station is to operate 90 percent of the time (10-percent idle time). In addition, the mean time a unit spends waiting to be packaged should not exceed 0.001 hr. Find the service rate and conveyor feed rate necessary to meet these conditions.

The mean proportion of time the service channel is idle is equivalent to the probability that the system is empty, P_0. Hence in this case:

$$P_0 = 1 - \rho = 0.10$$

If mean waiting time for service is to be 0.001 hr, then

$$W_q = \frac{\rho^2}{2\lambda(1 - \rho)} = 0.001$$

Since $P_0 = 0.10$, $\rho = 0.90$, and

$$\begin{aligned}\lambda &= \frac{\rho^2}{2(1 - \rho)W_q} \\ &= \frac{(0.90)^2}{2(0.10)(0.001)} \\ &= 4050 \text{ units/hr}\end{aligned}$$

and

$$\begin{aligned}\mu &= \frac{\lambda}{\rho} \\ &= \frac{4050}{0.90} \\ &= 4500 \text{ units/hr}\end{aligned}$$ ■

Exponential Service Time and Unlimited Waiting-Line Length

The situation considered here is the same as that treated above except that here we assume service time to be an exponentially distributed random with mean service rate μ. For this case, we have

$$P_0 = 1 - \rho \tag{9.21}$$

$$L = \frac{\rho}{1 - \rho} \tag{9.22}$$

$$L_q = \frac{\rho^2}{(1 - \rho)} \tag{9.23}$$

$$W = \frac{\rho}{[\lambda(1 - \rho)]} \tag{9.24}$$

$$W_q = \frac{\rho^2}{[\lambda(1 - \rho)]} \tag{9.25}$$

As in the case of constant service $0 < \rho < 1$. In this case the system is subject to two sources of random variation: that associated with the input process and that due to the service mechanism. The effect of the introduction of exponential random variation in service time can be observed by comparing the equations for L, L_q, W, and W_q in Eqs. (9.22) to (9.25) with the corresponding relationships for constant service in Eqs. (9.10) to (9.13). Let L_c, L_{qc}, W_c, and W_{qc} denote the mean number of units in the system, the mean number of units in the queue, the mean time a unit spends in the system, and the mean time a unit spends in the queue, respectively, for the case of constant service time and L_e, L_{qe}, W_e, and W_{qe} the corresponding quantities for exponential service. Then

$$\frac{L_c}{L_e} = \frac{2 - \rho}{2} < 1 \tag{9.26}$$

$$\frac{L_{qc}}{L_{qe}} = 0.50 \tag{9.27}$$

$$\frac{W_c}{W_e} = \frac{2 - \rho}{2} < 1 \tag{9.28}$$

$$\frac{W_{qc}}{W_{qe}} = 0.50 \tag{9.29}$$

and L, L_q, W, and W_q are greater in the case of exponential service time than in the case of constant service time.

Example 9.8

Jobs arrive at a computer terminal for processing in an exponential fashion at a mean rate of 120/hr. Processing time is exponentially distributed with a mean rate of 140/hr. Determine the mean idle time per hour for the terminal, the mean number of jobs awaiting processing, and the mean waiting time per job.

The terminal is idle whenever the system is empty. Thus mean idle time per hour is given by P_0, or

$$\begin{aligned} P_0 &= 1 - \rho \\ &= 1 - \frac{120}{140} \\ &= 0.143 \text{ hr} \end{aligned}$$

From Eq. (9.23) the mean number of jobs waiting for service is given by

$$\begin{aligned} L_q &= \frac{\rho^2}{1 - \rho} \\ &= \frac{(0.857)^2}{1.000 - 0.857} \\ &= 5.136 \text{ jobs} \end{aligned}$$

Mean waiting time per job is

$$\begin{aligned} W_q &= \frac{\rho^2}{\lambda(1 - \rho)} \\ &= \frac{5.136}{\lambda} \\ &= 0.043 \text{ hr} \end{aligned}$$

■

Example 9.9

Solve the problem in Example 9.6 if service time is exponentially distributed with mean rate 500 units per hour.

From Example 9.6:

$$\begin{aligned} C_T(\lambda) &= C_1 W_q + \frac{C_2}{\lambda} \\ &= \frac{C_1 \lambda}{\mu(\mu - \lambda)} + \frac{C_2}{\lambda} \end{aligned}$$

The stationary value of λ must satisfy

$$\frac{d}{d\lambda} C_T(\lambda) = \frac{C_1}{\mu(\mu - \lambda)} + \frac{C_1 \lambda}{\mu(\mu - \lambda)^2} - \frac{C_2}{\lambda^2} = 0$$

Solving for λ^* yields

$$\lambda^* = \frac{-C_2\mu + \mu\sqrt{C_1 C_2}}{(C_1 - C_2)}$$

Since $0 < \rho < 1$, for $C_1 > C_2$:

$$\lambda^* = \frac{-C_2\mu + \mu\sqrt{C_1 C_2}}{C_1 - C_2}$$

and for $C_1 < C_2$:

$$\lambda^* = \frac{C_2\mu - \mu\sqrt{C_1 C_2}}{C_2 - C_1}$$

For C_1 = \$10.00 and C_2 = \$4.00:

$$\lambda^* = 193.71 \text{ units/hr}$$

Since

$$\frac{d^2}{d\lambda^2} C_T(\lambda^*) = \frac{2C_1}{\mu(\mu - \lambda^*)^2} + \frac{2C_1\lambda^*}{\mu(\mu - \lambda^*)^3} + \frac{2C_2}{\lambda^{*3}}$$
$$= 1.797 \times 10^{-6} > 0$$

$\lambda^* = 193.71$ units/hr is the desired minimum yielding a cost per unit of

$$C_T(\lambda^*) = \$0.033$$ ■

Exponential Service Time and Limited Waiting-Line Length

The waiting-line system considered here is the same as that discussed in the last section except that there is room in the waiting line for only $M - 1$ units. If a unit arrives and finds $M - 1$ units in the waiting line (M units in the system), it is rejected from the system.

The expressions used to estimate mean total time in the system W and mean time waiting in the queue W_q depend upon the perspective taken in analyzing the system. If a customer arrives and finds the system full (M units in the system), he or she is rejected and spends zero time in the system. On the other hand, if the system is not full the arrival will be in the system for a positive period of time, although he or she may or may not have to wait for service in the waiting line. In most cases we are concerned only with arrivals that join the system in defining W and W_q. However, W and W_q may be defined to reflect the time spent in the system and the waiting line by customers rejected from the system as well as those who join the system. We restrict our attention here to consideration of customers who join the waiting-line system.

If $\lambda \neq \mu$ we have the following:

$$P_0 = \frac{1 - \rho}{1 - \rho^{M+1}} \tag{9.30}$$

$$L = \frac{\rho\,[(1 - \rho^M) - M(1 - \rho)\rho^M]}{(1 - \rho)(1 - \rho^{M+1})} \tag{9.31}$$

$$L_q = \frac{\rho^2(1 - \rho^M) - M(1 - \rho)\rho^{M+1}}{(1 - \rho)(1 - \rho^{M+1})} \tag{9.32}$$

$$W = \frac{\rho[(1 - \rho^M) - M(1 - \rho)\rho^M]}{\lambda(1 - \rho)(1 - \rho^{M+1})} \tag{9.33}$$

$$W_q = \frac{\rho^2(1 - \rho^M) - M(1 - \rho)\rho^{M+1}}{\lambda(1 - \rho)(1 - \rho^{M+1})} \tag{9.34}$$

P_M = probability system is full (M units in the system)

$$= \frac{(1 - \rho)\rho^M}{1 - \rho^{M+1}} \tag{9.35}$$

Expressions for Eqs. (9.30) to (9.35) can be developed for $\mu = \lambda$ but are not presented here. For a discussion of the case where $\lambda = \mu$, see Reference 8. In this case we do not require that $0 < \rho < 1$. As long as $\rho > 0$ and M is finite, the system will remain stable and achieve a steady state. However it should be noted that for $0 < \rho < 1$, Eqs. (9.30) to (9.34) approach Eqs. (9.21) to (9.25) as M approaches infinity (the case of an unlimited waiting line).

Example 9.10

Customers arrive at a carwash at a mean rate of 20/hr, where interarrival time is exponentially distributed. The time to wash a car is exponentially distributed with mean rate 22/hr. Cars are washed on a first-come, first-served basis. Space is to be provided for waiting cars such that the probability of an arrival being rejected (because no waiting space is available) will be 0.05 or less. How many spaces should be available to meet the criterion?

An arriving customer is rejected if the system is full. Since $\lambda \neq \mu$, the probability an arriving customer cannot wait for service is P_M, where

$$P_M = \frac{(1 - \rho)\rho^M}{1 - \rho^{M+1}}$$
$$= 0.05$$

where $\rho = 0.9091$. The solution for M is given in Table 9.1. Because $P_{11} < 0.05 < P_{10}$, $M = 11$ is the required solution and 10 positions should be provided for waiting customers. ■

Example 9.11

In Example 9.5, $M - 1$ spaces are provided for units waiting for packaging. Units rejected are lost at a cost C_1, and each space provided costs C_2 per hour. The total hourly cost of operation of the system is

$$C_T(M) = C_1\lambda P_M + C_2M$$

where λ is the hourly feed rate. Determine the number of spaces provided for waiting units that will minimize hourly cost of operation, where

$$\lambda = 400 \qquad C_1 = \$1.00$$
$$\mu = 500 \qquad C_2 = \$13.00$$

Table 9.1 Solution for *M* for the Problem in Example 9.10

M	P_M	M	P_M
1	0.4762	6	0.1054
2	0.3021	7	0.0875
3	0.2155	8	0.0736
4	0.1638	9	0.0628
5	0.1296	10	0.0540
		11	0.0468

$C_T(M)$ is given by

$$C_T(M) = C_1\lambda \frac{(1 - \rho)\rho^M}{(1 - \rho^{M+1})} + C_2M$$

For $\rho = 0.80$,

$$C_T(M) = 400 \left[\frac{0.20}{(1 - 0.80^{M+1})}\right](0.80)^M + 13M$$

The tabular solution for the optimum value of M is given in Table 9.2. Because the optimum value of M is 5, the required number of spaces for waiting units is 4. ■

9.4.2 Parallel-Channel Model

More often than not, a waiting line system is composed of more than one service channel. Several channels may be necessary to provide identical service simply because a single service channel cannot handle the volume of demand placed upon the system by arriving customers; that is, $\lambda > \mu$. This situation arises in banks and at air terminal ticket counters, rental car agencies, and supermarket checkout counters.

Let the number of channels providing an identical service in parallel be s. This arrangement of channels is shown in Figure 9.6. In some cases a single waiting line is formed before each of the s service channels; in others, all s channels are fed by a single waiting line. The former usually occurs in supermarkets and the latter at air terminal ticket counters. In this section we treat a situation in which all s channels are fed by a single waiting line, interarrival and service times are exponentially distributed random variables where the service rate in each channel is μ, and waiting-line length is unlimited. The parallel configuration of service channels is a common occurrence, but the assumption of an identical mean service rate in all channels is probably unrealistic in practice. However, an analysis based upon this assumption may be useful as a first approximation in modeling the behavior of the system.

Systems with s service channels in parallel where each channel is fed by an independent waiting line can be treated as individual waiting-line systems in restricted cases. However, this approach assumes an independent input process for each waiting line and server and that customers do not jockey among the waiting lines.

Table 9.2 Solution for the Optimum Value of M for the Problem in Example 9.11

M	$C_1\lambda[(1 - \rho)\rho^M/(1 - \rho^{M+1})]$	C_2M	$C_T(M)$
1	\$177.78	\$13.00	\$190.78
2	104.92	26.00	130.92
3	69.38	39.00	108.38
4	48.74	52.00	100.74
5	35.53	65.00	100.53
6	26.54	78.00	104.54

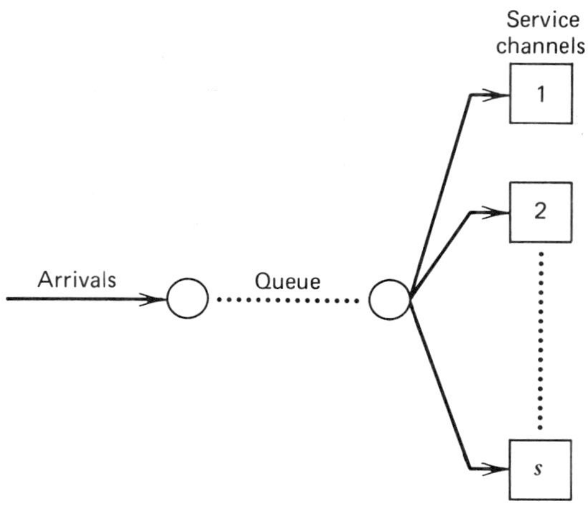

Figure 9.6 Parallel Arrangement of Service Channels

Exponential Service Time and Unlimited Waiting-Line Length

Since s service channels are available, the overall rate of service is $n\mu$ if there are n units in the system (all n in service) and $n < s$; for $n > s$ the overall service rate is $s\mu$. Thus, for system stability we require that $\lambda < s\mu$. If $\lambda > s\mu$ arrivals occur at a mean rate that is greater than the maximum mean service rate and queue length will tend to increase without limit. Letting $\rho = \lambda/\mu$:

$$P_0 = \frac{1}{\sum_{n=0}^{s-1} \rho^n/n! + \rho^s/[s!(1 - \rho/s)]} \tag{9.36}$$

$$L = \frac{P_0 \lambda \mu s \rho^s}{[s!(s\mu - \lambda)^2]} + \rho \tag{9.37}$$

$$L_q = \frac{\lambda \mu s \rho^s P_0}{s!\,(s\mu - \lambda)^2} \tag{9.38}$$

$$W = \frac{\mu s \rho^s P_0}{[s!(s\mu - \lambda)^2]} + \frac{1}{\mu} \tag{9.39}$$

$$W_q = \frac{\mu s \rho^s P_0}{[s!(s\mu - \lambda)^2]} \tag{9.40}$$

$$P_s = \text{Probability all servers are busy}$$

$$= \rho^s/s! \tag{9.41}$$

Note that if $s = 1$ the parallel-channel system reduces to a single-channel system and Eqs. (9.36) to (9.40) are equivalent to Eqs. (9.21) to (9.25).

Example 9.12

Three airline ticket counters are available for ticketing passengers and baggage handling. Each counter can handle passengers at a mean rate of 20/hr, and arrivals occur at a mean rate of 58/hr. Both interarrival time and service time at all three counters are exponentially distributed. Compute the mean waiting time per customer and the proportion of time all three counters are busy. How many additional identical servers would be necessary if mean waiting time were to be no more than 1 min?

From Eq. (9.40):

$$W_q = \frac{20(3)(2.90)^3 P_0}{3![3(20) - 58]^2}$$

where $\rho = 2.90$. P_0 is given by

$$P_0 = \frac{1}{1 + 2.90 + (2.90)^2/2 + (2.90)^3/[3!(1 - 2.90/3)]}$$
$$= 0.0077$$

and

$$W_q = 0.4695 \text{ hr} = 28.17 \text{ min}$$

Since $W_q > 1.0$ min, additional servers are necessary to reduce W_q to a value less than or equal to 1.0 min. The trial-and-error solution for s such that $W_q < 1.0$ min is summarized in Table 9.3, where the required value of s is found to be 5. ■

Example 9.13

A maintenance department is responsible for the repair of equipment in a machine center. Equipment failures are exponentially distributed with a mean failure rate of 12 failures per day. A single repair crew can service failing equipment at a mean rate of 6 per day where repair time is exponentially distributed. The average cost of downtime per machine is $1000/hr, and each repair crew costs $400.00 per 8-hr day. The equipment operates and must be maintained 24 hours per day, and the population of machines is large enough to be considered infinite. How many repair crews should be on duty each 8-hr shift to minimize the sum of mean downtime cost and mean repair crew cost?

Mean downtime per failing machine is W, and the mean number of failures per

Table 9.3 Solution for the Value of s such that $W_q < 1.5$ min in Example 9.12

s	P_0	$\mu s \rho^s/[s!(s\mu - \lambda)^2]$	W_q (min)
3	0.0077	60.9725	28.17
4	0.0437	0.4871	1.28
5	0.0521	0.0970	0.30

Table 9.4 Solution for *s* for Example 9.13

s	P_0	W	$C_T(s)$
3	0.0476	2.38	$20,246.40
4	0.1304	0.54	18,990.93
5	0.1343	0.51	18,320.00
6	0.1351	0.50	18,472.05

8-hr shift is $\lambda(4)$. Therefore, the mean cost of downtime per shift is

$$\text{Mean downtime cost/shift} = 8000\ \lambda W$$

The mean cost of repair crews per shift is $400s$. Therefore mean total cost is given by

$$C_T(s) = 8000\ \lambda W + 400s$$

$$= 8000\ \lambda\mu s\rho^s P_0[s!(s\mu - \lambda)^2] + 400s$$

Because $\lambda < s\mu$, for system stability $s > 2$. The trial-and-error solution for s is given in Table 9.4, where $\rho = 2.00$. As indicated in Table 9.4 the optimum number of repair crews is 5 at a cost of $18,320.00 per shift. ■

9.5 SUMMARY

In this chapter we present a fundamental overview of the basic concepts of waiting-line or queueing theory followed by results obtained for very basic queueing systems. The discussion in Section 9.2 gives the reader an appreciation of the variety and complexity of waiting-line systems that exist in practice. The purpose of the material in Section 9.3 is to place the concepts in section 9.2 in a modeling framework. Section 9.4 presents results for simple waiting-line systems and illustrates the use of these results in analyzing practical problems related to these systems.

The scope of this textbook is such that a detailed discussion of the development of basic queueing models is not possible; for this reason, only results are presented. These results are based upon the assumptions stated, and the reader should exercise caution in applying them to systems where the associated assumptions may not apply. However, the results contained in Section 9.4 may frequently be applied as a first approximation in analyzing the performance of simple systems even though the assumptions stated are at variance with reality. Of course, the performance of any model based upon the results presented here should be compared with the performance of the system itself to determine whether or not the model is a valid representation of the system even in the sense of a first approximation.

Current research on the mathematical analysis and modeling of waiting-line systems is extensive and growing. Of particular interest from a practical point of view is the research ongoing in the analysis of networks of queues, as this is the type of situation encountered in most applications [3, 4]. The reader should be aware, however, that a complete understanding of this research requires a strong background in probability, stochastic processes, and mathematics.

REFERENCES AND SUGGESTED READING

1. R. B. Cooper, *Introduction to Queueing Theory,* Macmillan, New York, 1972.

2. Frederick S. Hillier and Gerald J. Lieberman, *Introduction to Operations Research,* Prentice-Hall, Englewood Cliffs, N.J., 1974.

3. Leonard Kleinrock, *Queueing Systems, Volume 1: Theory,* John Wiley, New York, 1975.

4. Leonard Kleinrock, *Queueing Systems, Volume 2: Computer Applications,* John Wiley, New York, 1976.

5. T. L. Saaty, *Elements of Queueing Theory,* McGraw-Hill, New York, 1961.

6. James E. Shamblin and G. T. Stevens, Jr., *Operations Research: A Fundamental Approach,* McGraw-Hill, New York, 1974.

7. H. A. Taha, *Operations Research: An Introduction,* Macmillan, New York, 1971.

8. Harvey M. Wagner, *Principles of Operations Research,* Prentice-Hall, Englewood Cliffs, N.J., 1969.

9. John A. White, J. William Schmidt, and G. K. Bennett, *Analysis of Queueing Systems,* Academic Press, New York, 1975.

REVIEW QUESTIONS AND PROBLEMS

9.1 Give an example of a real-world waiting-line system with the following combination of characteristics.

Input process: single infinite population with random interarrival times, units arrive one at a time, and all arriving units join the system.

Waiting-line discipline: first-come, first served.

Waiting-line characteristics: single waiting line with unlimited capacity and customers do not reneg.

Service mechanism: multiple channels in parallel, units served one at a time, and service time is a random variable.

Do not use an example already given in the text.

9.2 Give an example of a real-world waiting-line system with the combination of characteristics given in Problem 9.1 with the following exceptions.

Input process: single finite population.

Waiting-line discipline: priority service.

Do not use an example already given in the text.

9.3 Give an example of a real-world waiting-line system with the combination of characteristics given in Problem 9.1 with the following exceptions.

Input process: bulk arrivals

Waiting-line characteristics: one waiting line before each channel with jockeying.

Do not use an example already given in the text.

9.4 Give an example of a real-world waiting-line system with the combination of characteristics given in Problem 9.1 with the following exceptions.

Input process: several finite populations.

Waiting-line characteristics: limited capacity.

Service mechanism: service channels in parallel.

Do not use an example already given in the text.

9.5 Give an example of a real-world waiting-line system with the combination of characteristics given in Problem 9.1 with the following exceptions.

Input process: several finite populations.

Service mechanism: single service channel.

Do not use an example already given in the text.

9.6 Give an example of a real-world waiting-line system that is composed of a network of service channels. Do not use an example already given in the text.

9.7 Define the characteristics of the waiting-line system that serves customers demanding gasoline at a service station.

9.8 Define the characteristics of the waiting-line system encountered at a movie theater.

9.9 Define the characteristics of the waiting-line system associated with the operation of a restaurant.

9.10 Define the characteristics of the waiting-line system associated with a toll booth for traffic leaving an interstate highway.

9.11 Define the characteristics of the waiting-line system associated with a traffic signal at the intersection of two two-way streets.

9.12 Give an example of a waiting-line system where a customer's service may be interrupted until another customer is served.

9.13 Give an example of a waiting-line system with bulk arrivals and bulk service.

9.14 Solve the problem in Example 9.6 if $\mu = 800$ units/hr.

9.15 Solve the problem in Example 9.11 if $\lambda = 450$.

9.16 For the problem in Example 9.12 determine L, L_q, and W for $s = 5$.

9.17 For the optimum value of λ in Example 9.6 compute P_0, L, L_q, W, and W_q.

9.18 For the optimum value of M in Example 9.10 determine P_0, L, L_q, W, and W_q.

9.19 The time between arrivals to a single service channel is exponentially distributed with $\lambda = 100$. The mean service rate is $\mu = 110$. Determine P_0, L, L_q, W, and W_q if service time is constant.

9.20 Solve Problem 9.19 if service time is exponentially distributed.

9.21 Solve Problem 9.19 if the capacity of the system is 3 ($M = 3$) and service time is exponentially distributed.

9.22 Solve Problem 9.19 if 3 service channels operate in parallel ($s = 3$) and service time in each is exponentially distributed with $\mu = 55$.

9.23 In Example 9.11 how much would λ have to change before the optimum value of M would increase by one?

9.24 In Example 9.13 what change in λ would justify an increase in the number of parallel channels to 6?

9.25 Solve the problem in Example 9.12 if the mean service rate is 22 per hour.

9.26 Given the conditions assumed in Section 9.4.1, suppose that mean waiting time in the queue is to be 15 min or less and the mean waiting line length is to be 2 or less. Find the minimum service rate μ that will satisfy these conditions if $\lambda = 100$ units per hour.

9.27 Solve Problem 9.26 assuming the conditions given in Section 9.4.1.2.

9.28 Solve Problem 9.26 assuming the conditions given in Section 9.4.2.1 if there are two service channels in parallel.

9.29 In Example 9.9 a second service channel can be added to the system providing parallel service. The service rate in the second channel would be $\mu = 500$ units/hr. The cost of an additional channel is estimated to be \$20.00/hr. Find the optimum conveyor feed rate λ if two channels operate in parallel. Should a second channel be added to the system?

9.30 In Example 9.10 suppose that the carwash has room for no more than five waiting customers. What should the mean service rate μ be if P_0 is to be 0.05?

9.31 In Example 9.13 suppose a single repair crew can service failing equipment at a mean rate of 3 per day. How many repair crews are necessary each shift to minimize total cost per shift?

9.32 An automated inspection station inspects units of product at a mean rate of 6000/hr, where inspection time is exponentially distributed. Units arrive exponentially for inspection at a mean rate of 5990/hr. Additional identical inspection stations may be added in parallel to the system at a cost of \$600.00/hr each. The mean rate of inspection for the existing station may be increased at a cost (\$0.10 $\Delta\mu$)/hr, where $\Delta\mu$ is the incremental increase in the service rate. However, $\Delta\mu$ is restricted to increments of 100. Each hour a unit spends waiting costs \$10.00. Find the number of inspection stations and the service rate that will minimize the cost of operation per hour. The mean inspection rate for all inspection stations must be the same.

9.33 In Problem 9.32 suppose that the mean service rate cannot be varied and all service channels operate at a mean rate of 6000/hr. What is the optimum number of service channels?

9.34 In Problem 9.33 let s^* be the optimum number of parallel servers. What change in λ would justify a reduction in the number of servers to $s^* - 1$? What change in λ would justify an increase in the number of servers to $s^* + 1$?

9.35 The cost per hour of providing service for a single channel is given by $C_T(M)$, where:

$$C_T(M) = C_1\lambda P_M + C_2W_q + C_3(M - 1)$$

and M is the capacity of the system. Service and interarrival times are exponentially distributed with mean rates 500/hr and 450/hr, respectively. If

$$C_1 = \$100.00 \qquad C_2 = \$40.00/\text{hr} \qquad C_3 = \$1000.00$$

determine the number of positions that will minimize the cost of operations per hour.

9.36 In Problem 9.35 let M^* be the optimum capacity for the system. What change in λ would justify a reduction in capacity to $M^* - 1$? What change in λ would justify an increase in capacity to $M^* + 1$?

9.37 For the situation in Problem 9.35 suppose that units arriving when the system is full are rejected from the system and do not return for service. If the probability of losing a customer should not exceed 0.10, what is the required value of M?

9.38 In Problem 9.32 suppose that the mean service rate cannot be varied and all channels operate at the same rate, $\mu = 6000$/hr. However, the mean arrival rate varies by day of the week as follows:

Day	λ
Monday	8000
Tuesday	5000
Wednesday	6000
Thursday	7000
Friday	4000

How many inspection stations should be scheduled for use each day?

9.39 In Problem 9.38 suppose that the same number of inspection stations must be used each day. How many inspection stations should be operated?

9.40 The cost per day of operating s service channels in parallel is given by

$$C_T(s) = C_1s + C_2L_q + C_3W_q + C_4\rho$$

where interarrival and service times per channel are exponentially distributed with mean rates $\lambda = 1000$/day and $\mu = 500$/day/channel, respectively. If

$$C_1 = \$1000 \qquad C_2 = \$1000/\text{unit} \qquad C_3 = \$1000/\text{hr} \qquad C_4 = 2000$$

find the optimum number of channels.

Chapter 10

System Simulation

10.1 THE MEANING OF SIMULATION

In the preceding chapters we discussed the use of mathematical models to describe and analyze the behavior of physical systems. In those cases where such models can be developed and manipulated in a fashion appropriate for the analysis desired, mathematical models are to be recommended. However, the complexity of many organizations in today's society is such that the totality of the effects of the interacting elements included sometimes cannot be captured adequately through a mathematical model. This is particularly evident when many of the factors affecting the performance of the organization cannot be predicted with certainty. To illustrate, consider the problem of attempting to develop a model for estimation of the annual profit for a manufacturer. To estimate profit realistically, the analyst must capture the effects of sales, salary and wages, taxes, production scheduling and control, machine failure, machine maintenance, inventory control, quality control, shipping and receiving, raw materials sources, and labor training and turnover. Although the organization may be reasonably small and the product may be manufactured through one of several simple production lines, the organization is faced with the problems of dealing with the often capricious behavior of a multitude of customers and raw materials suppliers, maintaining production schedules in the face of unpredictable machine failures, minimizing the occurrence of many different defects that may appear in any single unit of product, salvaging raw material from substandard production lots, meeting shipping deadlines that may change at the whim of the customer, and maintaining an adequate inventory of product to meet shipping deadlines without exceeding warehouse capacity. To estimate annual profit adequately, the model developed must not only capture the interacting effects of each possible state of these variables but must also account for the probabilistic nature of changes in these variables.

For the most part, mathematical modeling has proved successful when applied to rather small subsystems or components of the total organization, such as the design of an inventory or quality-control system, the scheduling of jobs through a series of machine centers, scheduling of preventive maintenance, and manpower allocation, to mention but a few. As the complexity of the system analyzed is increased by including several subsystems, however, the difficulties associated with the development of an adequate mathematical model will certainly increase. It is in cases such as this that the analyst often turns to simulation modeling as an alternative approach to the problem.

The foregoing discussion would probably lead one to the conclusion that simulation is a last-resort technique: "When all else fails, simulate." This attitude is not without merit. When the analyst has a choice between the use of a mathematical and a simulation model—both equally adequate representations of reality—he will usually choose to model the system mathematically. Although the mathematical model may take longer to develop and therefore cost more in the developmental stage, simulation models usually require a much heavier investment in computer time for execution. In addition, the output of a simulation model usually contains a random component, clouding the analyst's interpretation of the effectiveness of the performance of the system.

The analyst usually turns to simulation to model the system under study for one of the following three reasons.

1. The system is so complex that it will not yield to mathematical analysis.
2. The analyst does not possess the level of mathematical sophistication required for development of an adequate mathematical model.
3. Management is more receptive to simulation modeling than to mathematical modeling.

The first case implies that the analyst has thoroughly investigated a variety of approaches to the problem, perhaps seeking the assistance of outside authorities, and has concluded that simulation modeling is the only adequate means available for analysis of the system under study. In the second case, the analyst chooses to apply simulation because it seems to be the simplest approach to the problem although mathematical analysis may be feasible. Finally, it is not unusual to find that upper levels of management, where projects such as those discussed here are approved, are more receptive to problem analysis through simulation than through the more sophisticated and abstract techniques of mathematical modeling. This is particularly true within those organizations which have experienced a successful history of problem analysis through simulation. Although this bias is often unjustified, management personnel with only a layman's understanding of operations research tend to feel "more comfortable" with a simulation model because they have the feeling that they can see how interacting elements within the system are coordinated within a simulation model.

Today simulation is broadly accepted as a technique for the analysis of engineering, business, and behavioral systems. This technique has been applied successfully to problems such as:

The analysis of air traffic control systems

The analysis of large-scale military operations

The analysis of the traffic-handling capability of telephone switchboards

Job-shop scheduling

The analysis of forestry operations
The analysis of the U.S. economy
The analysis of multiproduct inventory systems
Determination of manpower requirements
Instructional modeling for higher education
Competitive market analysis
Transportation planning
Man–machine interface
Corporate planning

10.2 SYSTEM MODELING THROUGH SIMULATION

A simulation model attempts to trace the behavior of a system as it is influenced by events that occur during the period of time over which the system is studied. To introduce the basic concepts of the simulation process, consider a single-channel waiting-line system with exponentially distributed interarrival times with $\lambda = 20/\text{hr}$ and constant service rate $\mu = 60/\text{hr}$. We assume that the system has unlimited capacity and units are served on a first-come, first-served basis. Finally all arriving units join the system and do not leave until they are served. Before discussing the simulation of this system we examine the behavior of the system based upon a period of observation of 2 hr. Suppose that during this 2-hr period the arrival data in Table 10.1 were recorded. Since service time is a constant of 0.0167 hr we have little trouble tracing the behavior of the system throughout the period of study. The behavior of the waiting-line length q and the number in the system n are summarized in Table 10.2, where we assume $q = 0$, $n = 0$ at the start of the period of observation. The reader will note that q and n change only at the time of a service or an arrival.

A simulation model imitates the event-by-event behavior of the system studied. In this sense the output of a simulation model of the waiting-line system discussed above would be similar in form to that given in Table 10.2. In the example above, the time at which arrivals occur cannot be predicted with certainty. The simulation model must then capture the random nature of arrival times to properly reflect the behavior of the system. Since random variation is a significant component of most systems, methods for incorporating or generating random variation in the simulation model are essential to systems analysis through simulation. Before addressing methods for generating values of random variables, we discuss the development of a simulation model for a waiting-line system where events occur in a predictable manner.

Example 10.1

The time between successive arrivals to a single server is a constant of 0.100 hr. The time to serve the first, third, fifth, and so on, arrivals is 0.060 hr and is 0.150 hr for the second, fourth, sixth, and so on, arrivals. The system has unlimited capacity, and

Table 10.1 Times of Occurrence of Arrivals to a Single-Channel Waiting-Line System, Where Interarrival Time Is Exponentially Distributed with λ = 20.0

Arrival No.	Arrival Time	Arrival No.	Arrival Time
1	0.044	20	1.057
2	0.081	21	1.229
3	0.172	22	1.251
4	0.193	23	1.277
5	0.235	24	1.295
6	0.247	25	1.297
7	0.265	26	1.351
8	0.331	27	1.384
9	0.399	28	1.593
10	0.493	29	1.631
11	0.519	30	1.651
12	0.607	31	1.662
13	0.624	32	1.674
14	0.636	33	1.704
15	0.721	34	1.820
16	0.847	35	1.835
17	0.856	36	1.869
18	0.914	37	1.908
19	0.969		

units are served in the order in which they arrive. Develop an algorithm for the simulation model, and summarize the behavior of the simulated system by determining the waiting-line length q and the number of units in the system n at each event (arrival or service) during the first 21.000 hr of the simulation. Assume $q = 0$, $n = 0$ at the start of the simulation, 0.000 hr.

Let a, b, and s be the points in time at which the ith arrival, $(i + 1)$-th arrival and the jth service occur measured from time 0.000. Because arrivals occur at constant intervals of 0.100 hr:

$$b = a + 0.100 \tag{10.1}$$

Let T be the time at which the service of the jth unit begins. Then

$$\begin{aligned} s &= T + 0.060 \qquad \text{odd } j \\ &= T + 0.150 \qquad \text{even } j \end{aligned} \tag{10.2}$$

We define the length of the simulation to be t hr.

Events in the course of the simulation must occur at times defined by a, s, and t. Thus, the time at which the next event will occur is the least of a, s, and t, and the time of the next event may be defined by comparing a, s, and t and choosing the least of the three. When the system is empty, no unit is in the service channel and the next

Table 10.2 Summary of the Behavior of a Single-Channel Waiting-Line System for a Period of 1.00 hr. Interarrival Times Are Exponentially Distributed, $\lambda = 20$, and Service Time Is Constant, $\mu = 60$. Queue Length q Is 0 and Number in the System n is 0 at the Start of the Period of Observation

Time	Event	q	n
0.000	Start	0	0
0.044	Arrival 1	0	1
0.061	Service 1	0	0
0.081	Arrival 2	0	1
0.098	Service 2	0	0
0.172	Arrival 3	0	1
0.188	Service 3	0	0
0.193	Arrival 4	0	1
0.210	Service 4	0	0
0.235	Arrival 5	0	1
0.248	Arrival 6	1	2
0.251	Service 5	0	1
0.265	Arrival 7	1	2
0.268	Service 6	0	1
0.285	Service 7	0	0
0.331	Arrival 8	0	1
0.348	Service 8	0	0
0.399	Arrival 9	0	1
0.416	Service 9	0	0
0.493	Arrival 10	0	1
0.510	Service 10	0	0
0.519	Arrival 11	0	1
0.536	Service 11	0	0
0.607	Arrival 12	0	1
0.624	Arrival 13	1	2
0.624	Service 12	0	1
0.636	Arrival 14	1	2
0.641	Service 13	0	1
0.657	Service 14	0	0
0.721	Arrival 15	0	1
0.738	Service 15	0	0
0.847	Arrival 16	0	1
0.856	Arrival 17	1	2
0.864	Service 16	0	1
0.881	Service 17	0	0
0.914	Arrival 18	0	1
0.931	Service 18	0	0
0.969	Arrival 19	0	1
0.986	Service 19	0	0

event cannot be a service. In this case we define the time of the next service as

$$s = t + 1.000 \tag{10.3}$$

Since the last event in the simulation always occurs at t, if $s = t + 1.000$ when the system is empty, the next event must occur either at a or t.

An arrival to the system at time T enters either the service channel or the waiting line. If, on the one hand, the number in the system *after* the arrival occurs is 1 ($n = 1$), the unit enters the service channel, service begins, and s is defined by Eq. (10.2). On the other hand, if $n > 1$ the unit enters the waiting line and the waiting-line length q is increased by 1. In either case, the time of the next arrival is defined by

$$a = T + 0.100 \tag{10.4}$$

When a service occurs at T, the number in the system n is reduced by 1. If after the service $n > 0$ the waiting-line length q is reduced by 1, a unit enters the service channel at the time of service completion, and s is defined by Eq. (10.2). If $n = 0$ after the service, the system is empty and s is defined by Eq. (10.3).

The algorithm for the simulation model is as follows, where j is the number of services initiated at any point in the simulation run.

Algorithm

1. $j = 0$, $q = 0$, $n = 0$, define t.
2. $a = 0.100$, $s = t + 1.000$.
3. If $t < s$ and $t < a$ go to step 7; otherwise, go to step 4.
4. If $s < a$ and $s < t$ go to step 6; otherwise, go to step 5.
5. Next event is an arrival.
   (a) $n = n + 1$.
   (b) If $n > 1$, $q = q + 1$.
   (c) If $n = 1$, $j = j + 1$, $s = a + 0.060$ for j odd, $s = a + 0.150$ for j even.
   (d) $a = a + 0.100$.
   (e) Go to step 3.
6. Next event is a service.
   (a) $n = n - 1$.
   (b) If $n > 0$, $q = q - 1$, $j = j + 1$, $s = s + 0.060$ for j odd, $s = s + 0.150$ for j even.
   (c) If $n = 0$, $s = t + 1.000$.
   (d) Go to step 3.
7. Next event is the end of the simulation.
   (a) Terminate the simulation.

Table 10.3 Summary of the Behavior of the Simulated System Described in Example 10.1

Time	Event	*q*	*n*
0.000	Start	0	0
0.100	Arrival 1	0	1
0.160	Service 1	0	0
0.200	Arrival 2	0	1
0.300	Arrival 3	1	2
0.350	Service 2	0	1
0.400	Arrival 4	1	2
0.410	Service 3	0	1
0.500	Arrival 5	1	2
0.560	Service 4	0	1
0.600	Arrival 6	1	2
0.620	Service 5	0	1
0.700	Arrival 7	1	2
0.770	Service 6	0	1
0.800	Arrival 8	1	2
0.830	Service 7	0	1
0.900	Arrival 9	1	2
0.980	Service 8	0	1
1.000	Arrival 10	1	2
1.040	Service 9	0	1
1.100	Arrival 11	1	2
1.190	Service 10	0	1
1.200	Arrival 12	1	2
1.250	Service 11	0	1
1.300	Arrival 13	1	2
1.400	Arrival 14	2	3
1.400	Service 12	1	2
1.460	Service 13	0	1
1.500	Arrival 15	1	2
1.600	Arrival 16	2	3
1.610	Service 14	1	2
1.670	Service 15	0	1
1.700	Arrival 17	1	2
1.800	Arrival 18	2	3
1.820	Service 16	1	2
1.880	Service 17	0	1
1.900	Arrival 19	1	2
2.000	Arrival 20	2	3

A summary of the behavior of the simulated system for the first 2.000 hours is given in Table 10.3. ■

The system simulated in Example 10.1 assumes that arrival and service times can be predicted with certainty. Such is rarely the case, as event times usually possess a random component. To define event times that possess a random component requires a method for generating values of the associated random variable.

10.2.1 Generation of Random Variables

The basic concepts of random variable generation will be introduced by example. Suppose a continuous manufacturing process produces units that may be either good or bad. If a unit is drawn at random from the production process, there is a 95-percent chance that it will be good and a 5-percent chance that it will be bad. A bad unit is assigned the number 0 and a good unit the number 1. Unit quality is then a random variable x that may assume the values 0 or 1 such that the probability of finding a 0 is 0.05 and the probability of a 1 is 0.95, expressed as

$$P(x = 0) = 0.05 \tag{10.5}$$

$$P(x = 1) = 0.95 \tag{10.6}$$

Now suppose we wish to generate values of x. To accomplish this we must first define the cumulative distribution function of x, $F(x)$. $F(x)$ may be thought of as an accumulator of probability. For x as defined here:

$$\begin{aligned} F(0) &= P(x \leq 0) \\ &= P(x = 0) \\ &= 0.05 \end{aligned} \tag{10.7}$$

$$\begin{aligned} F(1) &= P(x \leq 1) \\ &= P(x = 0) + P(x = 1) \\ &= 1.00 \end{aligned} \tag{10.8}$$

$F(x)$ is shown graphically in Figure 10.1. From Eqs. (10.7) and (10.8), $0 \leq F(x) \leq 1$.

Generation of values of x requires a mechanism for selecting values of $F(x)$ at random, where each value of $F(x)$ has an equal and independent chance of occurring. Since $0 \leq F(x) \leq 1$, the desired generative mechanism must produce values on the interval 0 to 1 such that each value has an equal chance of occurring at a given selection, and the value realized at one selection does not influence the value realized at any subsequent selection. Random numbers have these properties. A table of random numbers is given in Table 10.4. Each four-digit number in Table 10.4 must be preceded by a decimal point to yield a value of the random number r on the interval 0 to 1. Since Table 10.4 contains only a limited number of values of r, a more extensive set of values may be necessary for practical purposes. However, since most simulation models are executed on a digital computer, the analyst usually generates values of r through routines available on most digital computers.

To generate a value of x, we choose a value of r from Table 10.4. This selection should be at random and can be accomplished by closing your eyes and pointing to a position in the table. Successive values can be drawn by continuing down the column from the first selection or across the row, as successive values of r in either direction are independent. The value of r selected defines a value of $F(x)$. To illustrate, if $r = 0.1256$ then $F(x) = 0.1256$. The value of x is defined by entering the graph of $F(x)$

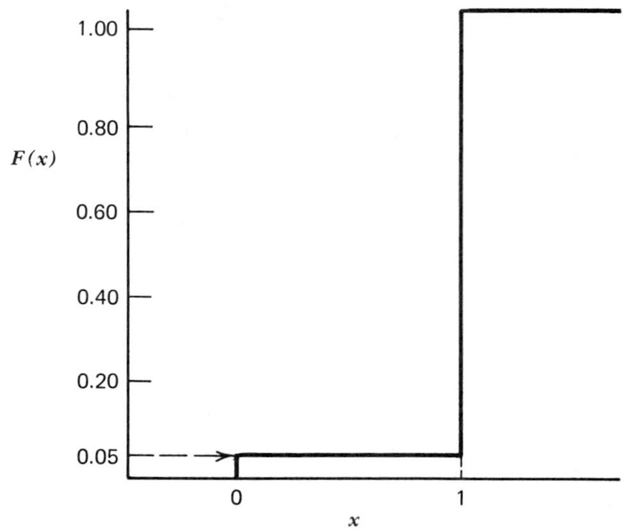

Figure 10.1 Cumulative Distribution Function $F(x)$ of the Random Variable x

in Figure 10.1 at $F(x) = r$ and continuing horizontally until the graph of $F(x)$ is intersected. The value of x at this intersection is that associated with $F(x) = r$ and is the generated value desired. The graphical selection of values of x for $r = 0.025$ and $r = 0.350$ is shown in Figure 10.2.

Mathematically, the process for generating values of x may be defined as follows:

$$\begin{aligned} x &= 0 \qquad 0.00 \leq r \leq 0.05 \\ &= 1 \qquad 0.05 < r \leq 1.00 \end{aligned} \tag{10.9}$$

Since each value on the interval 0 to 1 has an equal and independent chance of being defined by r at a given selection, in the long run we would expect to find 5 percent of the random numbers with values $0.00 \leq r \leq 0.05$ and 95 percent such that $0.05 < r \leq 1.00$, and thus generate values of x of 0 5 percent of the time and values of x of 1 95 percent of the time. This is precisely the property we desire if the generator for x is to have the property defined in Eqs. (10.5) and (10.6).

Example 10.2

A simulation model is to be developed to analyze the flow of traffic on a particular section of interstate highway. The analysis conducted is to distinguish between automobile and truck traffic. Data collected on traffic flow indicate that 68 percent of the traffic on the section is composed of automobiles and 32 percent is composed of trucks, the mixture being random. Develop a random variable generator that will determine whether any vehicle is a truck or an automobile.

Table 10.4 Four-Digit Random Numbers

1687	0121	5545	2178	3159	9348	7653	1787
1844	4978	3270	4816	9462	3422	5375	1447
0307	8813	0117	1380	7229	0953	0651	5333
6133	8803	7615	6462	0239	3273	7486	5456
5300	3057	0093	3049	7450	7261	6509	3707
3659	8586	8586	4239	8159	0803	1383	1073
3988	4272	9735	9962	2154	3266	0208	1855
9256	8834	9701	8697	4873	0966	1930	2891
9970	3800	3066	4193	7562	7634	7746	7769
6900	1474	6742	7184	2425	9889	7509	6051
8728	7905	8876	2113	2786	7702	1135	7488
4708	0856	2758	8844	8245	9871	5013	1240
2323	2772	5722	9389	4831	4481	3405	0097
9937	8750	3061	9619	0159	4382	4855	9693
4457	9507	6926	5987	3590	7651	3598	2729
3985	9350	0233	7246	1380	3065	5962	8189
5475	9141	5573	1168	6846	0566	1776	5563
7390	4271	9116	6261	5513	6733	0777	4058
7358	7624	9518	8489	5276	5250	4013	6825
4830	7549	1823	2994	1554	2374	0256	0174
8732	0828	6378	0808	7445	7402	7401	7790
0126	0644	2729	0575	8887	8144	8876	9958
9862	9550	8537	5274	4809	1389	5047	7776
1238	7436	3476	3930	2291	8374	9624	2379
7653	4504	8145	8338	6716	5253	1071	9152
5266	9228	7972	4780	6931	8565	9007	6952
0651	1329	2118	0745	5402	5706	5614	2330
3452	9743	7389	6645	3367	0392	2044	8741
4042	5587	7137	2538	0996	3132	9827	0771
6178	0129	5167	9842	2546	6698	7273	3353
4655	7758	4647	8057	6517	6589	0881	5983
7967	3952	2009	6482	0808	6503	1745	1944
5958	8247	5861	0937	2874	8804	6960	2521
2482	2204	0887	5477	4882	9994	6028	6219
3058	2381	6757	9113	3867	1178	2264	2985
7527	8297	2034	7533	6888	3534	9205	3425
7706	5406	3078	9815	1187	8782	2011	3024
0047	3063	7953	0153	9339	4654	3875	1361
3291	7492	5333	4569	9415	5361	7430	6333
1126	9758	8410	2638	0134	7060	1152	3370
9843	8731	3798	4201	1023	8329	0770	9654
0991	9060	5438	1086	7567	5629	5668	3341
9034	4137	3511	3829	1373	3776	0294	7776
4008	4069	8336	3390	5321	1409	0566	0712
9176	8645	9280	7873	3720	1460	5279	8534
3685	5305	8665	4243	7468	6621	2510	5468
0221	2111	0675	5047	4206	9813	1021	7811
7668	5712	5257	0131	3473	9656	6675	3139
8762	4315	7034	3362	6866	0938	3830	4536
2747	5655	9206	4341	3191	0080	1754	9804

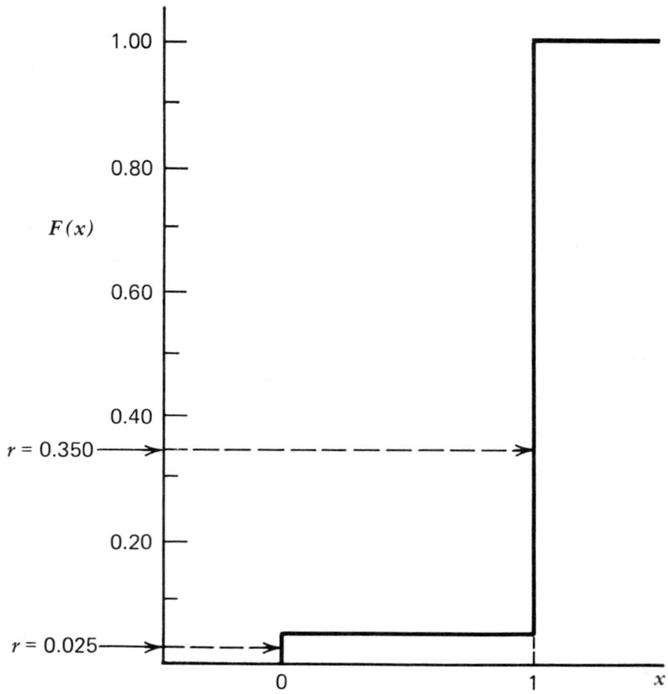

Figure 10.2 Values of x Associated with $r = 0.025$ and $r = 0.350$

Let the random variable x assume the value 0 if the passing vehicle is a truck and 1 if it is a car. Thus the random variable generator for x should possess the property that:

$$P(x = 0) = 0.32$$
$$P(x = 1) = 0.68$$

The cumulative distribution function of x is then defined by

$$F(x) = 0.32 \qquad x = 0$$
$$= 1.00 \qquad x = 1$$

To generate values of x we choose a random number r and

$$x = 0 \qquad 0.00 \le r \le 0.32$$
$$= 1 \qquad 0.32 < r \le 1.00$$

■

Example 10.3

For the problem in Example 10.2, classify the first 50 vehicles crossing the section of interstate highway.

The classification of the first 50 vehicles is summarized in Table 10.5. ■

Table 10.5 Summary of the Classification of the First 50 Vehicles Crossing the Section of Interstate Highway Discussed in Examples 10.2 and 10.3

Random Number	Vehicle Type
0.9962	Car
0.8697	Car
0.4193	Car
0.7184	Car
0.2113	Truck
0.8844	Car
0.9389	Car
0.9619	Car
0.5987	Car
0.7246	Car
0.1168	Truck
0.6261	Car
0.8489	Car
0.2994	Truck
0.0808	Truck
0.0575	Truck
0.5274	Car
0.3930	Car
0.8338	Car
0.4780	Car
0.0745	Truck
0.6645	Car
0.2538	Truck
0.9842	Car
0.8057	Car
0.6482	Car
0.0937	Truck
0.5477	Car
0.9113	Car
0.7533	Car
0.9815	Car
0.0153	Truck
0.4569	Car
0.2638	Truck
0.4201	Car
0.1086	Truck
0.3829	Car
0.3390	Car
0.7873	Car
0.4243	Car
0.5047	Car
0.0131	Truck
0.3362	Car
0.4341	Car
0.3159	Truck
0.9462	Car
0.7229	Car
0.0239	Truck
0.7450	Car
0.8159	Car

The procedure first outlined can be easily generalized to deal with discrete random variables that may assume more than two values. To illustrate, consider the simulation of the tossing of a set of dice. The number of points showing after any toss must lie between 2 and 12 inclusive with the probabilities shown in Table 10.6.

As before, we first define the cumulative distribution function for the random variable x, where x is the number of points showing at a single toss of the dice. Let $p(x)$ be the probability that x is the number of points showing on the dice. Then the cumulative distribution function of x is given by

$$\begin{aligned} F(x) &= p(2) + p(3) + \cdots + p(x-1) + p(x) \\ &= \sum_{y=2}^{x} p(y) \end{aligned} \tag{10.10}$$

$F(x)$ is tabulated in Table 10.7. We choose a random number r. If $0.0000 \leq r \leq 0.0278$, x assumes the value 2. If $0.0278 < r \leq 0.0834$, x assumes the value 3. Continuing in this manner, the value of x is defined by

$$\begin{aligned} x &= 2 & 0.0000 &\leq r \leq 0.0278 \\ &= 3 & 0.0278 &< r \leq 0.0834 \\ &= 4 & 0.0834 &< r \leq 0.1667 \\ &= 5 & 0.1667 &< r \leq 0.2778 \\ &= 6 & 0.2778 &< r \leq 0.4167 \\ &= 7 & 0.4167 &< r \leq 0.5833 \\ &= 8 & 0.5833 &< r \leq 0.7222 \\ &= 9 & 0.7222 &< r \leq 0.8333 \\ &= 10 & 0.8333 &< r \leq 0.9166 \\ &= 11 & 0.9166 &< r \leq 0.9722 \\ &= 12 & 0.9722 &< r \leq 1.0000 \end{aligned} \tag{10.11}$$

Table 10.6 Probability Distribution of the Number of Points Showing at an Arbitrary Toss of a Set of Dice

Points	Probability
2	0.0278
3	0.0556
4	0.0833
5	0.1111
6	0.1389
7	0.1666
8	0.1389
9	0.1111
10	0.0833
11	0.0556
12	0.0278

Table 10.7 Cumulative Distribution Function for the Number of Points Showing at the Toss of a Set of Dice

x	$p(x)$	$F(x)$
2	0.0278	0.0278
3	0.0556	0.0834
4	0.0833	0.1667
5	0.1111	0.2778
6	0.1389	0.4167
7	0.1666	0.5833
8	0.1389	0.7222
9	0.1111	0.8333
10	0.0833	0.9166
11	0.0556	0.9722
12	0.0278	1.0000

Since r has an equal and independent chance of assuming each value on the interval 0.0000 to 1.0000, we would expect to find values of between 0.0000 and 0.0278 with probability 0.0278; therefore, $x = 2$ should occur with probability 0.0278. Similarly, values of r between 0.0278 and 0.0834 occur with probability 0.0556; hence, $x = 3$ should occur with probability 0.0556. In general, r will lie between $F(x - 1)$ and $F(x)$ with probability $p(x) = F(x) - F(x - 1)$, and x will therefore occur with probability $p(x)$. The results of 24 simulated tosses of a set of dice are summarized in Table 10.8.

Example 10.4

Table 10.9 presents data on daily demand for rental cars, taken over a 200-day period. Develop a method for generating daily demand for rental cars.

In Table 10.9, f_x represents the number of days for which x demands occurred out of the 200 days analyzed. Thus the best estimate of the probability of x demands per day, $p(x)$, is given by:

$$p(x) = \frac{f_x}{200} \tag{10.12}$$

Table 10.8 Simulation of 24 Tosses of a Set of Dice

Random Number	Point
0.5475	7
0.7390	9
0.7358	9
0.4830	7
0.8732	10
0.0256	2
0.7401	9
0.8876	10
0.5047	7
0.9624	11
0.1071	4
0.9007	10
0.1168	4
0.6261	8
0.8489	10
0.2994	6
0.0808	3
0.0575	3
0.5274	7
0.3930	6
0.8338	10
0.4780	7
0.0745	3
0.6645	8

Table 10.9 Frequency Distribution of Daily Demand for Rental Cars

Cars Demanded x	Frequency of Occurrence f_x
25 or less	0
26	2
27	12
28	10
29	12
30	18
31	20
32	24
33	21
34	16
35	15
36	22
37	10
38	9
39	7
40	2
41 or more	0
Total	200

Applying Eq. (10.10) the cumulative distribution function of x is given in Table 10.10, and the random variable generator for x is:

$$\begin{aligned}
x &= 26 \quad 0.0000 \leq r \leq 0.0100 \\
&= 27 \quad 0.0100 < r \leq 0.0700 \\
&= 28 \quad 0.0700 < r \leq 0.1200 \\
&= 29 \quad 0.1200 < r \leq 0.1800 \\
&= 30 \quad 0.1800 < r \leq 0.2700 \\
&= 31 \quad 0.2700 < r \leq 0.3700 \\
&= 32 \quad 0.3700 < r \leq 0.4900 \\
&= 33 \quad 0.4900 < r \leq 0.5950 \\
&= 34 \quad 0.5950 < r \leq 0.6750 \\
&= 35 \quad 0.6750 < r \leq 0.7500 \\
&= 36 \quad 0.7500 < r \leq 0.8600 \\
&= 37 \quad 0.8600 < r \leq 0.9100 \\
&= 38 \quad 0.9100 < r \leq 0.9550 \\
&= 39 \quad 0.9550 < r \leq 0.9900 \\
&= 40 \quad 0.9900 < r \leq 1.0000
\end{aligned} \tag{10.13}$$

■

Table 10.10 Cumulative Distribution Function for Daily Car Rental Demand

x	$p(x)$	$F(x)$
25 or less	0.0000	0.0000
26	0.0100	0.0100
27	0.0600	0.0700
28	0.0500	0.1200
29	0.0600	0.1800
30	0.0900	0.2700
31	0.1000	0.3700
32	0.1200	0.4900
33	0.1050	0.5950
34	0.0800	0.6750
35	0.0750	0.7500
36	0.1100	0.8600
37	0.0500	0.9100
38	0.0450	0.9550
39	0.0350	0.9900
40	0.0100	1.0000
41 or more	0.0000	1.0000

The methods presented for generation of discrete random variables may be extended to continuous random variables with slight modification. We introduce this approach by example. Soft-drink bottles are to be filled with 16 fluid ounces. A sample of 1000 bottles has been drawn and the contents of each, x, have been measured with the results in Table 10.11. For simplicity, in Table 10.11 bottle contents are summarized in intervals of width 0.20 oz. In the table f_{a_i,b_i} represents the frequency of occurrence of observations for which $a_i < x \leq b_i$ and $p(a_i < x \leq b_i)$ is the observed probability that $a_i < x \leq b_i$ and is defined by

$$p(a_i < x \leq b_i) = \frac{f_{a_i,b_i}}{1000} \tag{10.14}$$

For continuous random variables we define the cumulative distribution function of x at the interval end points b_i, where

$$\begin{aligned} F(b_i) &= p(-\infty < x \leq b_i) \\ &= \frac{1}{1000}\sum_{j=1}^{i} f_{a_j,b_j} \end{aligned} \tag{10.15}$$

Based upon Eq. (10.15) the cumulative distribution function of x is shown in Figure 10.3.

If we select a random number r to define a value of $F(x)$, as shown in Figure 10.3, we will generate only those values of x corresponding to the interval end points in Table 10.11. That is, applying the random-variable-generation procedure defined for discrete random variables will yield a discrete set of values (although not integer values) for the continuous random variable to be generated. Although this is only a gross approximation to reality, it may be satisfactory for practical purposes.

Table 10.11 Contents of a Sample of 1000 Soft-Drink Bottles

Interval Number i	Interval for Contents (oz) $a_i < x \leq b_i$	Frequency f_{a_i, b_i}	$P(a_i < x \leq b_i)$
1	$-\infty < x \leq 15.00$	0	0.0000
2	$15.00 < x \leq 15.20$	20	0.0200
3	$15.20 < x \leq 15.40$	30	0.0300
4	$15.40 < x \leq 15.60$	50	0.0500
5	$15.60 < x \leq 15.80$	80	0.0800
6	$15.80 < x \leq 16.00$	380	0.3800
7	$16.00 < x \leq 16.20$	420	0.4200
8	$16.20 < x \leq 16.40$	20	0.0200
9	$16.40 < x < \infty$	0	0.0000
	Total	1000	1.0000

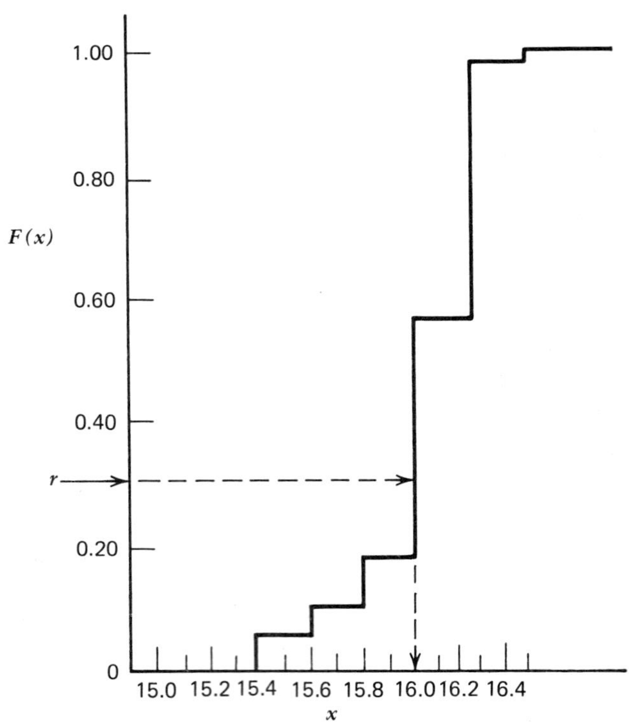

Figure 10.3 Cumulative Distribution Function of Bottle Contents

To provide for generation of a true continuum of values for a continuous random variable we may connect successive points of discontinuity of the "discrete" cumulative distribution function by a straight line. This process is illustrated in Figure 10.4 for the data in Table 10.11. As Figure 10.4 shows, the procedure just outlined yields a continuous cumulative distribution function for bottle contents x with the result that all values of x between 15.00 and 16.40 may be generated.

The technique we have presented for generation of values of discrete and continuous random variables is often called the tabulation method. Other methods available for this purpose are sometimes more convenient and efficient. (For a discussion of these techniques, see References 5, 9, 14, and 18.) Despite the availability of other methods, the tabulation procedure remains one of the most commonly used techniques for random variable generation owing to its simplicity and general applicability.

10.2.2 Random Numbers

As the above discussion indicates, random-number generation is an important component of every simulation model. An essential property of every random number generator is the ability to generate random variables that are uniformly distributed on

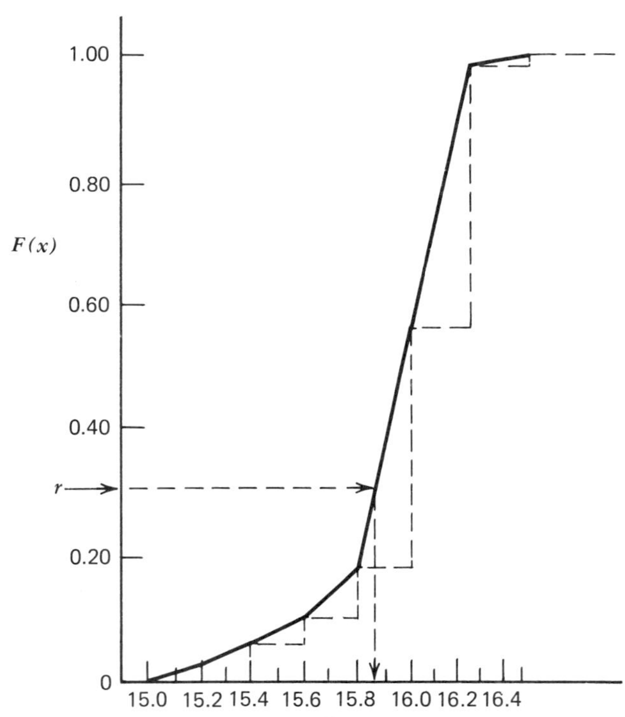

Figure 10.4 Continuous Cumulative Distribution Function of Bottle Contents

the interval (0, 1). Actually, digital methods for generating random numbers are algorithmic; therefore, the numbers resulting are generally termed pseudorandom numbers. That is, because the numbers are generated algorithmically they are not actually random. However, were one to compare a set of numbers derived from a reliable digital generator with numbers that are truly random, the distinction between the two sets of numbers would not be apparent. A discussion of algorithmic methods for the generation of random numbers is beyond the scope of this text; however, random-number generators are available for most high-speed digital computers.

10.2.3 Simulation Model Development

Development of a valid simulation model depends upon a thorough understanding of the operation of the system modeled and the ability to generate values of the random variables that influence the behavior of the system. The development of simulation models is illustrated in the following simple examples. In each case, an algorithm or step-by-step procedure for execution of the simulation model is presented along with sample results.

Example 10.5
A dice game operates according to the following rules.

1. A player bets \$1.00.
2. The player rolls the dice.
3. If a 7 or 11 appears on the first toss, the player wins \$1.00.
4. If a 7 or 11 does not appear on the toss, the player must match the point appearing on the first roll.
5. The player tosses the dice successively until he or she matches the point on the first toss or until a 7 or 11 appears. If a 7 or 11 appears before the point is matched, the player loses \$1.00. If the point is matched before a 7 or 11 appears, he or she wins \$1.00.

Develop an algorithm for simulation of the play of the game N times. The model should record the average winnings (losses) for the player per play of the game. Simulate 10 plays of the game.

The simulation of a single play of the game starts with the initial toss of the dice or the generation of the number of points showing on the dice at this toss. The random variable generator in Eq. (10.11) may be used for this purpose where x is the number of points recorded at a single toss. Let x_0 be the value of x at the first toss. If $x_0 = 7$ or $x_0 = 11$, the player wins \$1.00. If $x_0 \neq 7$ and $x_0 \neq 11$, the player rolls again. Let x_1 be the value recorded at a toss after the first. If $x_1 = x_0$, the player wins \$1.00. If $x_1 = 7$ or $x_1 = 11$, the player loses \$1.00. If $x_1 \neq x_0$, $x_1 \neq 7$, and $x_1 \neq 11$, the entire process is repeated until the player either wins or loses \$1.00.

The algorithm for N plays of the game is as follows, where i is used to count the number of times the game is played and W is the profit accumulated by the player.

1. Define N, $i = 1$, $W = 0$.
2. Select a random number r, and generate x_0 using Eq. (10.11).
3. If $x_0 = 7$ or $x_0 = 11$, $W = W + 1.00$ and go to step 7; otherwise, go to step 4.
4. Select a random number r, and generate x_1 using Eq. (10.11).
5. If $x_1 = x_0$, $W = W + 1.00$ and go to step 7; otherwise, go to step 6.
6. If $x_1 = 7$ or $x_1 = 11$, $W = W - 1.00$ and go to step 7; otherwise, go to step 4.
7. $i = i + 1$.
8. If $i \leq N$, go to step 2; otherwise, go to step 9.
9. Average profit for play is given by W/N.
10. Terminate the simulation.

A summary of 10 plays of the game is shown in Table 10.12. The profit accumulated is \$2.00 for an average gain of \$0.20 per play. ■

Table 10.12 Summary of the Simulation of 10 (*N*) Plays of the Dice Game in Example 10.5

i	r	x_0	x_1	W	Remarks
1				\$0.00	
1	0.9389	11		1.00	Player wins
2	0.3452	6			
2	0.2874		6	2.00	Player wins
3	0.6716	8			
3	0.0651		3		
3	0.5513		7	1.00	Player loses
4	0.1329	4			
4	0.7402		9		
4	0.2374		5		
4	0.0174		2		
4	0.7358		9		
4	0.8750		10		
4	0.4382		7	0.00	Player loses
5	0.6900	8			
5	0.7358		9		
5	0.5922		7	−1.00	Player loses
6	0.7746	9			
6	0.7634		9	0.00	Player wins
7	0.4873	7		1.00	Player wins
8	0.9415	11		2.00	Player wins
9	0.9507	11		3.00	Player wins
10	0.0134	2			
10	0.4654		7	2.00	Player loses

Example 10.6

A rental care agency maintains a fleet of 33 cars for daily rental. It costs the agency \$14.00 per day to keep a car on hand, and income per day for each rental is \$20.00. Daily demand for rental cars is given in Table 10.9. Develop an algorithm for a simulation model that will estimate daily profit. Simulate 20 days of operation of the agency.

The demand for rental cars for each day may be generated using Eq. (10.13). The cost of maintaining the fleet of cars is \$462.00 per day, no matter how many cars are rented. If x is the demand for a given day, then profit is given by

$$\text{Profit} = 20x - 462, \qquad \text{for } x \leq 33$$

The algorithm for simulation of N days of operation of the agency is as follows, where P is accumulated profit and i counts the number of days simulated.

1. Define N, $i = 1$, $P = 0$.
2. Generate r and define x using Eq. (10.13).
3. If $x \leq 33$, $P = P + 20x - 462$ and go to step 5; otherwise, go to step 4.
4. $P = P + 20(33) - 462$ and go to step 5.
5. $i = i + 1$.
6. If $i \leq N$ go to step 2; otherwise, go to step 7.
7. Profit per day is P/N.
8. Terminate the simulation.

A summary of 20 days of operation of the car rental agency is given in Table 10.13. Based upon this analysis, average daily profit is \$178.00 per day. ■

Example 10.7

A manufacturer maintains a training program that a large portion of its labor force must successfully complete. The training program comprises two phases. The first phase must be completed successfully before the second is attempted, each phase requiring one week of training. At the end of each phase, each trainee is given a test. If he or she fails the test, he must retake that phase. Past experience indicates that the probability that an individual will fail the first phase on the first attempt is p_1. If he or she fails the first attempt, the probability that he or she will fail the first phase on his second attempt is p_1^2. To generalize, if an individual fails to successfully complete the first phase on his or her first $(n - 1)$ attempts, the probability that he or she will fail it on the nth attempt is p_1^n. The probability that an individual will fail the second phase on his or her first attempt is p_2. In a fashion similar to the first phase, if the trainee fails the second phase on the first $(n - 1)$ attempts, the probability that he or she will fail that phase on the nth attempt is p_2^n.

A simulation model is to be developed that will estimate the average number of weeks of training per trainee required before the program is successfully completed if $p_1 = 0.60$ and $p_2 = 0.40$.

Table 10.13 Simulation of 20 (*N*) Days of Operation of the Car Rental Agency in Example 10.6

i	r	x	Income	Profit/Day	P
1	0.4239	32	$640	$178	$ 178
2	0.9962	40	660	198	376
3	0.8697	37	660	198	574
4	0.4193	32	640	178	752
5	0.7184	35	660	198	950
6	0.2113	30	600	138	1088
7	0.8844	37	660	198	1286
8	0.9389	38	660	198	1484
9	0.9619	39	660	198	1682
10	0.5987	34	660	198	1880
11	0.0129	27	540	78	1958
12	0.7758	36	660	198	2156
13	0.3952	32	640	178	2334
14	0.8247	36	660	198	2532
15	0.2204	30	600	138	2670
16	0.2381	30	600	138	2808
17	0.8297	36	660	198	3006
18	0.5406	33	660	198	3204
19	0.3063	31	620	158	3362
20	0.5706	33	660	198	3560

The following notation will be used in this simulation model.

N = total number of trainees simulated

j = counter indicating the number of trainees simulated thus far

n = number of weeks of training for the first phase

r = random number

m = number of weeks of training for the second phase

M = cumulative number of weeks of training over all trainees simulated.

The logical organization of the simulation model is summarized in the flowchart in Figure 10.5. For each trainee simulated, the model must determine the number of weeks of training required in each of the two phases. The probability that the first phase is completed on the first attempt is $1 - p_1$. To simulate the outcome of the first attempt at the first phase we generate or select a random number r. If $r \leq p_1$, then we conclude that the phase has not been passed, and we increase the number of attempts at this phase n to 2. The probability of passing the first phase on the second attempt is $1 - p_1^2$. Thus we generate a second random number r, and if $r \leq p_1^2$, we again conclude that the first phase has been failed again and n is increased to 3. This process is repeated until $r > p_1^n$, where a new random number r is generated for each attempt. The entire process is then carried out in an identical fashion for phase II

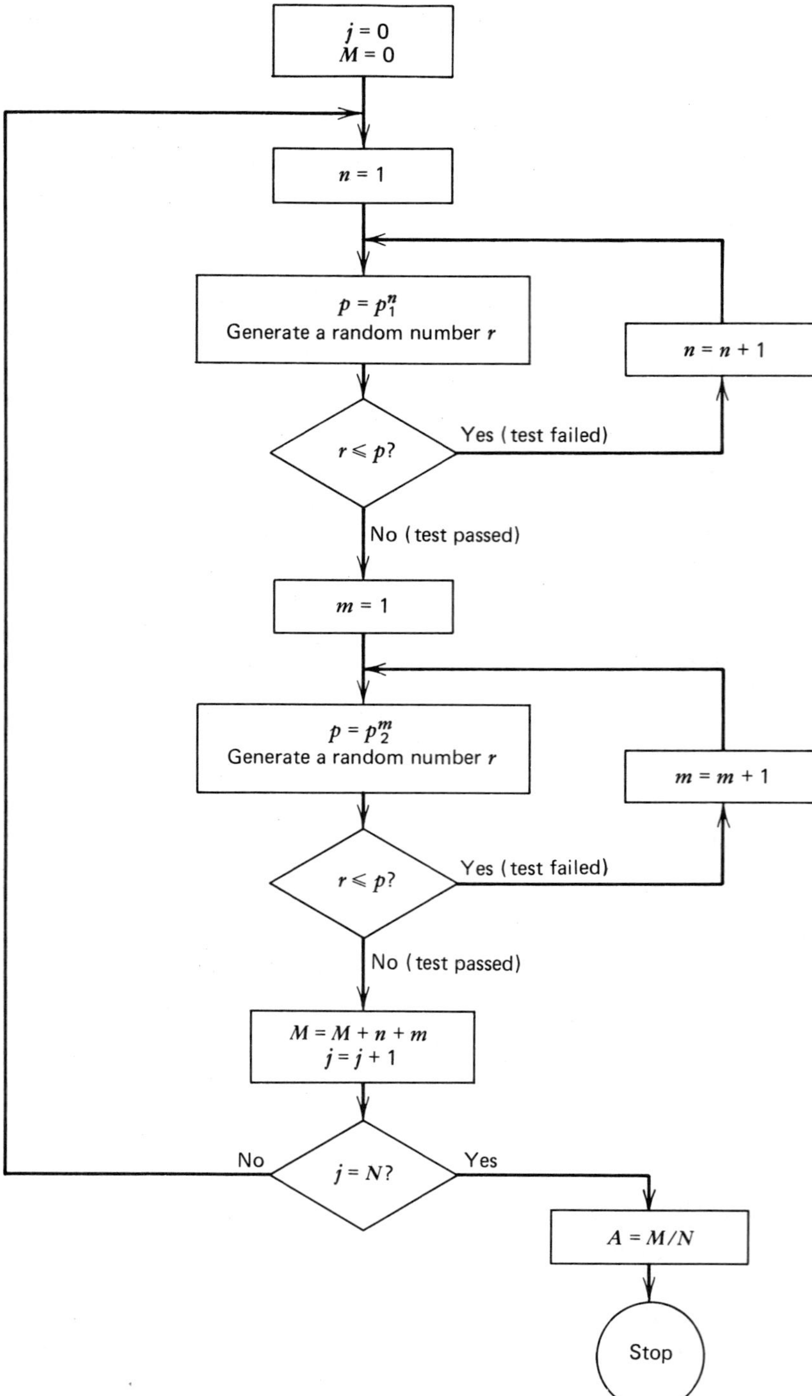

Figure 10.5 Flowchart for the Simulation Model in Example 10.7

comparing successively generated random numbers r with p_2^m until a point is reached where $r > p_2^m$.

The total number of weeks of training for the two phases is given by $n + m$ for each trainee simulated. If n_j and m_j are the total weeks of training for phases I and II, respectively, for the jth trainee, and if the number of trainees simulated is N, then the average total weeks of training per trainee A is given by:

$$A = \frac{1}{N} \sum_{j=1}^{N} (n_j + m_j) \tag{10.16}$$

The results of the simulation of 30 trainees are summarized in Table 10.14. As a result of this analysis, the average number of weeks of training per trainee is 3.20. ■

10.3 SYSTEMS ANALYSIS THROUGH SIMULATION

Thus far, our discussion of simulation has dealt primarily with the mechanics of the development of simulation models. We have said very little about how such models can be used as an analytic tool and as an aid to the decision-making process. An in-depth discussion of these topics is far beyond the scope of a single chapter on simulation; however, there are certain critical aspects with respect to systems analysis through simulation of which the reader should be aware. We simply introduce the reader to the most important of these problem areas without attempting to deal with their solution.

In attempting to develop a simulation model, one of the first difficulties to be dealt with is that of model validation. Before a simulation model can be used for systems analysis, the analyst must first verify that the model is an adequate reflection of reality. This can sometimes be achieved by comparing model results with those of the real-world system. However, even though a model realistically reflects the performance of a system in the past and today, there is no guarantee it will do so under conditions that may prevail in the future. All too often the analyst has no objective basis for ascertaining model validity but must rely upon his or her judgment and that of others familiar with the system under study.

As the examples in the preceding section indicate, simulation models are often used to estimate a measure or measures of performance of the system studied. However, unlike mathematical models, the output of a simulation model is one or more random variables. Hence, there is error in an estimate of system performance through simulation. This is both an advantage and a disadvantage. Since the output of the real-world system is also a random variable, the random nature of the output of the simulation model might be considered a realistic reflection of true system performance, making no attempt to hide the difficulties that exist in assessing the value of that measure. But suppose that we wish to compare several systems based upon the mean value of an appropriate measure of system performance. In this case we require an estimate of system performance for each system that is precise enough to validly carry out the comparison indicated. The precision with which such measures of performance

Table 10.14 Output of the Simulation of 30 Trainees for the Problem in Example 10.7.

Trial j	Phase I Attempt n_j	Phase I $p_1^{n_j}$	Phase I r_1	Phase II Attempt m_j	Phase II $p_2^{m_j}$	Phase II r_2	Total Weeks Training $n_j + m_j$
1	1	0.60	0.89	1	0.40	0.41	2
2	1	0.60	0.35	1	0.40	0.84	3
	2	0.36	0.76				
3	1	0.60	0.40	1	0.40	0.71	3
	2	0.36	0.83				
4	1	0.60	0.17	1	0.40	0.29	4
	2	0.36	0.43	2	0.16	0.74	
5	1	0.60	0.07	1	0.40	0.62	4
	2	0.36	0.03				
	3	0.22	0.63				
6	1	0.60	0.09	1	0.40	0.87	3
	2	0.36	0.38				
7	1	0.60	0.66	1	0.40	0.29	3
				2	0.16	0.69	
8	1	0.60	0.62	1	0.40	0.53	2
9	1	0.60	0.78	1	0.40	0.52	2
10	1	0.60	0.27	1	0.40	0.41	4
	2	0.36	0.21				
	3	0.22	0.89				
11	1	0.60	0.47	1	0.40	0.54	3
	2	0.36	0.65				
12	1	0.60	0.64	1	0.40	0.17	3
				2	0.16	0.52	
13	1	0.60	0.82	1	0.40	0.43	2
14	1	0.60	0.34	1	0.40	0.03	5
	2	0.36	0.11	2	0.16	0.59	
	3	0.22	0.30				
15	1	0.60	0.19	1	0.40	0.55	4
	2	0.36	0.09				
	3	0.22	0.84				
16	1	0.60	0.25	1	0.40	0.31	4
	2	0.36	0.43	2	0.16	0.83	
17	1	0.60	0.89	1	0.40	0.50	2
18	1	0.60	0.96	1	0.40	0.49	2
19	1	0.60	0.08	1	0.40	0.91	4
	2	0.36	0.08				
	3	0.22	0.83				

Table 10.14 (Continued)

Trial j	Phase I Attempt n_j	$p_1^{n_j}$	r_1	Phase II Attempt m_j	$p_2^{m_j}$	r_2	Total Weeks Training $n_j + m_j$
20	1	0.60	0.48	1	0.40	0.86	3
	2	0.36	0.91				
21	1	0.60	0.86	1	0.40	0.87	2
22	1	0.60	0.19	1	0.40	0.53	4
	2	0.36	0.08				
	3	0.22	0.37				
23	1	0.60	0.09	1	0.40	0.67	3
	2	0.36	0.60				
24	1	0.60	0.70	1	0.40	0.19	3
				2	0.16	0.73	
25	1	0.60	0.04	1	0.40	0.39	4
	2	0.36	0.43	2	0.16	0.87	
26	1	0.60	0.18	1	0.40	0.56	3
	2	0.36	0.69				
27	1	0.60	0.26	1	0.40	0.34	5
	2	0.36	0.25	2	0.16	0.60	
	3	0.22	0.34				
28	1	0.60	0.43	1	0.40	0.22	4
	2	0.36	0.59	2	0.16	0.97	
29	1	0.60	0.23	1	0.40	0.71	3
	2	0.36	0.56				
30	1	0.60	0.57	1	0.40	0.72	3
	2	0.36	0.97				

are estimated can usually be improved by increasing the length of the simulation run. As the simulation run length is increased, however, the cost of the analysis also increases. Thus, the analyst is faced with the problem of balancing the cost of the analysis against the precision of estimation to be achieved.

Finally, the reader should realize that the results obtained from a simulation experiment may be strongly influenced by the starting conditions assumed. To illustrate this point, consider the simulation of a single-channel waiting-line system where interarrival time is exponentially distributed and service time is constant. Assume that the system starts out empty. We would expect a short waiting line for early arrivals to the system with longer waiting lines occurring at later stages. Eventually the system will reach a steady or equilibrium state. However, for simulation runs of short duration, estimates of quantities such as the mean waiting-line length and the mean number of units in the system may be biased by the initial conditions assumed. Thus, if the analyst is concerned with estimation of steady-state conditions for the system modeled, simulation runs of considerable duration may be required.

10.4 SIMULATION LANGUAGES

The complexity of the systems usually analyzed through simulation generally requires that the analysis be carried out on a digital computer. In such cases the analyst must translate the simulation model into a medium that can be interpreted by the computer. This translation is accomplished through a programming language. Translation of the simulation model can be accomplished through a general-purpose language or a special-purpose language. On the one hand, general-purpose languages such as FORTRAN, BASIC, and PL/I provide the programmer with a tool for the analysis of a virtually limitless number of problems, of which simulation is only one. On the other hand, special-purpose simulation languages are designed to address problems to be analyzed through simulation, although the variety of simulation problems that can be handled by these languages is quite broad. Included in the category of special-purpose simulation languages are GPSS, SIMSCRIPT, SLAM, DYNAMO, and SIMULA.

Perhaps the principle advantage of general-purpose languages lies in the fact that one of these languages is probably already known to the programmer. In addition, these languages provide the analyst with a maximum of flexibility in the design of his analysis. However, because special-purpose languages are oriented toward the specific application of simulation, the programming time required for translation of the model is generally less than that required in the case of general-purpose languages, as the time-keeping mechanism and many of the subroutines normally required in any simulation model are built in. In addition, the structure of special-purpose languages will often help the analyst to formulate his model. In using a special-purpose language, however, the analyst is restricted to a prescribed output format and increased computer running time.

10.5 SUMMARY

Simulation has proved to be an effective and versatile modeling technique for the analysis of complex interactive systems in both the private and public sectors. In addition, as compared with mathematical modeling, simulation offers the advantage of relative simplicity in model construction, although the analysis of simulated results may require an advanced knowledge of probability theory and statistical methodology.

Although the advantages of simulation are substantial, it should not be used indiscriminately. In general, if a system can be modeled mathematically, it should be. A mathematical model can usually be evaluated more quickly than a simulation model because replication of a given system condition is normally required for reliable estimation of the measure of system performance when the system is modeled through simulation. Thus, the cost of the analysis is likely to be greater when the system is modeled through simulation than through a mathematical model.

REFERENCES AND SUGGESTED READING

1. W. G. Bulgren, *Discrete System Simulation,* Prentice-Hall, Englewood Cliffs, NJ, 1982.
2. F. Cellier (Ed.), *Progress in Modelling and Simulation,* Academic, New York, 1982.
3. D. N. Chorafas, *Systems and Simulation,* Academic, New York, 1965.
4. J. Clark and S. Cole, *Global Simulation Models: A Comparative Study,* John Wiley, New York, 1975.
5. G. S. Fishman, *Principles of Discrete Event Simulation,* John Wiley, New York, 1978.
6. J. W. Forrester, *Industrial Dynamics,* M.I.T. Press, Cambridge, Mass., 1961.
7. G. Gordon, *System Simulation,* 2nd Ed., Prentice-Hall, Englewood Cliffs, NJ, 1978.
8. S. H. Hollingdale (Ed.), *Digital Simulation In Operational Research,* American Elsevier, New York, 1967.
9. A. M. Law and W. D. Kelton, *Simulation Modeling and Analysis,* McGraw-Hill, New York, 1982.
10. F. F. Martin, *Computer Modeling and Simulation,* John Wiley, New York, 1968.
11. C. McMillan and R. F. Gonzalez, *Systems Analysis*, Homewood, IL: Irwin, 1965.
12. A. G. Mihram, *Simulation: Statistical Foundations and Methodology,* Academic, New York, 1972.
13. J. H. Mize and J. G. Cox, *Essentials of Simulation,* Prentice-Hall, Englewood Cliffs, NJ, 1968.
14. T. H. Naylor, J. L. Balintfy, D. S. Burdick, and K. Chu, *Computer Simulation Techniques,* Wiley, New York, 1968.
15. A. A. B. Pritsker and C. D. Pegdon, *Introduction to Simulation and SLAM*, Systems, West Lafayette, IN, 1979.
16. J. Reitman, *Computer Simulation Applications*, John Wiley, New York, 1971.
17. N. Roberts, D. Anderson, R. Deal, M. Garet, and W. Shaffer, *Introduction to Computer Simulation: The System Dynamics Approach*, Addison-Wesley, Reading, MA, 1983.
18. J. W. Schmidt and R. E. Taylor, *Simulation and Analysis of Industrial Systems,* Irwin, Homewood, IL, 1970.

REVIEW QUESTIONS AND PROBLEMS

10.1 A manufacturing process produces units that are packaged in lots of 1000. To monitor the quality of manufactured lots an inspector draws a sample of 5 units from each lot of 1000 and inspects each unit in the sample, classifying each unit inspected as either good or bad. The process produces defective (bad) items at an average rate of 10/lot under normal operating conditions. By simulation, estimate the average number of defective units per lot found by the inspector if the process is operating normally. Simulate the inspection of 50 lots.

10.2 The following dice game has been proposed. The player pays \$8.00 to roll the dice. If x is the number of points showing on the dice when they come to rest, the player is paid x dollars. The probability distribution of the number of points showing on the dice, $P(x)$, is given in Table 10.6. Using the random variable generator in Eq. (10.11) and random numbers, simulate 100 plays in the game and determine the average gain to the player per play of the game. The gain to the player at a given play of the game is x − \$8.00.

10.3 A game such as that described in Problem 10.2 is said to be "fair" if the true mean gain per play of the game is \$0.00. This implies that in order for the game described in Problem 10.2 to be fair, a player must pay \$7.00 to play the game (that is, \$7.00 is his break-even price). By simulating 200 plays of the game, determine whether or not this assertion seems to be true.

10.4 In Problem 10.3, if the assertion that the game is fair is true, would you expect the simulated average gain per play of the game to be exactly \$0.00? If so, why? If not, why not?

10.5 The data in the following table represent runway occupation time for landings by commercial aircraft. Define the cumulative distribution function of runway occupation time, and develop a procedure for generating simulated values.

Interval for Occupation Time, t	Frequency of Occurence
$0.00 < t \leq 1.10$	24
$1.10 < t \leq 1.80$	136
$1.80 < t \leq 2.50$	346
$2.50 < t \leq 3.20$	342
$3.20 < t \leq 3.90$	130
$3.90 < t \leq 4.60$	22
$4.60 < t$	0

10.6 The data in the following table have been collected on the life of a certain brand of calculator under normal use. Define the cumulative distribution function of calculator life, and develop a method for generating simulated values of this random variable.

Interval for Life, t (yrs)	Frequency of Occurence
$0.00 \leq t \leq 1.00$	185
$1.00 < t \leq 2.00$	130
$2.00 < t \leq 3.00$	59
$3.00 < t \leq 4.00$	56
$4.00 < t \leq 5.00$	32
$5.00 < t \leq 6.00$	14
$6.00 < t \leq 7.00$	10
$7.00 < t \leq 8.00$	5
$8.00 < t \leq 9.00$	4
$9.00 < t \leq 10.00$	2
$10.00 < t \leq 11.00$	3

10.7 The following data have been collected on the number of commercial aircraft delayed at the departure gate at a major air terminal. The data were taken during peak hour traffic under normal operating conditions.

Number of Aircraft Delayed	Frequency of Occurrence
0	0
1	2
2	8
3	15
4	19
5	26
6	32
7	35
8	42
9	30
10	22
11	18
12	13
13	11
14	8
15	4
16	5
17	4
18	4
19	2
20	0

Define the cumulative distribution function and a method of generating simulated values of aircraft delays during the peak hour.

10.8 In Example 10.1 suppose that the time between the occurrence of successive arrivals is a random variable. The data summarized in the following table describe the frequency distribution of interarrival time resulting from the observation of 800 arrivals.

Interval Time, t (hr)	Frequency of Occurrence
$0.00 \leq t \leq 0.03$	30
$0.03 < t \leq 0.06$	70
$0.06 < t \leq 0.09$	100
$0.09 < t \leq 0.12$	250
$0.12 < t \leq 0.15$	250
$0.15 < t \leq 0.18$	70
$0.18 < t \leq 0.21$	20
$0.21 < t \leq 0.24$	5
$0.24 < t \leq 0.27$	4
$0.27 < t \leq 0.30$	1

a. Define the cumulative distribution function for interarrival time.
b. Develop a method for generating simulated values of interarrival times.

10.9 The following data have been collected on the number of units of product x demanded per day.

Demand, x	Frequency of Occurrence
0	0
1	10
2	31
3	37
4	16
5	6
6	0

Define the cumulative distribution function for x based on the observed data and the associated random variable generator.

10.10 A building contains 1000 light fixtures. The following data on time until failure x have been collected on these fixtures.

Time until Failure in Months, x	Frequency of Occurrence
$0.00 \leq x \leq 4.00$	27
$4.00 < x \leq 8.00$	35
$8.00 < x \leq 12.00$	130
$12.00 < x \leq 16.00$	151
$16.00 < x \leq 20.00$	349
$20.00 < x \leq 24.00$	308

Define the cumulative distribution function for x and the associated random variable generator.

10.11 Continue the simulation in Example 10.6 for an additional 80 days beyond the period summarized in Table 10.13. Based upon the total sample of 100 days, estimate average daily profit.

10.12 The rental car agency in Example 10.6 hopes to increase its profit by making more cars available. Specifically, it intends to purchase an additional 7 cars, increasing its fleet to 40. The agency anticipates that this increase in its fleet will reduce its daily maintenance cost to \$13.00. It is assumed that daily demand for rental cars will continue at the same levels experienced in the past. By simulating the operation of the agency for a period of 100 days, estimate average profit per day. Based upon a comparison of the results obtained in Problem 10.11, should the rental car agency increase the size of its fleet?

10.13 Solve Problem 10.12 using the same random numbers used in Problem 10.11. What advantage is gained in comparing alternative systems by using a common stream of random numbers in simulating the behavior of each alternative? Are there any disadvantages to this procedure?

10.14 Continue the simulation described in Example 10.7 for an additional 30 trainees. Based upon the simulation of 60 trainees, estimate the average number of weeks required to train an employee.

10.15 In Example 10.7 the average number of weeks of training per trainee was found to be 3.20 based upon the simulated training of 30 new employees. In an effort to reduce the training period, a revision of the training period has been proposed. The revision requires 1.5 weeks for the first phase and 1 week for the second. However, the revision of the first phase is predicted to reduce the probability of failure of the first phase on the nth attempt from $(0.60)^n$ to $(0.20)^n$. In addition, the probability of failing the second phase on the nth attempt will be reduced from $(0.40)^n$ to $(0.30)^n$. By simulating the training of 30 new employees estimate the reduction in the average time to complete the training program.

10.16 The distribution of the contents of soft-drink bottles is given in Table 10.11. The soft drinks are actually sold in six-packs. Simulate the total contents of 100 six-packs, and estimate the proportion of six-packs that will contain a total of 95.10 oz or less.

10.17 Simulate the arrival and service of 100 customers for Example 10.1 if interarrival time is a random variable with the probability distribution given in Problem 10.8. Estimate the average time an arrival spends waiting to get into the service channel and the average time an arrival spends in the system (waiting time + service time) based upon the simulation of the service of 100 arrivals.

10.18 Carry out the exercise in Problem 10.17 if service time is a random variable with cumulative distribution function given in the following table.

Service Time, t (hr)	Cumulative Probability, P (Service Time $\leq t$)
0.03	0.10
0.06	0.30
0.09	0.50
0.12	0.80
0.15	0.90
0.18	1.00

10.19 Carry out the analysis in Problem 10.17 if an arriving customer refuses to wait for service if there is already one unit in the waiting line. What percentage of customer business will be lost?

10.20 In Problem 10.18 the service facility charges \$100.00 to each customer it serves. The fixed cost of operating the facility is \$2000.00/day, and it costs an additional \$20.00 to serve each customer. Furthermore, if a customer is required to wait more than 0.10 hr before his or her service begins, the service is provided free of charge. The service facility operates 8 hr/day, and all customers arriving before closing time are served the same day. By simulating 2 days of activity, estimate the average total profit per customer.

10.21 In Problem 10.10, suppose that it costs \$4.00 to replace each light fixture when it burns out. The building manager is considering implementing a replacement policy calling for replacement of all fixtures every 12 months, as bulk replacement costs only \$1.50 per fixture. By simulating this replacement policy, estimate the average monthly cost of fixture replacement.

SECTION FOUR

THE USE OF IE AND OR TOOLS IN TRADITIONAL APPLICATION AREAS

Many of the design and operational problems addressed by industrial engineers since Fredrick Taylor's era are amenable to analysis by operations research techniques as well as by traditional IE tools. For example, assembly lines have been "balanced" (that is, a smooth flow with no bottlenecks has been created) since Henry Ford's Model T production through the application of traditional work methods procedures. This approach involves altering the design of one or more jobs along the assembly line through the methods procedures discussed in Chapter 4. Assembly lines can also be balanced through the application of several of the OR techniques discussed in the previous section. For instance, an integer programming model can be formulated to represent the design problem of which work elements to assign to which jobs along the line so as to minimize bottlenecks. Problem areas such as assembly-line balancing are reflective of the overlap that exists between traditional IE and mathematical OR. As a result of this overlap, these two areas have become fused into a single scientific discipline.

In this section, we discuss several of these overlap problem areas. Each topic is described from a functional standpoint, emphasizing the underlying problem first and then discussing the various ways the problem can be analyzed. For example, Chapter 11 is concerned with facility planning and design. This chapter first presents a discussion of the meaning of facility planning and facility design—that is, the types of decision problems involved and their nature. This desciption is followed by a discussion of the various classical or traditional ways of analyzing these decision issues, as well as more recent mathematical analytical approaches.

In this facility planning and design chapter, the first decision issue discussed is the determination of the number of machines to utilize in the shop. As pointed out, the classical way to answer this question is to solve a relatively simple descriptive mathematical model or equation, rounding the results to yield the desired number of machines. The discussion also points out that recently several OR techniques have been applied as alternative ways to answer this question. These include a mathematical programming model similar to those discussed in Chapter 8, a queueing model along the lines of those described in Chapter 9, and a simulation analysis like that mentioned in Chapter 10.

A second decision issue discussed in Chapter 11 is where to position equipment inside a plant or office. The traditional way to generate a desired layout is to use graphical models called templates and to move these templates around until a desirable ''eyeball'' solution is found. As mentioned in Chapter 10, an alternative layout approach is to use mathematical programming models or to use simulation. However, both the traditional approach and the OR approach have a common base—they both require the same fundamental data, such as what equipment is to be located or the frequency of movement between pieces of equipment. These data are obtained through the use of the traditional IE techniques and procedures discussed in Section Two, including time studies, work-sampling studies, and flow process charts. This highlights the fact that traditional IE tools are used as a direct means of analyzing facility design problems as well as of supplying data for OR analyses of these problems.

The same type of overlap between traditional IE and OR exists in other problem areas, such as production planning and inventory control. In Chapter 12, several common production-planning decision issues are discussed, including the problem of how best to schedule jobs on various pieces of equipment, and the determination of the best overall output rate to use in running a plant. Traditional methods of looking at such problems include descriptive analogue models such as the Gantt Chart. The data required to build a Gantt Chart for use in, say, scheduling the work through a job shop, include the standard times and flow process diagrams described in the work methods chapter. The same can be said of the data needed to apply OR techniques to these problems. The same process diagrams and standard times used in applying a Gantt Chart are used in applying an integer programming model or simulation to the scheduling problem.

Chapter 13 is concerned with inventory control. The chapter discussion includes a description of the decision problems underlying the control of inventories, as well as the various mathematical models that are commonly used to address these problems. Inventories have been controlled since Taylor's era, using physical devices such as the ''two-bin'' system as well as mathematical models. For example, the Wilson Lot Size Model was introduced in 1934 as a mathematical means of determining the most economical order quantity and is still perhaps the most common approach to inventory control. However, as Chapter 13 indicates, more accurate models have been introduced recently, and they require more sophisticated analytical procedures such as nonlinear programming. As can be seen from the discussion in Chapter 13, inventory control is a good example of a problem area in which a primary distinction between a traditional IE approach and an OR approach is the level of mathematical rigor involved.

This same conclusion can be drawn for the problem area discussed in Chapter 14, quality control. Statistical techniques have been applied for years to the problem of proper control of the quality of a process. The same type of techniques used in work measurement for determining how many cycles to time in a stopwatch study or for determining how many observations to take in a work-sampling study, are also used in quality control. However, more involved statistical and mathematical procedures are also used, such as in cost-based quality control. The main difference is that the more involved OR procedures are more powerful, yielding control information that is more precise and more accurate. However, they are also more difficult to derive and to convey to management.

The chapters included in this section are all concerned with application areas or decision issues faced by the IE/OR analyst. Obviously, this section is not all-inclusive. Many application areas common to IE/OR analysts in various organizations are not mentioned here. These include material handling, assembly-line balancing, product mix decisions, project planning, vehicle routing, capital budgeting, and reliability planning. However, the survey given here should give the reader a good idea of the types of problem areas faced by the IE/OR analyst, as well as the overlap between the traditional IE tools and OR procedures used in analyzing these problems.

Chapter 11

Facility Design

11.1 THE MEANING OF FACILITY DESIGN

Facility design is concerned with planning the arrangement of physical facilities such as machines and buildings. It involves the specification of (1) what *type* of facilities to utilize, (2) the *number* of each type of facility to have, and (3) the *location* of each facility. Hence, a wide variety of decision problems can be classified as facility design problems, as illustrated by the following examples.

1. Deciding on the site to use in locating a new sales warehouse.
2. Determining what type of process to use to manufacture a vice.
3. Specifying the configuration or relative positions of the machines used in a job shop.
4. Determining where to relocate one of the existing fire stations at an airport after a new runway is added.
5. Deciding on the path that a new power line should take through a region.

Traditionally, facility design has been taken to mean "plant layout" and "plant location." However, as can be seen from the above list, its scope goes well beyond the plant environment. The concept of a "facility" includes computer equipment, ambulances, and fire stations just as it does a drill press or a plant. Therefore, a facility design problem can arise in service organizations and in the public sector just as readily as it can in a plant context.

In this chapter, we survey several of the more common areas of facility design. Specifically, we discuss three topics: determination of the number of machines to have in a manufacturing plant, layout of the internal activities or facilities inside a structure, and location of structures. As you will see, each of these areas lends itself both to traditional industrial engineering and to operation research. Traditional IE tools are used to investigate and understand the problem as well as to provide insight into good solutions. Operations research techniques are also used to derive potential solutions, sometimes as an alternative approach to a traditional analysis and sometimes as a supplemental tool.

11.2 DETERMINING THE NUMBER OF MACHINES

In designing a manufacturing system, one of the first decisions that must be made is the choice of what types of basic production processes to use, such as metal casting versus forging, and what types of equipment and machines to have to implement these processes. This sort of decision making is usually called *process planning*. The primary end result of process planning is a route sheet that lists each step in the production of a particular part, and the type of machine that is to be used in that step.[1] An important follow-on decision that must be made is the determination of exactly how many of each of the types of machines specified in the route sheet to have in the plant or shop. This decision is called the *machine requirements* problem.

Deciding on the number of machines to have in a shop goes beyond merely calculating the number necessary to meet a specific production output requirement. Because of the relative costs involved, it may be best not to have the necessary number of machines on hand but, rather, to subcontract out some of the production requirements. Or we may want to use a fewer number of machines and work them on an overtime basis in order to meet these requirements. In addition, the availability of production resources, such as manpower, space, and investment funds, must be considered in this decision. For example, it may be that we need four automatic drill presses and seven lathes to meet our projected monthly requirements for our products. However, these machines may require more physical space in our plant than we now have or can afford in expansion. Therefore, we have to consider the alternative trade-offs that are available, such as subcontracting some of our monthly requirements or working overtime.

Determining the number of machines to use in a shop is also dependent on the configuration of the manufacturing processes in use in the shop, and on the characteristics of the machines themselves. For example, it is generally true that fewer machines are needed if the shop is organized in a "job shop" fashion with similar machines grouped together in work centers than if it is organized like an "assembly line" with one machine after another. Also, the speed and efficiency of each type of machine plays an obvious role in dictating whether more or fewer machines are needed.

A graphical illustration of how these type of effects can affect machine requirements is given in Figure 11.1 for a typical job shop process. Each work center contains a specific type of machine; all machines in a center are somewhat homogeneous with respect to operating characteristics such as speed, scrap rate, and cost. Because of the serial organization of work centers, the output rate of any particular center is dependent on the input it receives from the previous center. Therefore, in order to prevent

[1]Process Planning was discussed in Chapter 3.

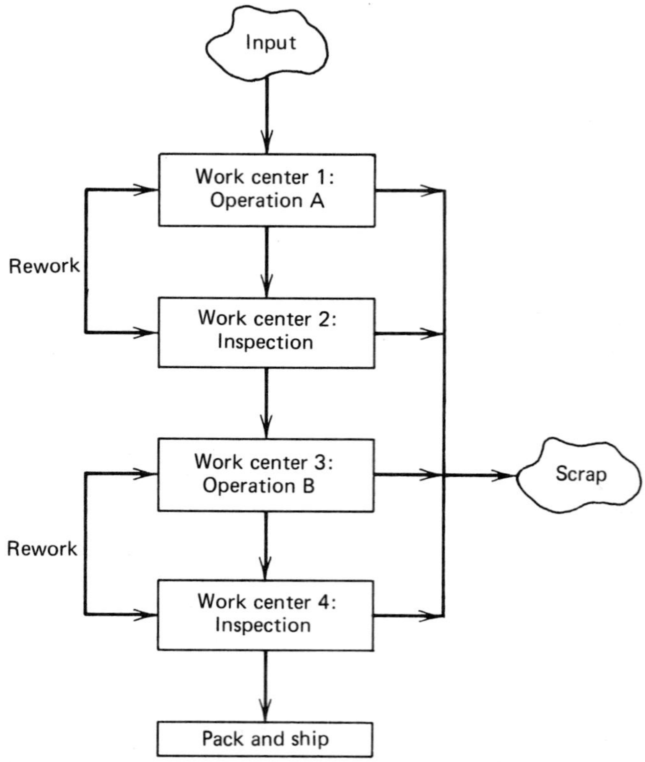

Figure 11.1 Flow of Products through a Job Shop Process

bottlenecks and to have a smoother flow of products, the number of machines required at one work center must be based on the total input rate required at the next center as well as on the output rates of a machine in that center.

Analysis of machine requirements problems has been a basic IE activity since Fredrick Taylor's era. Therefore, a classical or traditional approach to this problem has been developed and in use for some time. Recently, there have been attempts to apply OR tools to this same problem. Specifically, the use of simulation, statistical analysis, and mathematical programming has been advocated (see Reference 5 for an overview of approaches to the machine requirements problem). In the remainder of this section, we describe these two basic approaches to the machine requirements problem.

11.2.1 Traditional IE Approach

Most traditional methods of determining how many machines to have on hand are calculations using a descriptive model relating the requirements for finished products to various operating factors for a given work center. For example, Shubin and Madeheim [7] suggest using the following formula for calculating the desired number of machines for a single work center, with only one product passing through the center:

$$x = \frac{t}{60} \times \frac{p}{hu} \qquad (11.1)$$

where x is the desired number of machines to have, t is the standard time required to process one unit on a machine at the center, p is the total production of units required per day, h is the standard number of hours available per day per machine, and u is the percentage efficiency or scrap factor.

Models similar to Eq. (11.1) are the essence of most traditional approaches to the machine requirements problem. However, more comprehensive approaches explicitly consider the transfer of products from one center to another (see, for example, Reference 1). The procedure is to calculate the input requirements in "good" units needed at the final work center in order to meet the prespecified process production requirements. Then, the input requirement at the next-to-last center is calculated, using as an output requirement of this center the input required at the final center. This backwards analysis is continued until the initial center is included. The output rates at all centers are then inflated by a plant-wide efficiency factor. Finally, the number of machines at each center is found using a calculation like model (11.1).

Example 11.1

In order to illustrate the traditional approach to the machine requirements problem, consider the decision facing the GAT Company.[1] GAT is building a new wing on its existing plant in order to produce a component part it has previously purchased. An assembly chart for the part, called a "powrarm," is shown in Figure 11.2; its route sheet is given in Figure 11.3. GAT wishes to know how many men and machines it needs to acquire for this new operation in view of its estimated internal needs for 134,000 powrarms per year.

Based on 2000 hr available for production during the year, the output of work center 7 must be (134,000/2000) = 67 *good* units per hour. Based on data from the route sheet and scrap factors supplied by equipment vendors, the backward analysis is performed to determine the required output rate at each center (see Table 11.1). As an example of the calculations in Table 11.1, note that to supply the required input of 67 units/hr to work center 6, center 5 must have 67/(1.0 − 0.02) = 68.4/hr as its average input, as 2 percent of these units will be lost as scrap. This, in turn, means that center 4 must produce 68.4 good units per hour to keep center 5 stocked with working inventory.

Based on past records, GAT tends to operate at 90 percent of its total production capacity. That is, 10 percent of its capability to turn out products is lost due to absenteeism, downtime of equipment, poor work scheduling, and the like. This means that the number of machines to be planned on for each work center must be capable

[1]This illustration is based on an example given in Reference 1, pp. 41–45].

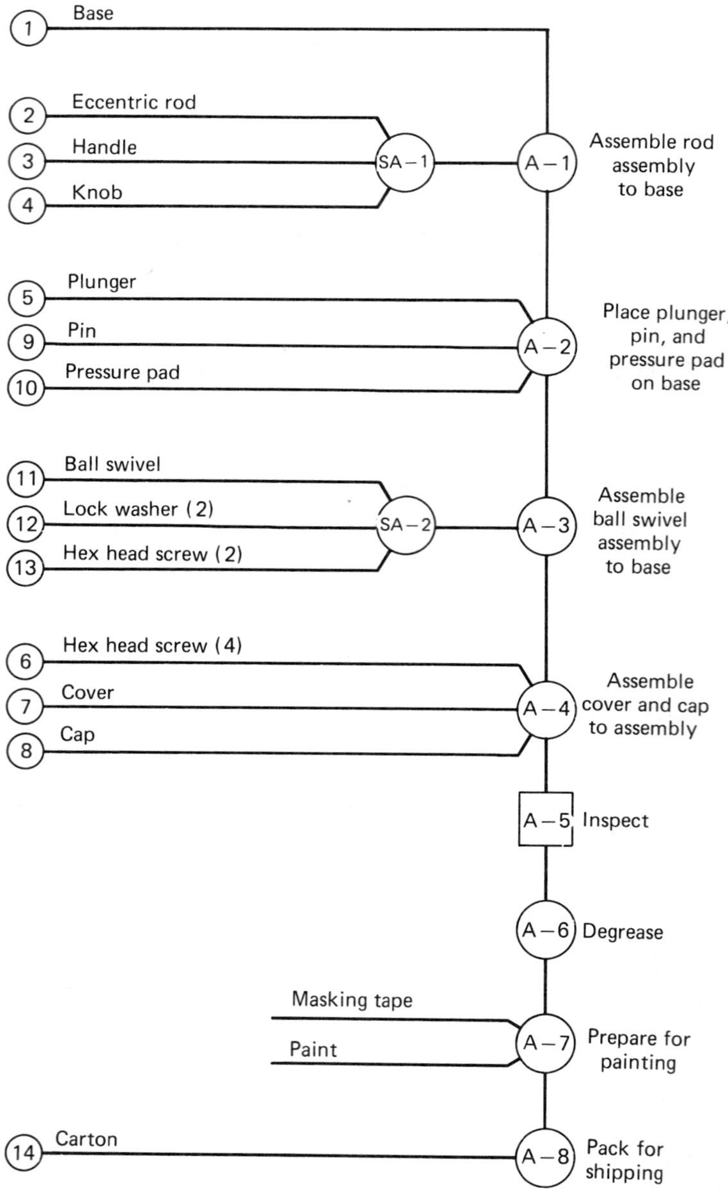

Figure 11.2 Assembly Chart for Powrarm Example (By permission of Ronald Press)

PRODUCTION ROUTING

DRAWING NO. C - 285

PART NAME Powrarm Base

PART NO. 1

Oper. No.	Operation Description	Machine Name	Jigs, Fixtures, Tools, etc.	Dept. No.	Std. Time (hrs.)	Hourly Machine Capacity	No. Machine	No. Operators	Floor Space Required
1.	Face bottom.	14″ LeBlond Engine lathe	Special chuck Turning tool		0.0167	60			
2.	Face top, turn O.D., neck, drill & ream $\frac{5}{8}$″ hole.	Warner & Swasey turret lathe	Special chuck Face Turn-neck Drill Ream		0.042	23.8			
3.	Drill 3 bolt holes.	21″ Cleereman drill press	Box jig 3-spindle head		0.0120	83.4			
4.	Drill pin hole.	Delta drill press	Plate jig		0.0042	238.0			
5.	Drill & ream $\frac{3}{4}$″ eccentric hole.	#4 Fosdick 2 spindle drill press	Box jig		0.0153	65.4			
6.	Inspect				0.018	55.5			
7.	Degrease	Detrex			0.0070	143.0			

IE - 17

Figure 11.3 Route Sheet for Powrarm Example (By permission of Ronald Press)

Table 11.1 Work Center Output Requirements For Powrarm Example

Operation Number	Machine	Output Required (Good Units per Hour)	Expected Scrap (%)	Input Needs (Good Units per Hour)
7	Degrease	67.0	0	67.0
6	Inspect	67.0	0	67.0
5	#4 Fosdick D.P.	67.0	2	68.4
4	Delta D.P.	68.4	3	70.5
3	Cleereman D.P.	70.5	2	72.0
2	W & S Turret Lathe	72.0	5	75.8
1	LeBlond Eng. Lathe	75.8	4	78.9

of handling 10 percent more throughput than it would if the plant operated at 100-percent efficiency. Therefore, the output rates for each center in Table 11.1 must be inflated by 10 percent. After doing this, the theoretical number of machines required at each center can be calculated based on the standard time for that machine as given in the route sheet of Figure 11.2. These results are given in Table 11.2. As an illustration of the calculations involved, note that the number of machines needed at center 5 is $(68.4/0.90)/65.4 = 1.16$ drill presses. ■

This type of analysis often results in fractional numbers of machines. The usual procedure is to round these answers based on a subjective appraisal of the relative merits of rounding up or down. This appraisal is based on factors such as the sensitivity of the calculations to errors in estimates of scrap rates, standard times, and so on; effect of a breakdown in a single machine on the rest of the work center; and the availability and desirability of overtime as opposed to an extra machine. In addition, it may be the case that other processes in the plant can utilize the equipment being considered so that the fractional machine hours can be consumed by these other processes or products.

The traditional approach for planning machine requirements is a straightforward approach that is easy to use and to explain. However, it has several disadvantages.

Table 11.2 Required Number Machines for Powrarm Example

Operation Number	Machine	Effective Output Rate (At 90% Efficiency)	Hourly Machine Capacity (Units per Hour)	Number of Machines
1	LeBlond Eng. Lathe	87.7	60.0	1.46
2	W & S Turret Lathe	84.2	23.8	3.45
3	Cleereman D.P.	80.0	83.4	0.96
4	Delta D.P.	78.3	238.0	0.33
5	Fosdick D.P.	76.0	65.4	1.16
6	Inspect	74.4	55.5	1.34 (men)
7	Degrease	74.4	143.0	0.52 (men)

For instance, it does not represent or model the realities involved in the underlying decision very well. Specifically, there is no explicit accounting for factors such as dynamic production requirements, the probabilistic nature of factors such as scrap loss and machine output rate, or limitations on production resources such as floor space. By using operations research tools, more realistic analyses can usually be made for a machine requirements problem.

11.2.2 Operations Research Approach

The machine requirements problem can be treated as a mathematical programming problem, a simulation analysis, or as a queueing problem, depending on the assumption made regarding the underlying production process. As an illustration, in this section we present a mixed integer programming approach to the problem (see Reference 6 for details).

The basic philosophy underlying the math programming model to be described is that the machines at each work center are in competition for the use of scarce production resources such as floor space, capital, and overtime hours. That is, each additional machine added to one work center means that there is less floor space, less capital, and fewer overtime hours available to support more machines at the other centers. Explicit consideration of this competition between work centers requires that a *normative* modeling approach be used rather than a descriptive approach like that used in the traditional analysis.

There are several other major differences between the traditional approach and the use of a math programming model. The math programming approach allows the end results (that is, the number of machines desired at each center) to be found directly as an integer, rather than requiring a subjective rounding of a fractional answer. As was pointed out in Chapter 8, rounding fractional results can lead to answers that are infeasible or far from optimal. Also, the math programming model to be presented allows the dynamic aspects of the problem to be modeled. Specifically, the model accounts for fluctuations from one time period to the next in the expected value of final production requirements, cost parameters, and available production resources. Hence, the answer to the overall machine requirements problem is a period-to-period schedule of how many machines to have on hand in each center. For instance, a center may start with five machines, then at the end of the second month or period add two more machines, and at the end of the fourth month sell one machine so that six are on hand.

The math programming model is based on the following terminology:

x_{it} = the number of machines to have on hand in work center i at the beginning of and during period t ($i = 1, 2, \ldots, N$; $t = 1, 2, \ldots, T$).

M_{it} = the theoretical number of machines needed in work center i in period t to satisfy exactly the final production requirements in that period.

f_{it} = the first cost of a machine in center i when it is procurred at the beginning of period t.

s_{it} = the average net salvage value of a machine in center i when it is disposed of at the beginning of period t.

m_{it} = the annual operating and maintenance expense of a machine in center i during period t.

c_{it}^{+} = the cost per machine per period of an excess machine of type i carried during period t.

c_{it}^{-} = the overtime cost per machine per period for carrying fewer machines in center i during period t than is theoretically desired.

B_t = the funds budgeted for procurring new equipment in period t.

b_i = the number of hours consumed by a machine in center i when it operates on an overtime basis for one period.

L_t = the square footage of floor space available in period t for placing machinery.

a_i = the square footage required to place one machine in work center i.

U_{it} = the hours available in period t for machines in center i to be worked overtime.

In addition, define the mathematical operator $|\cdot|^{\pm}$ as follows (where a and b are dummy arguments):

$$|a - b|^{+} = a - b \qquad \text{if } a \geq b$$
$$= 0 \qquad \text{otherwise}$$

and similarly:

$$|a - b|^{-} = b - a \qquad \text{if } a \leq b$$
$$= 0 \qquad \text{otherwise}$$

Using the above notation, the math programming model can be written as

$$\max x = \sum_{t=1}^{T} \sum_{i=1}^{N} (f_{it}|x_{it} - x_{i,t-1}|^{+} - s_{it}|x_{it} - x_{i,t-1}|^{-}$$
$$+ m_{it}x_{it} + c_{it}^{+}|x_{it} - M_{it}|^{+} + c_{it}^{-}|x_{it} - M_{it}|^{-}) \qquad (12.2)$$

S.T.: (1) The shop's budget must not be exceeded:

$$\sum_{i=1}^{N} (f_{it}|x_{it} - x_{i,t-1}|^{+} - s_{it}|x_{it} - x_{i,t-1}|^{-}) \leq B_t \qquad \text{for all } t$$

(2) The shop's available floor space must not be exceeded:

$$\sum_{i=1}^{N} a_i x_{it} \leq L_t \qquad \text{for all } t$$

(3) The available overtime must not be exceeded:

$$b_i|x_{it} - M_{it}|^{-} \leq U_{it} \qquad \text{for all } i \text{ and all } t,$$

(4) The number of machines disposed of in a period must be no greater than the number of machines on hand at the end of the previous period:

$$|x_{it} - x_{i,t-1}|^{-} \leq x_{i,t-1} \qquad \text{for all } i \text{ and all } t$$

(5) The number of machines must not be negative:

$$x_{it} \geq 0 \qquad \text{for all } i \text{ and all } t$$

Model (11.2) is a nonlinear programming problem, as the absolute value operators $|\cdot|^{+}$ and $|\cdot|^{-}$ are nonlinear functions of their arguments. However, the problem can readily be converted into a linear model by use of the following auxillary decision variables:

$$\begin{aligned}
y_{it}^{+} &= |x_{it} - x_{i,t-1}|^{+} = (x_{it} - x_{i,t-1}) && \text{if } x_{it} \geq x_{i,t-1}, \text{ 0 otherwise} \\
y_{it}^{-} &= |x_{it} - x_{i,t-1}|^{-} = (x_{i,t-1} - x_{it}) && \text{if } x_{it} \leq x_{i,t-1}, \text{ 0 otherwise} \\
\mu_{it}^{+} &= |x_{iy} - M_{it}|^{+} = (x_{it} - M_{it}) && \text{if } x_{it} \geq M_{it}, \text{ 0 otherwise} \\
\mu_{it}^{-} &= |x_{it} - M_{it}|^{-} = (M_{it} - x_{it}) && \text{if } x_{it} \leq M_{it}, \text{ 0 otherwise}
\end{aligned} \qquad (11.3)$$

such that

$$y_{it}^{+} \cdot y_{it}^{-} = 0 \qquad \mu_{it}^{+} \cdot \mu_{it}^{-} = 0$$

$$x_{it} = y_{it}^{+} - y_{it}^{-} + x_{i,t-1}$$

and

$$x_{it} = \mu_{it}^{+} - \mu_{it}^{-} + M_{it}$$

After substituting these auxillary variables into model (11.2), an alternative form of the math programming model is obtained. It turns out to be a mixed integer, linear programming problem. Although it is more complex than the traditional approach, the use of a model such as (11.2) can lead to more economical and more feasible results. For example, in one illustration of the model as reported in Reference 6, the optimal number of machines was found and had an associated annual cost of \$64,600. On the other hand, the solution obtained by the traditional approach had an annual cost of \$95,400, or a 48 percent worse solution value. In addition, the rounded solution of the traditional approach required 20 percent more floor space than the company had available.

11.3 DESIGNING THE LAYOUT OF A FACILITY

Facility layout is the process of determining where each activity or piece of equipment is to be located inside a structure. It involves designing or specifying the internal configuration of the structure. The end result of this process is a layout, which is essentially a macrolevel blueprint of the interior of the structure. There is no standard format for the layout drawing; however, diagrams such as that shown in Figure 11.4 are often used.

The scope of facility layout problems is very broad. The following is a list of typical examples of layout problems.

1. Positioning all of the machinery and auxillary service areas such as tool cribs in a wire rope manufacturing plant.
2. Dividing the interior of a Social Securities headquarters into offices and work areas.
3. Allocating floor space in a warehouse to the products to be stored there.
4. Sizing and positioning the space to be used on a ship for various activities, such as crew quarters and a gallery.
5. Designing the length and shape of an assembly line for television sets, as well as specifying the placement for work areas and equipment along the line.

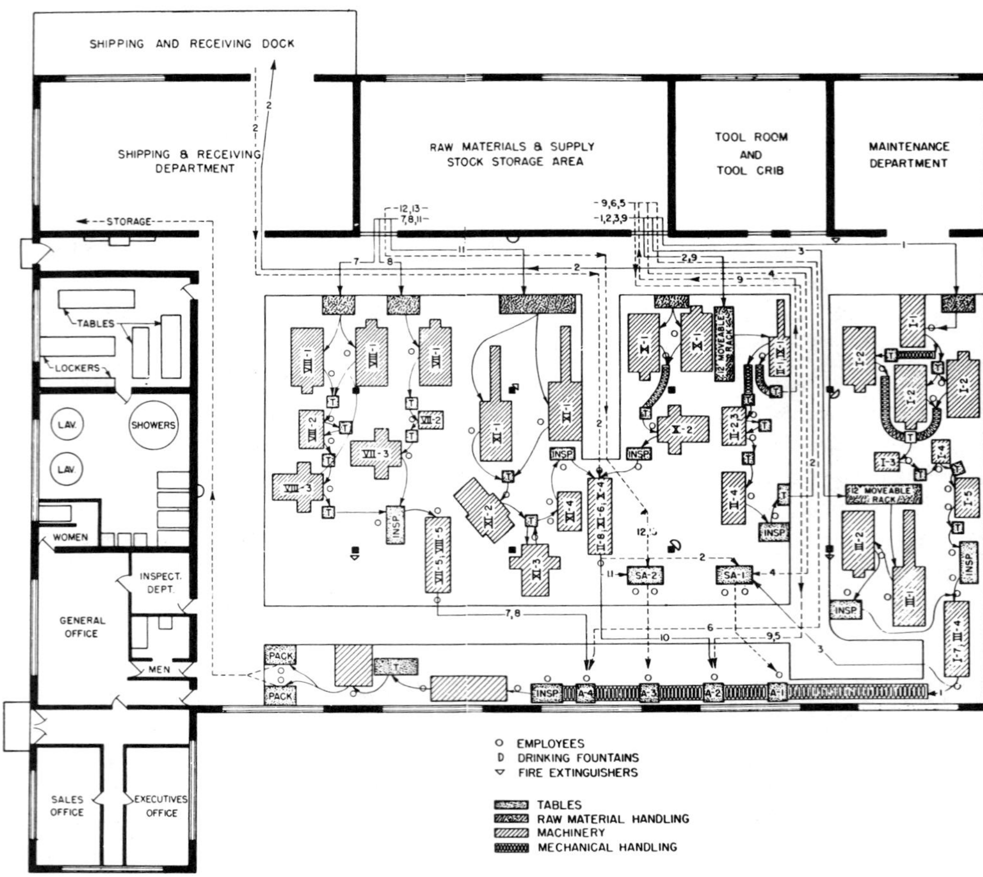

Figure 11.4 Illustration of a Layout Diagram (By permission of Ronald Press)

Regardless of its setting or nature, a facility layout problem is still a decision problem. Whether or not they are explicitly stated, there is an objective function, decision variables, and constraints or restrictions on these decision variables. For instance, in laying out a manufacturing plant the objective is usually to minimize the material-handling cost associated with moving goods between work stations. The decision variables would specify the exact placement of machines and work centers, as well as the size and shape of these centers. Assigning numerical values to these decision variables is equivalent to dictating a layout. The constraints in such a layout problem restrict the configuration of machines and work centers because of factors such as limited floor space, the size and shape of the existing building in which the layout is to be placed, and safety considerations. Notice that if the layout problem were modeled mathematically it would constitute a math programming problem (see Chapter 8).

The solution to the layout problem, that is, the ''layout'' itself, dictates the flow pattern that will take place within the facility. Depending on the application, this flow might refer to products as they are being manufactured, paper work, material-handling equipment, people, or some combination of these. For example, the layout of a warehouse determines the flow of material-handling equipment during the day as items are moved to and from storage and receiving–shipping areas. The desirability of this flow pattern is a major determinant of the quality of a layout. Good layouts tend to have efficient, short flow patterns without many points of congestion. Locating close together equipment or activities with a relatively high degree of interaction generally produces such patterns. And efficient flow patterns lead to reduced material-handling time and costs, increased safety performance, and reduced travel time for personnel and paper work. Development and use of a good layout can thus lead to significant dollar savings for an organization.

Three basic categories or types of layouts are generally used: process, product, and fixed position. Each of these represents a different approach or philosophy toward the basic organization of the equipment and activities to be positioned in a layout. In a process layout, all machinery and activities that are alike or have the same function are grouped together and positioned in the same general area (this area often constitutes the ''work center''). The most common example of this type of layout arrangement is the job shop, in which all engine lathes, say, are located in the same area, all drill presses are grouped together in another part of the shop, all assembly takes place in some other area, and so on. An illustration of a process layout is given in Figure 11.5, with the flow of two different products superimposed on the layout. Notice the amount of ''backtracking'' that takes place in the processing of the parts, and the intersection of flow paths that can cause material-handling congestion and safety problems.

The second class of layouts—product layout—derives its name from the fact that in this layout arrangement equipment and activities are positioned in accordance with their use in the development or manufacturing of the product. If the first step in making a particular part is cutting raw material to size on a band saw, then a band saw is located at the point in the shop where the processing of the part is to begin. The equipment required in the second processing step or operation is placed beside this band saw. An assembly line is a common example of this type of layout, as illustrated

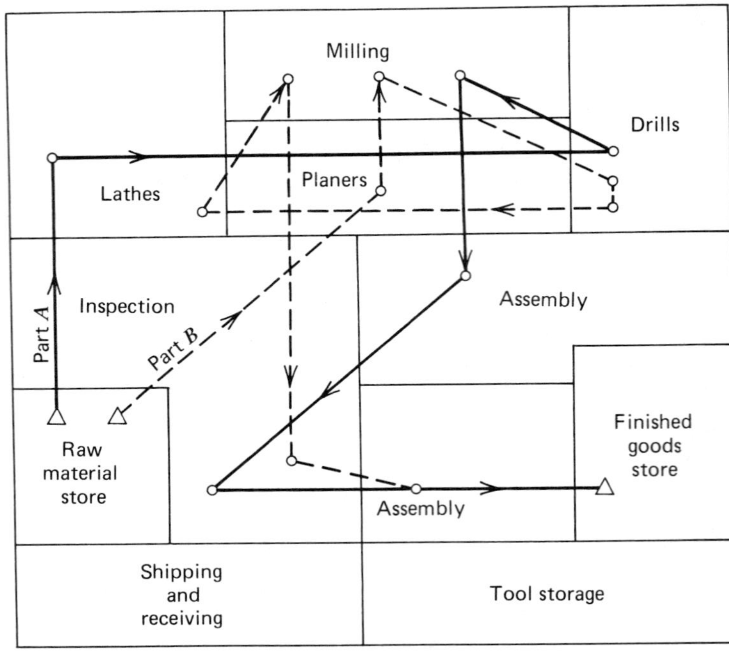

Figure 11.5 Illustration of a Process-Type Layout with Two Product Flows Shown

in Figure 11.6. Notice that in this type of layout duplication of equipment is often required, as the same machine may be needed in the second processing step and again, say, in the fifth step.

An important feature to keep in mind about a product layout arrangement is that the relative position of equipment and activities is fixed and not a decision variable as in the case of a process layout. The manner in which the product is to be manufactured, as reflected in its route sheet, dictates what equipment must be positioned in various places along the line. Therefore, the basic layout question concerns the shape of the line: straight, an "L" shape, a circular shape, or perhaps a serpentine or "S" shape.

In the third category of layouts—fixed position—the product remains stationary in a fixed location, and the machinery or work activities are moved to the product as required. A good example of this type of layout arrangement is a shipyard. The body of the ship stays in one place as it is being constructed, and all materials and processing equipment are moved to the ship as they are needed.

In developing the actual layout to be used in a particular organization, one of the first decisions to be made is the choice of the basic layout arrangement. If we decide on a product arrangement, then the remaining layout issue is the shape of the production line. If we decide on a process arrangement, then we must still determine where the individual machines or work centers are to be positioned in the shop, and the allocation of the shop's floor space to each machine or center as well as to isle ways, restrooms,

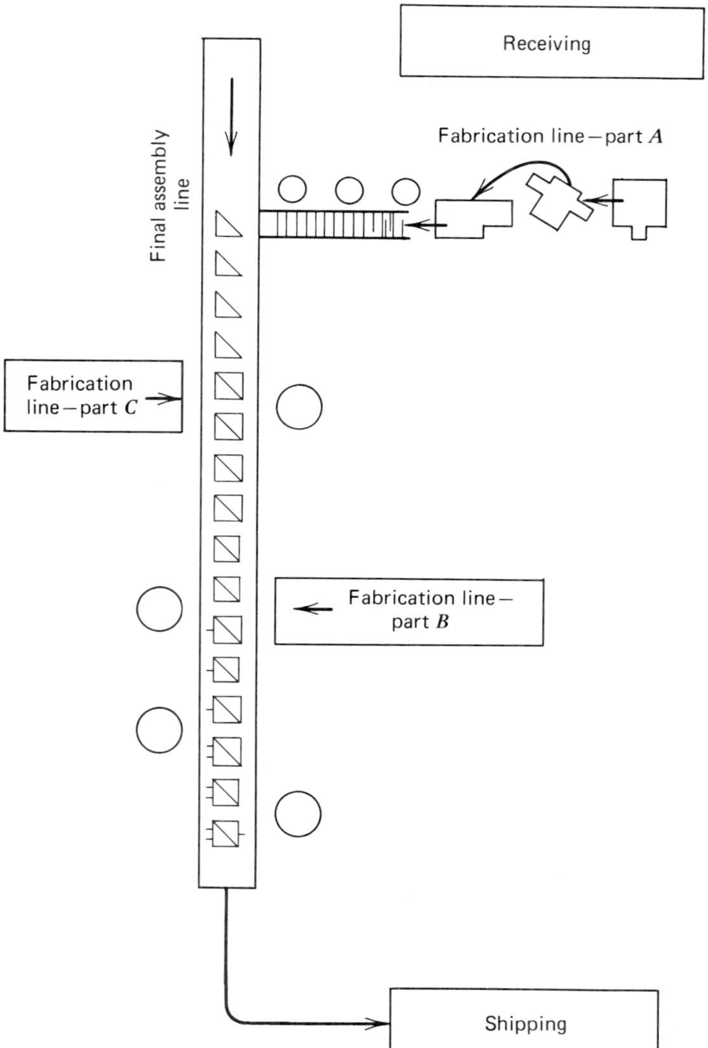

Figure 11.6 Illustration of a Product-Type Layout

and the like. In the remainder of this section we concentrate on this process layout problem. While our discussion will be oriented to a manufacturing environment, it should be kept in mind that the same concepts and techniques are applicable to layout problems in other settings such as banks, computer installations, and hospitals.

11.3.1 Basic Approach

Just as in the analysis of any other decision problem, determining the desired layout for an organization involves two fundamental steps: generating alternative solutions (that is, alternative layouts that could be used), and evaluating each alternative solution. There are traditional manual schemes for generating and evaluating layouts, as well as computerized methods that use mathematical models and search procedures. Regardless of the layout approach used, however, there is a need to gather and analyze fundamental information about the requirements of the layout problem. This information includes the following items.

1. *The scope of the layout problem.* For instance, are we interested in positioning individual machines, or only entire work centers or departments? Is the layout to go into an existing building with confining walls and columns, or are we free to specify an entirely new building size and shape?

2. *The internal facilities to be positioned in the layout.* This information comes from several sources. For instance, the route sheet for each part tells us what type of processing equipment and activities are required, and the results of our machine requirements analysis described in the previous section tells us the number of each type that is needed. There are also support activities and service facilities that must be positioned, such as tool cribs, storage rooms, and water fountains. Lists of such facilities required in a typical shop are available in the literature (see, for example, Reference 1, Chapter 8).

3. *The space requirements for each facility or activitiy to be positioned.* The floor space that will be taken up by each facility must be calculated as well as the space needed for walkways and other common areas. Rules of thumb are often used in arriving at these figures. For example, Apple [1] recommends that one foot of space be allowed for work room between two machines placed side to side, while three feet be allowed for machines placed front to back (with an operator between them).

4. *The interrelationships among the facilities or activities to be positioned.* Information is needed on the amount of interaction that will take place between each pair of facilities so that we can get an idea of how important it is to locate those facilities close together. This type of information can be obtained by developing various charts and diagrams that are simply descriptive models of the activities that will take place in the building once it has been laid out. Typical charts used for this purpose are the "multiproduct process chart" (see Figure 11.7 for an example), the "from-to chart" (see Figure 11.8), and the "activity relationship" or REL Chart (see Figure 11.9). The REL chart presents management's judgment as to relative desirability of having a pair of activities or facilities located together. Notice that every possible pair of activities is assigned a subjective rating.

11.3.2 Manual Layout Methods

The traditional approach to converting the basic layout input data such as that given above into one or more layouts for comparison is to develop *template* models. A template is nothing more than a scale model of a machine or other facility. An illustration of two-dimensional templates is given in Figure 11.10.

A template model of a layout could be developed by simply cutting out or

Operations	A Tin – base etched item	B Alum – base etched items	C Alum – base printed items	D Alum – base anodized items I	E Alum – base anodized items II	Business vol. each operation (%)
1. Cut to size	1	1	1	3		A – 18 B – 32 C – 28 D – 14 92
2. Polish	2					18
3. Wash out	3					18
4. Nickel – silver plate	4					18
5. Weld				1	1	D – 14 E – 8 22
6. Anodize				2	2	22
7. Colour				5	3	22
8. Print	5	2	2	4	4	100
9. Color etch					5	8
10. Dry spray	6	3				A – 18 B – 32 50
11. Retouch	7	4				50
12. Deep etch	8	5				50
13. Pickle	9					18
14. Rinse	10	7		6	6	72
15. Lacquer	11	8	3			78
16. Spray paint		6				32
17. Imbed colors (future consideration)	9 Alternate	7 Alternate				Future potential 50
Business volume (%)	18	32	28	14	8	100

Figure 11.7 Illustration of a Multiproduct Process Chart for Five Products (By permission of McGraw-Hill)

From \ To	R.M.	Saw	Lathe	Drill	Mill	Ins.	Pkg.	F.G.
R.M.		16	40	72	88	60	36	20
Saw	16		24	56	72	44	20	36
Lathe	40	24		32	48	20	44	60
Drill	72	56	32		16	36	60	76
Mill	88	72	48	16		52	76	92
Ins.	60	44	20	36	52		24	40
Pkg.	36	20	44	60	76	24		16
F.G.	20	36	60	76	92	40	16	

Figure 11.8 Illustration of a From-To Chart (The Number of Trips from a Given Department to Another Department are Shown in the Body of the Table) (By permission of Prentice-Hall)

constructing a template to represent each machine, operator, rest room, and so on, and placing them anywhere on a diagram of the shell of the building. The result is a trial layout. It may be a good layout, or it may be a poor layout; it may be a practical layout, or it may violate some constraints such as not placing the machine shop next to the drafting department. Nevertheless, it is a layout.

Usually, the placement of the template on the building diagram is based on an analysis of information contained in charts such as the from-to chart and the REL chart. For instance, the REL chart can be converted into a diagram like that shown in Figure 11.11. Notice that in this new chart the density of the lines between the various blocks (representing activities to be positioned) provides the analyst with a feel for where to position the activities. Obviously, those blocks having the most important mutual closeness rating (as represented by the line density) should be located near each other. When quantitative information, such as the number of trips made between machines per hour, is to be used in the analysis in lieu of the subjective closeness rating, the lines between the blocks in the REL diagram would be based on the numbers in a from-to chart such as the one given in Figure 11.8.

The REL diagram can be made even more useful by modifying each block to reflect the relative size and shape of the machine or activity represented by that block. The result is a "space relationship diagram." For instance, the REL diagram of Figure 11.11 has been converted into the space relationship diagram shown in Figure 11.12. This diagram becomes the basic tool with which the analyst works to obtain a layout. He or she shifts the blocks (or the associated templates) around to find positions for every internal facility which provide what appears to him or her to be a good, workable arrangement. Figure 11.4 is an illustration of a final layout derived in this fashion.

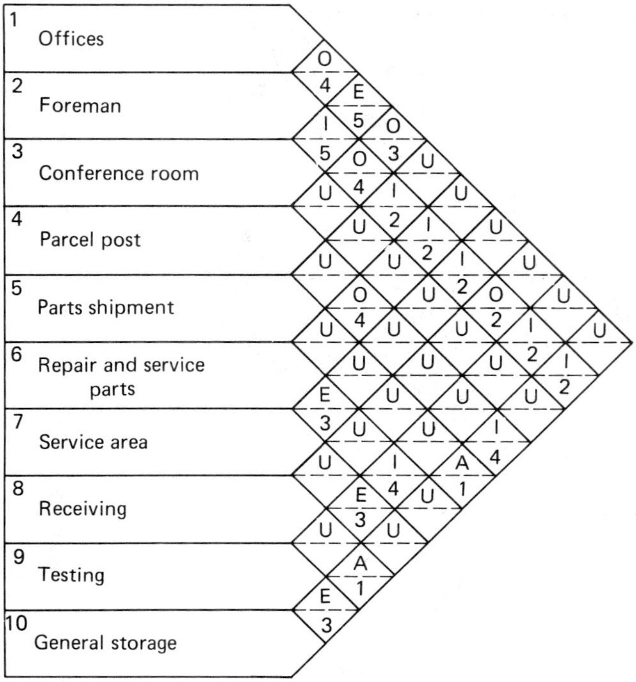

Code	Reason
1	Flow of materials
2	Ease of supervision
3	Common personnel
4	Contact necessary
5	Convenience
6	
7	
8	
9	
10	

Rating	Definition
A	Absolutely necessary
E	Especially important
I	Important
O	Ordinary closeness OK
U	Unimportant
X	Undesirable

Figure 11.9 Illustration of a REL Chart (By permission of Prentice-Hall)

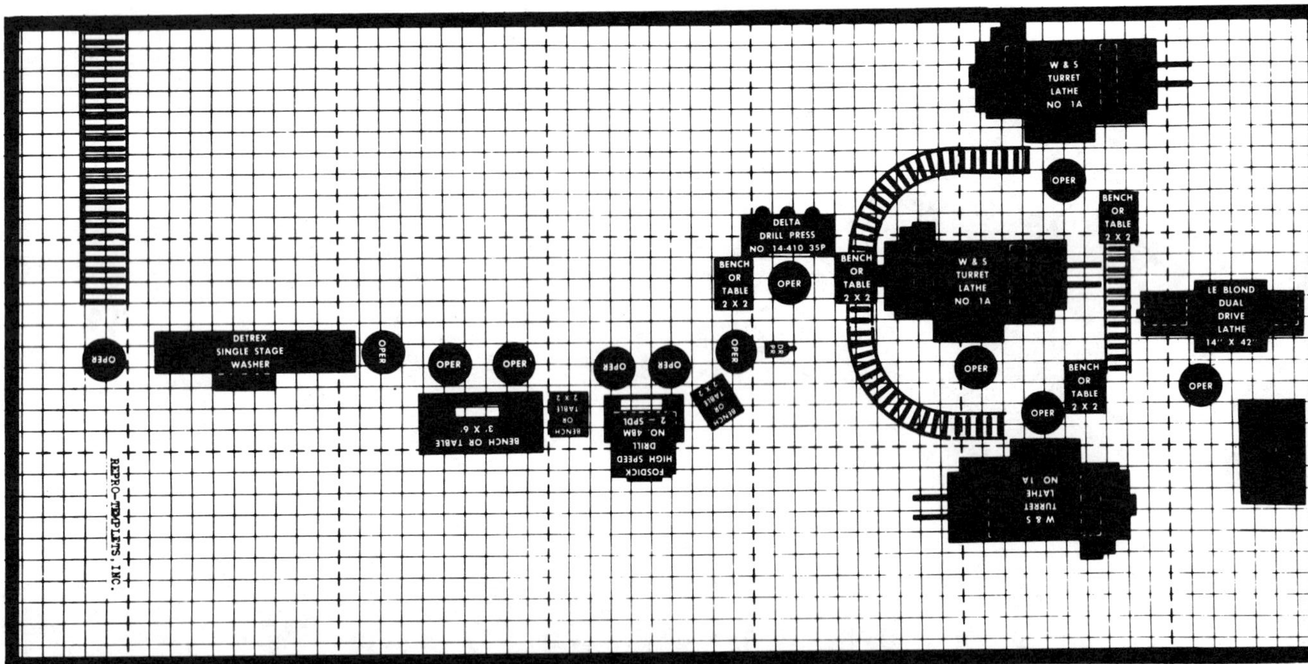

Figure 11.10 Illustration of Two-Dimensional Templates (By permission of Ronald Press)

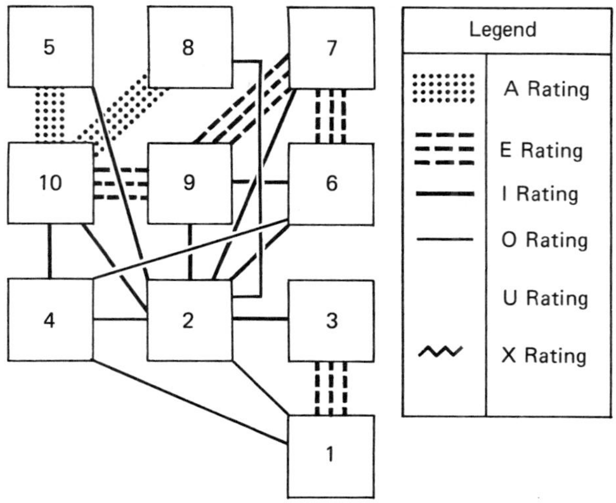

Figure 11.11 Illustration of an Activity Relationship Diagram (By permission of Prentice-Hall)

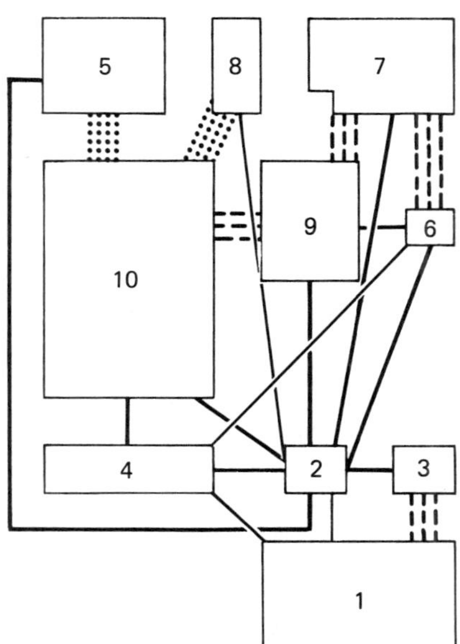

Figure 11.12 Space Activity Relationship Diagram (By permission of Prentice-Hall)

11.3.3 Computerized Layout Methods

Computer programs are available as an alternative to the ''template shuffling'' approach described above. These programs are essentially search procedures for solving mathematical models of the layout problem. They are computer-coded versions of algorithms similar to the mathematical programming algorithms discussed in Chapter 8. However, they use the same two basic steps as do the manual template shuffling methods—generate alternative solutions (layouts), and evaluate each one in an effort to find the best feasible layout. Whereas the manual methods rely on subjective evaluations and graphical models, the computerized or operation research methods rely on quantitative evaluations and mathematical models. Although there are pros and cons associated with both approaches, the computerized methods have a major advantage in their ability to generate and compare many more alternative layouts than can be done in a manual analysis.

As an illustration of the mathematical modeling orientation of the computerized layout methods, consider the following model that is addressed by one of the available computer programs:

$$\begin{aligned} \underset{\mathbf{x}}{\text{minimize}}\ x_0 &= \frac{1}{2}\sum_{i=1}^{n}\sum_{k=1}^{n}\sum_{j=1}^{n}\sum_{h=1}^{n} c_{ikjh}x_{ik}x_{jh} \\ \text{S.T.:}\quad \sum_{i=1}^{n} x_{ik} &= 1 \qquad \text{for } k = 1, 2, \ldots, n \\ \sum_{k=1}^{n} x_{ik} &= 1 \qquad \text{for } i = 1, 2, \ldots, n \\ x_{ik} &= 0 \text{ or } 1 \text{ for all } i \text{ and } k \end{aligned} \tag{11.4}$$

where $x_{ik} = 1$ if machine or activity i is positioned in the kth area or location site inside the shell of the building (the building's floor space is divided ahead of time into n different sites or areas, and a machine or activity is to be placed in each of the sites to produce a layout); x_{jh} is the same variable as x_{ik}, with the first index i and j referring to different machines to be positioned in the layout and the second index k and h referring to different location sites in the building; and c_{ikjh} is the annual cost of having machine i positioned at site k in the layout simultaneously with machine j being positioned at site h.

Recall from Chapter 8 that model (11.4) is a zero–one integer programming problem. Because of the nonlinear nature of the objective function it can be further classified as a nonlinear, zero–one integer programming problem. One of the most commonly used computerized layout programs, CRAFT, is basically an algorithm to solve this model.

There are a number of other ways in which the layout problem can be formulated and modeled. Because of this, a number of other computer programs have been developed as alternatives to CRAFT. These include CORELAP, ALDEP, PLANET, LSP, LAYOPT, and RUGR.

The algorithms used by these programs (and therefore the programs themselves) can be classified in terms of the type of search procedure they use in solving the layout problem. Two alternative approaches are generally used: a "construction" approach, or an "improvement" approach. A layout method that uses a construction approach begins with an empty floor plan of the building and places machines or activities in the layout one at a time until all machines are included in the layout. Construction methods, of which CORELAP and ALDEP are examples, differ in how the next machine or activity is selected to be added to a partial layout, and where that machine is to be placed after it is selected. Improvement layout programs begin with a complete layout—that is, with all machines or activities located somewhere in the building—and generate alternative layouts using this first trial solution as a basis. This is done by interchanging machines between their assigned positions in an effort to improve the desirability of the layout. Improvement programs differ in the manner in which machines are chosen for interchanging.

CRAFT is an example of an improvement method. It begins with a prespecified solution to model (11.4) and attempts to reduce the objective function value x_0 by changing the zero or one assignments made to the x_{ik} variables in such a way that an interchange between the location of two or more machines is achieved. For example, if the initial layout had machine A assigned to area I and machine B assigned to area II, then the decision variable values would be $x_{B,I} = x_{A,II} = 0$, $x_{A,I} = 1$, and $x_{B,II} = 1$. Then, if the first iteration of the CRAFT algorithm made an exchange of machines A and B, the decision variable values would be $x_{A,I} = 0$, $x_{A,II} = 1$, $x_{B,I} = 1$, and $x_{B,II} = 0$. A graphical illustration of this exchange is given in Figure 11.13.

Another way to classify computerized layout methods is in terms of the type of data used to evaluate alternative layouts. Some methods, such as CRAFT, use *quantitative* data such as material-handling costs as the objective function measure, whereas others, such as CORELAP and ALDEP, use *qualitative* assessments such as the REL chart ratings described in the previous section. The qualitative assessments are usually converted into intermediate numerical values through the use of some type of rating scale. For example, ALDEP internally converts the *A*, or "absolutely necessary," rating of the importance of locating two machines near each other as contained in a

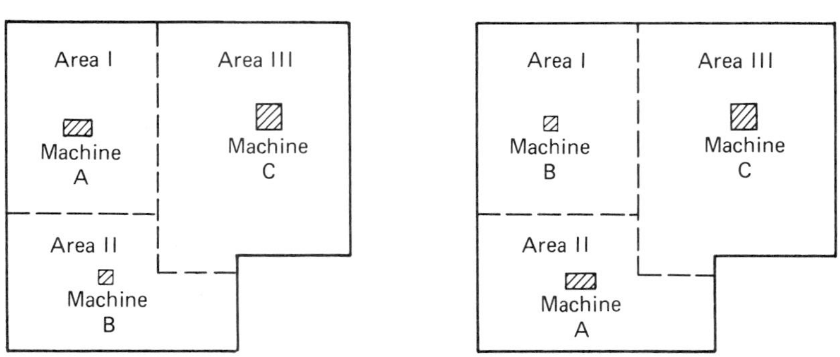

Figure 11.13 Illustration of an Interchange of Machines in CRAFT

REL chart into a numerical score of $(4)^3 = 64$, while an E is converted to $(4)^2 = 16$, $I = (4)^1 = 4$, $O = (4)^0 = 1$, $U = 0$, and $X = -(4)^5 = -1024$. Other qualitative layout methods use different conversion scales, but they all relay on the same REL chart data as input.

Construction approaches generally require that the analyst input the REL chart, along with information such as the length, width, and area requirements for the building, the number of machines or activities to be included in the layout, and the minimum allowable score, if any, for an acceptable layout. A typical printout of a solution based on such input data is shown in Figure 11.14. Note that the area of the building assigned by the program to a given machine or activity is represented by a collection of grid squares, each square represented by a number giving the numerical identification of the machine.

A useful feature incorporated by some computerized layout programs is the use of real-time interaction between the decision maker and the computer. A user, sitting at a computer terminal, makes changes directly in the trial layout generated by the basic layout program. These changes may be an attempt on the part of the user to improve the score of the layout (for example, the evaluation of the layout in terms of the REL chart data), or to improve the appearance of the shape of a work center. For example, a printout illustrating a typical exchange between the user and the computer in an interactive version of CORELAP is shown in Figure 11.15. The questions at the bottom of the printout are issued by the computer to guide the user through a series of possible changes. A change is basically a swap of grid squares that in effect alters the definition of the areas assigned to one or more machines in the partial layout at the time. For instance, notice in the layout shown in Figure 11.15 that the work centers for machine 16 and for machine 18 are irregular. Through the interchange of grid squares provided by the user in the lower right-hand part of the printout the centers

5	5	5	1	1	1	1			
5	5	5	1	1	1	1			
5	5	5	1	1	1	1			
2	2	2	1	1	1	1			
2	2	2	2	2	2	2			
2	2	2	2	2	2	2	4	4	4
0	3	3	3	6	6	6	4	4	4
0	3	3	3	6	6	6	7	7	7
0	3	3	3	6	6	6	7	7	7
0	0	0	0	0	0	0	0	0	0

Figure 11.14 Typical Layout Supplied by a Computer Program

Partial layout number 5

	1	2	3	4	5	6	7	8	9	10	11	12	13	14	15	16
1																
2																
3																
4																
5																
6	0	0	0	0	0	0	0	0	0	18	18	18	16	0	0	0
7	0	0	0	0	0	0	0	0	0	17	18	16	16	0	0	0
8	0	0	0	0	0	19	19	19	19	19	13	13	16	16	0	0
9	0	0	0	0	0	19	19	19	19	19	13	13	0	0	0	0
10	0	0	0	0	0	0	19	19	19	19	0	0	0	0	0	0
11																
12																
13																
14																
15																
16																

Do you wish a score for this layout? 1
The total score for this layout is 13
Do you wish to make changes? 1

Which change option do you desire 1, 2, or 3? 1
What are your option 1 changes? 8, 14, 0, 6, 12, 16

Option 0, 1, 2, or 3? 1
What are your option 1 changes? 7, 10, 18, 7, 8, 17

Option 0, 1, 2, or 3? 0

Figure 11.15 Typical Exchange in Interactive CORELAP [After *Industrial Engineering* (August 1971). Copyright American Institute of Industrial Engineers, Inc., 25 Technology Park/Atlanta, Norcross, GA 30071]

are made more compact. In fact, the work center for machine 18 is now square. Interactive computer methods generally have more utility than pure batch methods because the decision maker is involved in the development of the layout. He can also incorporate subjective criteria into the solution process that are not included in the mathematical model underlying the computerized layout method.

11.4 LOCATION OF FACILITIES

Another problem generally classified as a facility design issue is that of determining where to locate a facility. The layout problems discussed in the previous section are basically facility-location problems, with the machines or work centers being the facilities to be located. However, in the layout problem there is also the question of allocating the internal floor space to each machine or center, and designing the configuration of the area allocated to each center. In this section, we concentrate on "pure" location problems, that is, decision problems involving the selection of specific sites on which to locate one or more facilities. The discussion is oriented to locating structures such as a new plant or a new warehouse. However, as in the case of layout problems, it should be kept in mind that the scope of facility-location problems is very broad. For instance, the following list of typical location problems should indicate the wide variety of potential applications.

1. Choosing the city in which to locate a new regional sales office.
2. Selecting a set of cities to use as bulk mail processing centers for the U.S. postal service.
3. Choosing the area of the existing plant grounds on which to house the company's centralized fleet of material handling vehicles.
4. Selecting specific building sites within the city for construction of four new fire stations.
5. Determining where to locate each of 45 new trash containers for public use.
6. Choosing the exact position on the floor of the plant for a new numerically controlled drill press that will receive work and feed work to machines already positioned in the shop.

11.4.1 Classification of Facility-Location Problems

As can be seen from the above list, a wide variety of decision issues are essentially facility-location problems. However, all facility-location problems have certain characteristics that can be used as a basis for distinguishing one type of problem from another. For instance, a location problem is classified as a *single-facility* problem if it involves determining the location site for only one new facility. Alternatively, if the problem involves the simultaneous selection of location sites for several new facilities, it is called a *multifacility* location problem. Examples 1, 3, and 6 in the list of typical problems in the introduction to Section 11.3 are illustrations of single-facility problems, whereas examples 2, 4, and 5 are illustrations of multifacility problems.

Another way to classify location problems is in terms of the availability of alternative sites that are being considered as candidates for locating the new facilities. If there is a list of specific sites under considerations, the problem is called a *discrete-space* problem. On the other hand, if the new facilities can be placed *anywhere* within an area, the problem is classified as a *continuous-space* problem. For example, if we are trying to relocate a distribution center and management has indicated that we should move the center to either Atlanta, Mobile, New Orleans, Memphis, or Nashville, then we are dealing with a single-facility, discrete-space location problem. If management wished to consider all possible points in the Southeast instead of providing five specific sites, the problem would be a single-facility, continuous-space problem.

Identifying the type of location problem we are trying to solve is important in selecting an appropriate solution procedure. Methods that are designed to solve single-facility location problems are not necessarily useful in solving multifacility problems.

11.4.2 Plant Location

Still another way to classify location problems is in terms of the type of new facility to be located and the problem environment in general. For instance, if we are interested in locating (or relocating) a retail or a wholesale warehouse, we are dealing with a "warehouse location problem." If, instead, the facility were a fire station or ambulance center, we would be analyzing an emergency-facility location problem. One of the most important classes of location problems is the "plant location problem" in which the new facility is a manufacturing concern.

In a plant-location problem we are trying to determine the best site to use for one or more plants that are to be added to our existing system of manufacturing concerns (there may be no existing plants). In addition, we may be simultaneously trying to determine which customers to supply from which plants. The new plants will receive raw material from remote sources and will send their finished products to customers.

Normally the plant location problem is solved in a two-step fashion. First, the general area, such as a county or city, is selected from among alternative areas; then second, a specific plot of land or address within the selected area is chosen. While these two steps are usually performed in subsequent analyses, they are sometimes treated simultaneously.

There are many factors or criteria that are considered in selecting the general area for locating a plant. Some of these are quantitative in nature, whereas others are qualitative. Quantitative criteria for evaluating alternative location areas include transportation costs for bringing in raw material and delivering finished products to customers, income and property taxes in the area, cost of utilities in the area, construction and land costs, and the general labor costs in the area. Subjective or qualitative factors usually include the attitude of labor and government in the area toward the type of industry the plant belongs to, the availability of labor in the area, the quality of the standard of living in the area as represented by the number of churches and schools, the area's climate, and the availability of transportation facilities such as airports and railroad terminals. These subjective factors are important in their own right, but also can indicate the *future* values that will be realized by many of the quantitative factors.

The traditional manner of analyzing the quantitative and qualitative criteria is through a subjective comparison of the values of these criteria that are associated with each of a set of alternative location areas. A table is usually developed and subjectively reviewed. The area that appears to be the best overall is selected for the location for the plant.

Sometimes, however, techniques are used to convert the evaluation and comparison of all criteria to a quantitative basis. For example, the same type of "point system" scheme that is used in job evaluation (see Chapter 5) can be used to assess the relative worth of alternative location sites. The location criteria, such as labor availability, would be synonomous with the "job factors" of job evaluation, and would be subdivided into four or five "degrees." Rating points are then assigned to each degree of each location factor. For example, the factor "attitude of the area's community" might be divided into the following degrees: degree 1, very negative; degree 2, negative; degree 3, cooperative; and degree 4, very cooperative. The points would be assigned in the same manner as described in Chapter 5 and would result in a table similar to Table 5.5. Each alternative location site would be scored with this table, and the site with the largest total number of points would be the most desirable solution.

In situations in which cost criteria are deemed more important than subjective factors, an approach often used is to formulate a normative cost model in order to obtain the two or three best least-cost locations, then choose among these candidates based on a subjective comparison of secondary qualitative factors. The cost model is usually a mathematical programming problem (see Chapter 8). Several models that have been used in analyzing location problems are described in the next section. A more detailed discussion of location models can be found in Francis and White [3].

11.4.3 Location Models

The type of model used in the analysis depends on the characteristics of the location problem, such as the number of new plants to be located, the type of cost factors to be considered, and the existence of restrictions like unsuitable geographical regions. The concepts involved in formulating mathematical models in general were discussed in Chapter 2. In fact, an example of formulating a location model was given in Example 2.3. Three representative forms of location models often used are presented below.

As a first illustration, consider the commonly occurring location problem in which we are trying to find the site for a single new plant that will result in the minimum transportation cost in shipping goods from the plant to a set of customers. When it is true that (1) the transportation cost between the plant and customers is directly proportional to the distance between them, (2) this distance is adequately represented by a "rectilinear distance" model, and (3) the plant to be located can be positioned anywhere, then an optimization model that can be used is

$$\min_{x,y} x_0 = \sum_{i=1}^{n} c_i w_i(|x - a_i| + |y - b_i|) \tag{11.5}$$

where x and y are the unknown Cartesian coordinates of the position of the new plant, a_i and b_i are the coordinates of the position of the ith customer, c_i is the cost to move one unit of product one distance unit between the plant and the ith customer, and w_i is the forecast number of trips required between the plant and customer i. Note that the rectilinear distance metric employed in model (11.5) implies that movement between two points is adequately represented by the sum of horizontal and vertical movements, as opposed to, say, the straight-line distance between the two points.

Model (11.5) constitutes an unconstrained, nonlinear optimization model. It can be solved for the best location of the new plant (that is, for optimum values of x and y) by the methods described in Chapter 7. Alternatively, it can be shown that the solution (x^*, y^*) is the coordinate point having the following properties [3, p. 171].

1. x^* is the same value as a_i for some customer i, and y^* is the same value as b'_i for some customer i', where i' is not necessarily the same as i.
2. x^* and y^* are *median sum locations*; that is, no more than one half of the

item movement between the plant and all the customers is to the left of x^*, while no more than half the movement is to the right of x^*; similarly, no more than half the movement is below y^*, while no more than half is above y^*.

Example 11.2

As an illustration of the rectilinear model in plant-location problems, consider the decision that faced a nut-and-bolt manufacturer in a large industrial city. The company decided to build a new plant in the city in order to obtain the sales of four major customers who were also located in the city. The company decided to site the plant in as central a location as possible with respect to the four customers. They felt that this would lead to a location that minimized their shipping cost and, at the same time, could be used as a marketing tool in getting the new business.

After taking into account the potential share of each customer's total business the company could expect *after* building the new plant, it was predicted that the annual sales to the four customers would be as follows: customer 1, \$55,000 (or approximately 100 deliveries per year); customer 2, \$40,000 (or 70 deliveries); customer 3, \$60,000 (or 110 deliveries); and customer 4, \$80,000 (or 145 deliveries). A model along the lines of Eq. (11.5) was formulated in order to derive an "ideal" location site for the new plant. The formulation assumed that the unit cost of deliveries is the same for all customers and can therefore be ignored, and is proportional to the street distance traveled in making the deliveries. A rectilinear distance model was chosen as the most appropriate manner to represent plant-to-customer travel because of the perpendicular nature of the city's streets. To obtain the necessary position coordinates (a_i, b_i) for each of the four customers, a city map was used with an x–y grid placed on top of it. As shown in Figure 11.16, the customers were found to be located at (8, 26), (11, 15), (25, 18), and (30, 45), respectively. The model that was used is as follows.

$$\min_{x,\,y} x_0 = 100|x - 8| + 100|y - 26| + 70|x - 11| + 70|y - 15|$$
$$+ 110|x - 25| + 110|y - 18| + 145|x - 30| + 145|y - 45| \quad (11.6)$$

The solution to model (11.6) was obtained by applying the medium sum conditions. Calculations for finding x^* are summarized in the following table.

Customer	x Coordinate	Weight	Cumulative Weight	Comparison to Medium Sum
1	8	100	100	$100 < \frac{425}{2} = 212.5$
2	11	70	170	$170 < 212.5$
3	25	110	280	$280 > 212.5$
4	30	145	425	$425 > 212.5$

Since the medium sum of the weights (i.e., deliveries in this case) is first exceeded by the cumulative weights at customer 3, then $x^* = a_3 = 25$ is the medium sum solution. That is, $x^* = a_3$ since no more than half the deliveries are expected to take place to the left of $a_3 = 25$ and half to the right of a_3. Similarly, $y^* = a = 26$ as indicated in the following tabulation:

Customer	y coordinate	Weight	Cumulative Weight
2	15	70	70
3	18	110	180 < 212.5
1	26	100	280 > 212.5
4	45	145	425

■

The rectilinear distance model (11.5) can also be formulated as a linear programming problem. This requires use of extra, auxiliary variables so that the nonlinear absolute value functions can be transformed into linear functions. Specifically, consider the following notation.

$$\begin{aligned} r_i &= |x - a_i|^+ = x - a_i && \text{if } x \geq a_i \\ &= 0 && \text{if } x \leq a_i \\ s_i &= |x - a_i|^- = y - b_i && \text{if } y \geq b_i \\ &= 0 && \text{if } y \leq b_i \\ u_i &= |y - b_i|^+ = y - b_i && \text{if } y \geq b_i \\ &= 0 && \text{if } y \leq b_i \\ v_i &= |y - b_i|^- = 0 && \text{if } y \geq b_i \\ &= b_i - y && \text{if } y \leq b_i \end{aligned} \tag{11.6}$$

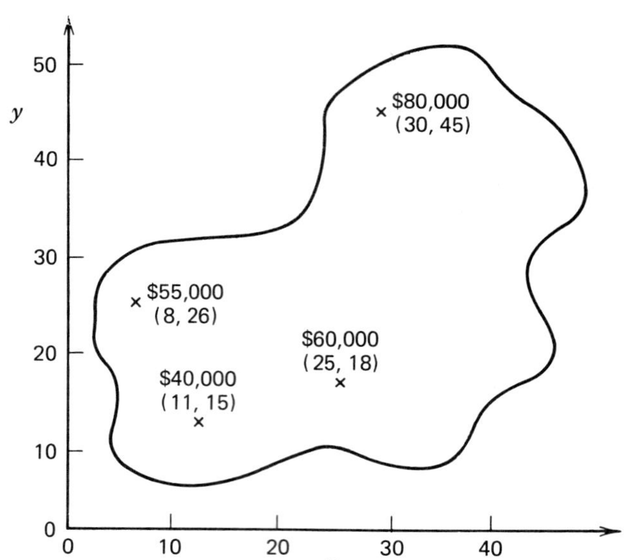

Figure 11.16 Location of Major Customers in the Nut-and-Bolt Problem

Note that these variables lead to

$$|x - a_i| = |x - a_i|^+ + |x - a_i|^- = r_i + s_i \tag{11.7}$$

and

$$|y - b_i| = |y - b_i|^+ + |y - b_i|^- = u_i + v_i$$

when we require that the following conditions hold:

$$(1) \quad r_i - s_i = x - a_i \quad \text{and} \quad u_i - v_i = y - b_i \tag{11.8}$$

and

$$(2) \quad r_i s_i = 0 \quad \text{and} \quad u_i v_i = 0$$

Then, model (11.5) can be rewritten as

$$\min_{x,y,\mathbf{r},\mathbf{s},\mathbf{u},\mathbf{v}} x_0 = \sum_i c_i w_i (r_i + s_i + u_i + v_i) \tag{11.9}$$

$$\begin{array}{lll} \text{S.T.:} & (1) \quad x - r_i + s_i = a_i & \text{for all } i \\ & (2) \quad y - u_i + v_i = b_i & \text{for all } i \\ & (3) \quad r_i s_i = 0 & \text{for all } i \\ & (4) \quad u_i v_i = 0 & \text{for all } i \\ & (5) \quad x, y, r_i, s_i, u_i, v_i > 0 & \text{for all } it \end{array}$$

Note that model (11.9) would be a linear programming problem if nonlinear constraints (3) and (4) were not present. As it turns out, they can be removed from the model because they are automatically met by any solution to the reduced problem (see Ref. 3, p. 241 for a detailed explanation). In addition, note that model (11.9) can be decomposed into two separate linear programming problems, one involving the decision variables x, r_i, and s_i (i.e., the "x-coordinate" problem), the other involving y, u_i, and v_i (i.e., the "y-coordinate" problem). For example, for the problem in Example 11.2, the x-coordinate linear programming model is:

$$\min_{x,r_i,s_i} x_0 = 100r_1 + 100s_1 + 70r_2 + 70s_2 + 110r_3 + 110s_3 + 145r_4 + 145s_4$$

$$\begin{array}{llr} \text{S.T.:} & x - r_1 + s_1 & = 8 \\ & x - r_2 + s_2 & = 11 \\ & x - r_3 + s_3 & = 25 \\ & x - r_4 + s_4 & = 30 \\ & x, r_1, s_1, r_2, s_2, r_3, s_3 & > 0 \end{array}$$

The optimal solution to this model is $x^* = 25$, $r_1^* = 17$, $s_1^* = 0$, $r_2^* = 14$, $s_2^* = 0$, $r_3^* = 0$, $s_3^* = 0$, $r_4^* = 0$, and $s_4^* = 5$.

A second type of mathematical programming model that is often used in analyzing plant-location problems is the single-facility "Euclidean distance" model, also called the "Steiner–Weber" model. It is similar to model (11.5), except that a straight-line distance metric is used rather than the rectilinear metric. The model is:

$$\min_{x,y} x_0 = \sum_i c_i w_i [(x - a_i)^2 + (y - b_i)^2]^{1/2} \qquad (11.10)$$

Note that unlike the rectilinear distance model, problem (11.10) is not separable into x and y problems. In addition, the solution to (11.10) is not as readily obtainable.

The basic approach used to solve (11.10) is to treat it as an unconstrained optimization problem (see Chapter 7). The "necessary conditions" for (11.10) are:

$$\frac{\partial x_0}{\partial x} = 0 = \sum_i \left\{ \frac{c_i w_i (x - a_i)}{[(x - a_i)^2 + (y - b_i)^2]^{1/2}} \right\} \qquad (11.11)$$

and

$$\frac{\partial x_0}{\partial y} = 0 = \sum_i \left\{ \frac{c_i w_i (y - b_i)}{[(x - a_i)^2 + (t - b_i)^2]^{1/2}} \right\} \qquad (11.12)$$

Equation (11.11) can be rewritten as

$$x = \frac{\Sigma_i \, a_i g_i(x, y)}{\Sigma_i \, g_i(x, y)} \qquad (11.13)$$

where

$$g_i(x, y) = \frac{c_i w_i}{[(x - a_i)^2 + (y - b_i)^2]^{1/2}}$$

Similarly, Eq. (11.12) can be expressed as

$$y = \frac{\Sigma_i \, b_i g_i(x, y)}{\Sigma_i \, g_i(x, y)} \qquad (11.14)$$

Equations (11.13) and (11.14) constitute the necessary (and sufficient) conditions we must solve in order to obtain the optimal location (x^*, y^*). That is, we must solve the two simultaneous, nonlinear equations (11.13) and (11.14) for (x^*, y^*). As long as $g_i(x, y)$ is defined, a solution can be obtained by the following algorithm:

Step 1. Choose an initial trial solution, $x^{(0)}$ and $y^{(0)}$. (The medium sum solution can be used for this starting point.)

Step 2. Find the next trial solution, $x^{(k)}$ and $y^{(k)}$, from the following equations:

$$x^{(k)} = \frac{\Sigma_1 \, a_i g_i(x^{(k-1)}, y^{(k-1)})}{\Sigma_i \, g_i(x^{(k-1)}, y^{(k-1)})}$$

and

$$y^{(k)} = \frac{\Sigma_1\, b_i g_i(x^{(k-1)}, y^{(k-1)})}{\Sigma_i\, g_i(x^{(k-1)}, y^{(k-1)})}$$

Step 3. If $x^{(k)} = x^{(k-1)}$ and $y^{(k)} = y^{(k-1)}$, stop; the optimal location is $x^* = k^{(k)}$ and $y^* = y^{(k)}$. If not, go to step 2.

Notice that step 2 cannot be executed whenever $x^{(k-1)} = a_i$ and simultaneously $y^{(k-1)} = b_i$. This implies that we have selected a trial solution that is the location of one of the customers. Such a selection leads to undefined necessary conditions; therefore, the optimal location for the new facility can never be the site of an existing customer.

An alternative method of solving the Euclidean distance problem is to use an analogue model of the problem rather than the mathematical model (11.10). Specifically, a location table like that described in Chapter 2 and shown in Figure 2.2 can be used to find the optimal location (x^*, y^*). If the table were free of friction, the solution obtained would be identical to that found by solving Eq. (11.10). Although solving the math model has certain advantages, the use of the table provides the analyst with an easily understood methodology that can be readily explained and sold to management.

A third example of a mathematical model often used in location problems is the "set-covering" model. This model deals with the problem of finding the locations for the fewest number of new facilities required to provide a prespecified minimum level of service or "coverage." Its mathematical representation is as follows:

$$\min_{\mathbf{x}} x_0 = \sum_{j=1}^{n} x_j \tag{11.15}$$

$$\text{S.T.:} \quad \sum_{j=1}^{n} a_{ij}x_j > 1 \qquad \text{for } i = 1, 2, \ldots, m$$

$$x_j = 1 \text{ or } 0 \qquad \text{for all } j$$

where $x_j = 1$ if a new facility is positioned at the jth alternative location site and 0 otherwise; $a_{ij} = 1$ if a facility located at site j can "cover" the ith customer, and 0 otherwise.

The cover coefficients a_{ij} are input parameters that are based on whether or not a new facility located at a particular site can meet a given service goal in satisfying a particular customer. For example, if we are locating new fire stations in a city, and if a station that was located at alternation site 3 (say, the corner of 5th Avenue and 12th Street) could respond to fire alarms in the 7th neighborhood (say, a particular subdivision) in *less than* the fire department's minimal acceptable time of 5 min, then the coefficient $a_{3,7}$ would be assigned a value of 1. If, on the other hand, this response took *more than* 5 min, $a_{3,7}$ would be set equal to 0.

Whereas the two previous models (11.5) and (11.10) were single-facility location problems, the set-covering model (11.15) is a multifacility location problem. It is also a discrete-location space problem, as there are a finite number of predefined alternative sites at which facilities can be located.

Model (11.15) constitutes a zero–one integer programming problem. It can be analyzed like any other integer programming model as discussed in Chapter 8. However, it can also be solved with a special-purpose algorithm designed specifically for dealing with set-covering problems. There are several such algorithms, with one straightforward method outlined below.

Step 1. Set up the "cover matrix" **A**, consisting of the cover coefficients a_{ij}. Note that there is one row in this matrix for each customer, and one column for each new facility site.

Step 2. Reduce the size of the A matrix by invoking the following rules (see Ref. 4 for details):

1. For each row that constitutes a unit vector, remove that row by crossing it out. Set $x_k = 1$, where k is the number of the column in which the one of the unit row vector is positioned. Cross out the kth column of **A**.
2. For any row r in the reduced matrix that is "dominated by" another row p, cross the dominated row out. Row r is dominated by row p if every element of row p is less than or equal to row r.
3. For any column j in the current reduced matrix that is "dominated by" another column k, remove the dominated column. Column j is dominated by column k if every element in column k is greater than or equal to column j.

Step 3. Using either direct observation or complete enumeration, find the combination of the fewest number of columns in the reduced **A** matrix that have a sum of at least one in each row. The x_j's for these columns are equal to one [along with any x_j's from Step 2(1)]. All other x_j's are equal to zero. This is the final solution.

Example 11.3

In order to reduce the customer waiting time for catalog sales, a national department store decided to implement a new policy, that they would fill and deliver all catalog orders placed through their retail stores within 3 days. In order to carry out this policy, the company's Southeastern sales manager realized he would have to open several new distribution warehouses from which catalog orders in his region were filled. The decisions with which the sales manager was faced were those of determining the fewest number of new warehouses needed to implement the company's new delivery time goal, and where each of these warehouses should be located.

There are 7 retail stores (stores A–G) in the Southeastern region, all located in different cities. Delivery of catalog orders to each store from a warehouse is to be made by commercial truck carriers such as UPS. After investigating factors like construction cost and local taxes, the sale manager selected 9 cities as alternative sites in which to open a warehouse. All 9 cities were considered by the manager to be virtually equivalent in terms of these preliminary factors.

On the basis of quotes of expected travel times from the truck carriers, the manager determined that it was possible to fill and deliver orders for retail stores A, E, and G from a warehouse located in city 1 within 3 days. Stores A, B, C, and F could be "covered" by a warehouse in city 2 within the 3-day goal; stores A, B, E, and F could be served by city 3; stores D and G by city 4; stores A, C, E, and G by city 5; stores C and E by city 6; stores A and F by city 7; stores A, B, C, and G by city 8; and stores C, E, F, and G by a warehouse in city 9.

Formulating this decision as a set-covering problem results in the following model:

$$\min_{\mathbf{x}} x_0 = x_1 + x_2 + x_3 + x_4 + x_5 + x_6 + x_7 + x_8 + x_9 \qquad (11.16)$$

$$\begin{aligned}
\text{S.T.:}\quad & x_1 + x_2 + x_3 + x_5 + x_7 + x + x_8 > 1 \\
& x_2 + x_3 + x_8 > 1 \\
& x_2 + x_5 + x_6 + x_8 + x_9 > 1 \\
& x_4 > 1 \\
& x_1 + x_3 + x_5 + x_6 + x_9 > 1 \\
& x_2 + x_3 + x_7 + x_9 > 1 \\
& x_1 + x_4 + x_5 + x_8 + x_9 > 1 \\
& \qquad x_j = 1 \text{ or } 0 \text{ for } j = 1, 2, \ldots, 9
\end{aligned}$$

Model (11.16) can be solved using the procedure outlined in this section as follows:

Step 1.

$$\mathbf{A} = \begin{array}{c} \\ A \\ B \\ C \\ D \\ E \\ F \\ G \end{array}
\begin{array}{c}
\begin{matrix} 1 & 2 & 3 & 4 & 5 & 6 & 7 & 8 & 9 \end{matrix} \\
\begin{bmatrix}
1 & 1 & 1 & 0 & 1 & 0 & 1 & 1 & 0 \\
0 & 1 & 1 & 0 & 0 & 0 & 0 & 1 & 0 \\
0 & 1 & 0 & 0 & 1 & 1 & 0 & 1 & 1 \\
0 & 0 & 0 & 1 & 0 & 0 & 0 & 0 & 0 \\
1 & 0 & 1 & 0 & 1 & 1 & 0 & 0 & 1 \\
0 & 1 & 1 & 0 & 0 & 0 & 1 & 0 & 1 \\
1 & 0 & 0 & 1 & 1 & 0 & 0 & 1 & 1
\end{bmatrix}
\end{array}$$

Step 2. The reduction principles are applied as follows:

(2) Row $r = G$ is dominated by row $p = D$; remove row G from the matrix.

(1) Row D constitutes a unit vector with the one in column $k = 4$; set $x_4 = 1$ and remove row D and column 4.

(2) Row A is dominated by row B; remove row A.

$$\begin{array}{c} \\ B \\ C \\ E \\ F \end{array} \begin{array}{c} \begin{array}{cccccccc} 1 & 2 & 3 & 5 & 6 & 7 & 8 & 9 \end{array} \\ \begin{bmatrix} 0 & 1 & 1 & 0 & 0 & 0 & 1 & 0 \\ 0 & 1 & 0 & 1 & 1 & 0 & 1 & 1 \\ 1 & 0 & 1 & 1 & 1 & 0 & 0 & 1 \\ 0 & 1 & 1 & 0 & 0 & 1 & 0 & 1 \end{bmatrix} \end{array}$$

(3) Column $j = 7$ is dominated by column $k = 2$; remove column 7.

(3) Column 1 is dominated by column 3; remove column 1.

(3) Column 6 is dominated by column 5; remove column 6.

(3) Column 5 is dominated by column 9; remove column 5.

(2) Row F is dominated by row E; remove row F (*after* all the above column deletions).

(3) Column 2 is dominated by column 8; remove column 2.

The reduced coefficient matrix at this point is

$$\begin{array}{c} \\ B \\ C \\ E \end{array} \begin{array}{c} \begin{array}{ccc} 3 & 8 & 9 \end{array} \\ \begin{bmatrix} 1 & 1 & 0 \\ 0 & 1 & 1 \\ 1 & 0 & 1 \end{bmatrix} \end{array}$$

Step 3. By direct observation, no further reductions in the cover matrix are apparent. From the reduced matrix it is evident that only two of the remaining three warehouses need be used since these two can cover all three cities. Arbitrarily select warehouses 8 and 9; therefore, the solution to the problem is to open a warehouse in cities 4, 8, and 9. (Notice that an alternative optimum solution is to use warehouses in cities 4, 3, and 8.) ■

11.5 SUMMARY

Facility design is a broad IE/OR activity that includes addressing decisions as to what type of facilities to have, how many of each type to use, and where to locate these facilities. The concept of a "facility" includes buildings, machines, vehicles, and activities such as letter filing. Therefore, problems that can be classified as facility-design problems go well beyond the classical "plant layout" and "plant location" problems.

In this chapter we address three major facility-design issues. The first is the machine requirements problem. The traditional approach to determining the number of machines to have on hand involves inflating a work center's requirements by the effects of scrap in subsequent centers as well as by the center's efficiency. Mathematical programming has also been applied to this problem. Although the OR approach tends to provide a more realistic treatment of the machine requirements problem than does the traditional approach, it is more complicated.

The second issue is facility layout, which concerns the arrangement of equipment and work centers inside a building. The classical approach used in laying out a structure is to develop template models of the internal facilities and to move these models around on a diagram of the building until a good layout is apparent. Schematic models such as the REL chart and the form-to chart are often used to provide insight as to the value of a particular layout in conjunction with this ''template shuffling.'' Computerized methods are also available for designing layouts. These programs, such as CRAFT and CORELAP, utilize the same information as the traditional ''template shuffling'' approach, but involve formalized searches over many alternative layouts in an attempt to find the optimal design.

The third issue discussed in this chapter is that of facility location. In selecting a general area or region in which to locate a facility, subjective factors such as community attitude and labor availability are often the dominating criteria. However, when quantitative factors are of primary concern, mathematical models such as the Stiener–Weber model or the set-covering model are frequently used to find the best location.

REFERENCES AND SUGGESTED READING

1. J. M. Apple, *Plant Layout and Material Handling*, Ronald Press, New York, 1963.
2. E. S. Buffa, *Modern Production Management*, Wiley, New York, 1973.
3. R. L. Francis, and J. A. White, *Facility Layout and Location*, Prentice-Hall, Englewood Cliffs, N.J., 1974.
4. R. S. Garfinkel, and G. L. Nemhauser, *Integer Programming*, Wiley, New York, 1972.
5. D. M. Miller, and R. P. Davis, ''The Machine Requirements Problem,'' *International Journal of Production Research*, **15,** No. 2 (1977).
6. D. M. Miller, and R. P. David, ''A Dynamic Resource Allocation Model for a Machine Requirements Problem,'' *AIIE Transactions*, **10,** No. 3 (Sept. 1978).
7. J. A. Shubin, and H. Madeheim, *Plant Layout*, Prentice-Hall, New York, 1951.

REVIEW QUESTIONS AND PROBLEMS

11.1 What is ''facility design''? Discuss its scope.

11.2 Define or explain the following terms:

(a) Template
(b) A machine requirements problem
(c) A facility layout
(d) REL Chart

(e) Space relationship diagram
(f) CRAFT
(g) A discrete-space location problem
(h) A medium sum location
(i) Steiner–Weber model
(j) Set-covering model

11.3 Describe the difference between a product-type and a process-type layout, and give their relative advantages and disadvantages.

11.4 For what production situations is a product-type layout better suited than a process layout? For what situations is the process layout better suited? Why?

11.5 Describe the basic decision problem involved in developing a process-type layout.

11.6 Compare the decision issue involved in designing a process layout with that involved in a product layout.

11.7 How are the decision problems involved in the machine requirements problem and process planning interrelated with those involved in a layout problem?

11.8 What type of information is needed in analyzing a layout problem?

11.9 Describe the traditional approach used to lay out a job shop.

11.10 Critique the following statement: "Facility design is as much an operations research topic as it is a traditional industrial engineering topic."

11.11 Discuss the relevance of each of the following in analyzing a machine requirements problem:

(a) Queueing theory
(b) Linear programming
(c) Mathematical programming
(d) Simulation
(e) Work measurement (time standards)
(f) Job evaluation
(g) Engineering economy
(h) Probability and statistics
(i) Operations process charts
(j) Route sheets

11.12 Discuss the relevance of each of the following in analyzing a plant location problem:

(a) Queueing theory
(b) Linear programming
(c) Mathematical programming
(d) Simulation
(e) Work measurement (time standards)
(f) Job evaluation
(g) Engineering economy
(h) Probability and statistics
(i) Operations process charts
(j) Route sheets

11.13 (a) Describe the "template shuffling" approach used in facility layout.
(b) Discuss the advantages and disadvantages of template shuffling relative to computerized layout methods.

11.14 Compare and contrast the solution approach used in "construction" type computerized layout programs with that used in "improvement" type programs.

11.15 Suggest a step-by-step algorithm that could be used to solve the model of the layout problem as assumed by the CRAFT computer program.

11.16 Write a computer program to lay out an office. Base your program logic on the template shuffling approach. Validate your code. Give the mathematical model you are implicitly solving.

11.17 Describe the layout model used by the CRAFT program in terms of the classification scheme used for facility-location problems.

11.18 Give four or five qualitative factors that are relevant in selecting the general area to locate each of the following:

(a) A beer brewery
(b) A new-product research center
(c) A steel mill
(d) A nuclear power plant
(e) A salt-dome oil-storage facility
(f) An airport

11.19 Consider the decision to locate a beer brewery in the New York City area as opposed to a small rural town in Louisiana on the Mississippi River. Develop a point system like that used in job evaluation to choose between these alternative sites.

11.20 Verify that the optimal location for the nut-and-bolt plant of Example 11.2 has a y coordinate of 26 by formulating and solving the associated linear programming model.

11.21 Verify that the optimal location for the nut-and-bolt plant of Example 11.2 has an x coordinate of 25 by solving the associated linear programming model.

11.22 Give a multifacility version of the location model (11.9).

11.23 The Freeland Metal Products Company makes U-bolts that function as clamps for wire rope. There are three main steps in making the U-bolts: cutting to length, bending, and threading. A route sheet for processing these bolts is given in Figure 11.17. If 500,000 U-bolts are expected to be sold next year, how many of each type of processing equipment does the company need to produce U-bolts? List all assumptions.

11.24 The ABC Wire Rope Company is building a new plant to produce three new types of wire rope. The process used to manufacture these ropes requires three work stations: drawing, annealing, and spooling. In addition, there are three auxiliary areas that are required to support these production operations: shipping and receiving, weighing, and storage. A proposed layout of the new plant is given in Figure 11.18. Using the REL chart also shown in Figure 11.18, evaluate the proposed layout. List any assumptions.

11.25 Using the same building shell and the same work station area, give an improved layout for the ABC Company of Problem 11.24. Use space relationship diagrams in deriving your proposed layout. Verify that your layout is in fact an improvement.

Operation Number	Work Center	Operation	Machine	Time (min) Set Up	Run
B-1	Saw	Cut stock to length	Bank saw	15	0.02
B-2	Bend	Bend stock	Avery bender	30	0.70
B-3	Thread	Turn threads	Engine lathe	25	0.30

Figure 11.17 Route Sheet for U-Bolts

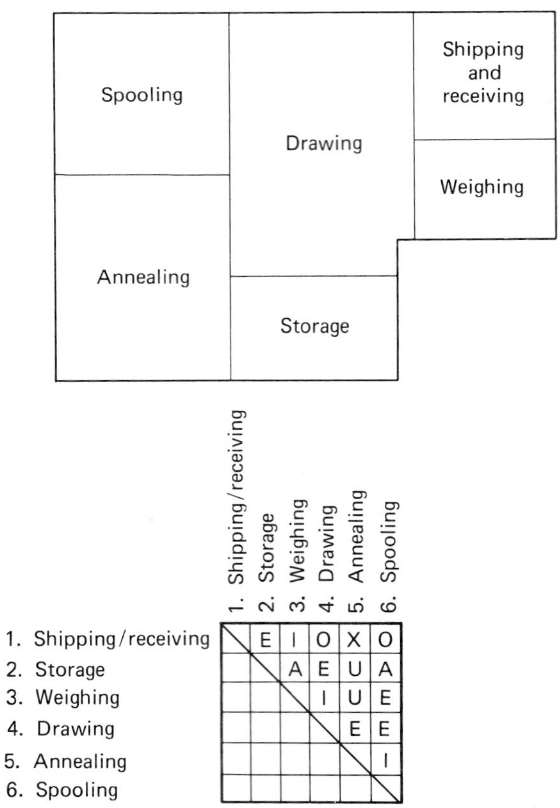

Figure 11.18 Layout and REL Chart for ABC Wire Rope Company

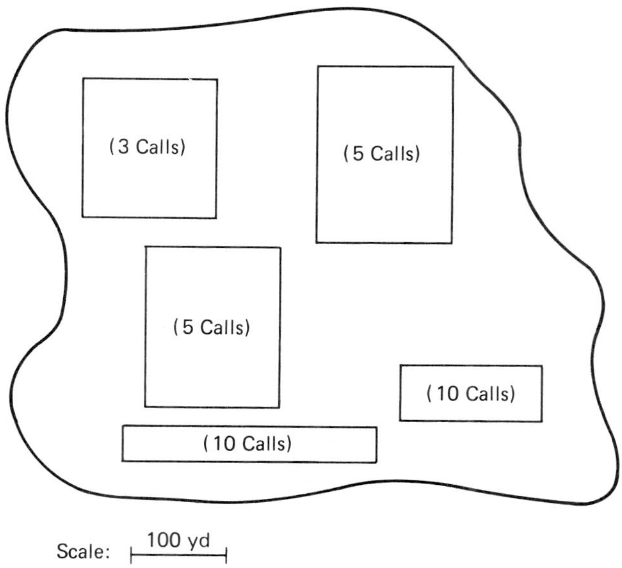

Figure 11.19 Layout of the Grounds of the World Wide Steel Company

11.26 The World Wide Steel Company has a plant in Opp, Alabama, that encompasses five buildings all located on the same grounds. The company plans to centralize their pool of material handling equipment and house it in a sixth building somewhere on the grounds. The locations of the existing five buildings are shown in Figure 11.19, along with the average number of calls per day for handling service occurring at each building.

(a) If material handling cost is assumed to be proportional to the total straight-line distance traveled, where should the new equipment pool be located so as to minimize material handling costs? List all assumptions.

(b) Where should the pool be located if cost is proportional to the square of the straight-line distance?

(c) Where should the pool be located if cost is proportional to the total rectilinear distance traveled?

(d) Suppose that management wishes that the new pool be located no more than 300 yards from any of the existing buildings, even if it means decentralizing the pool by having several substations scattered around the plant. What is the smallest number of such substations needed, and where should they be located if they can only be housed in an existing building?

Chapter 12

Production Planning and Control

12.1 THE MEANING OF PRODUCTION PLANNING AND CONTROL

Production planning and control (PP&C) refers to the process of determining the specific manner in which an organization is to operate in producing goods or providing services. It involves setting policies and strategies to govern the level and timing of production, as well as determining which facilities are to be used in various production operations. For example, the selection of the specific product mix to run next month in a paint factory is a PP&C problem, as is the choice of the exact batch or lot size of each type of paint to produce when that type is being made during the next month.

The scope of decisions that are part of PP&C is quite large. These range from a macrolevel involving overall production capacities and long time horizons, to a microlevel dealing with specific pieces of equipment and immediate operations. The product mix issue is an example of a PP&C decision at the macrolevel, as is the choice of the specific output level to be achieved by a plant in each of the next 12 months (production smoothing). Examples of microlevel PP&C decisions are the selection of a daily assignment of jobs to machines (scheduling), choice of the number of units to process at a particular work station once the equipment is set up (lot size planning), and determining the desired output rate of each work station on an assembly line to achieve a smooth flow (line balancing). Other examples of typical PP&C decisions and activities are expediting late customer orders through a plant, setting policies on the quantity and timing of reordering inventoried material (inventory control), monitoring the status of production jobs, setting policies to test and track the quality of production operations (quality control), and predicting output requirements or demand (forecasting).

Consider the operational structure of a specific company from the viewpoint of production planning and control activities. Figure 12.1 shows the operations as well as the information flow in a large manufacturing organization. Production planning in this situation begins after the product has been defined and orders have been received.

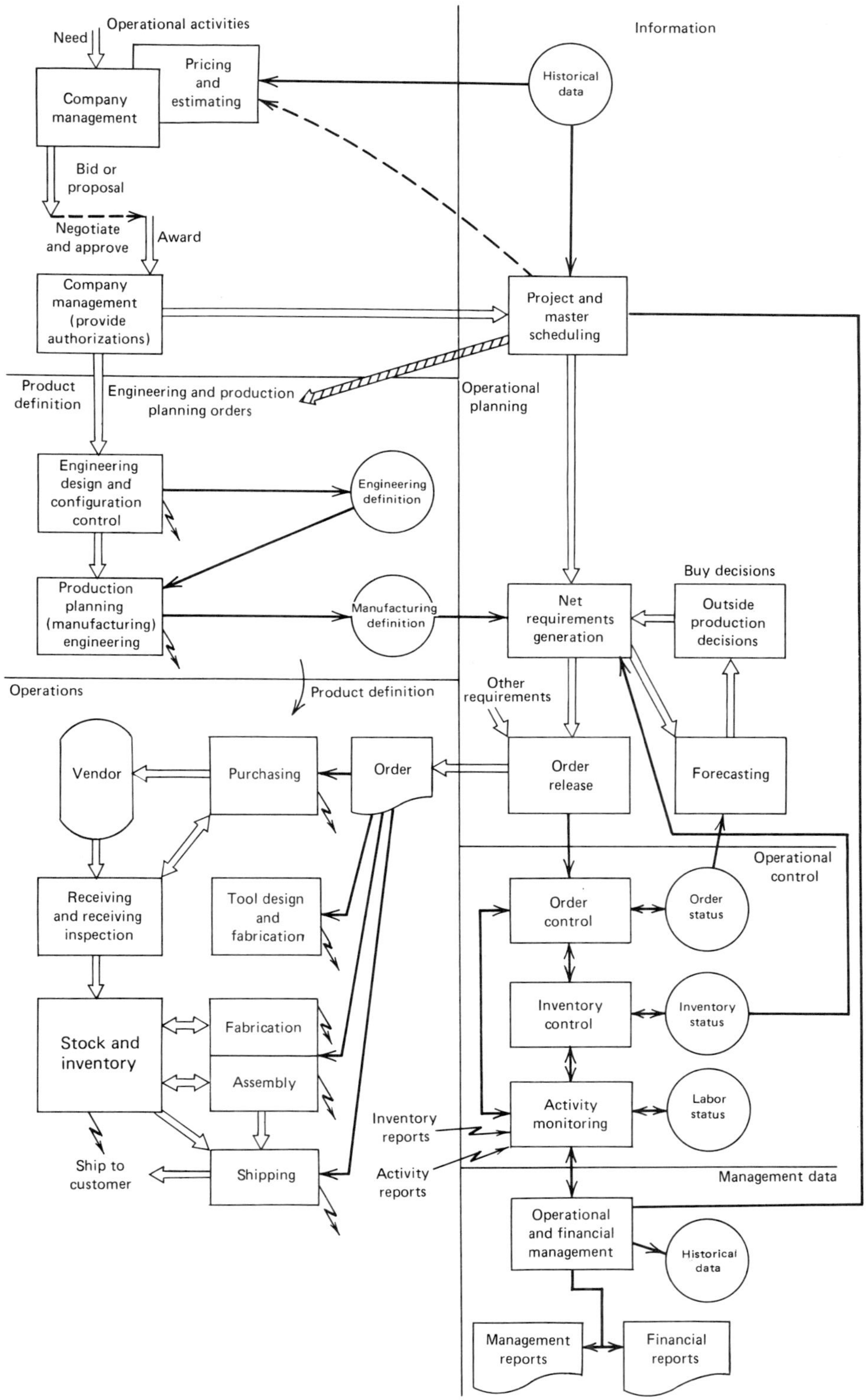

Figure 12.1 Operations and Information Flows in a Particular Manufacturing Organization (From A. A. Kusnick, "Management and Engineering Information," *Industrial Management Review,* Spring 1966, pp. 3–16)

In fact, all of the activities shown in boxes in Figure 12.1 in the "Operational Planning" and "Operational Control" segments of the diagram are PP&C functions.

PP&C decisions arise in service organizations as well as in manufacturing concerns. Banks, hospitals, local governments, and the military all have decisions to make as to the mix and timing of services to provide, and must also monitor and control the quality of these services. For example, a local fire department must periodically reassess the number of people assigned to various jobs such as building inspection and arson investigation. This is as much a PP&C decision as is an automobile manufacturer's periodic reassessment of the desired capacity of a painting or assembly work station. Scheduling policemen to beats, controlling the quality of care in a hospital, and forecasting cash-flow requirements in a bank are other examples of PP&C problems in service organizations.

While the specifics of the problem may differ, the same techniques used to make PP&C decisions in the manufacturing sector are used to analyze PP&C problems in the service sector. These techniques include many of the operations research tools discussed in Section Three, such as queueing theory, linear programming, and simulation, as well as the traditional industrial engineering tools discussed in Section Two, such as flow process charts and Gantt charts. The main difference in applying these techniques is that in the manufacturing sector, *output* is measured in units produced, whereas in the service sector it is measured in criteria such as people served per hour.

In this chapter we describe several PP&C problem areas and discuss the techniques commonly used in the analysis of these problems. Specifically, we examine the areas of aggregate production planning, operations scheduling, and forecasting. Because of the broad range of its applications outside PP&C decisions, inventory control is treated separately in Chapter 13. Likewise, the important PP&C topic of quality control is discussed in Chapter 14. More details on these as well as other PP&C areas can be found in several of the books referenced at the end of this chapter. Johnson and Montgomery [10] provide an excellent quantitative treatment of these topics, whereas Moore and Jablonski [13] present these topics from a more pragmatic, qualitative viewpoint.

12.2 FORECASTING

Most PP&C problems involve *demand* on operations and facilities. Depending on the particular problem in question, demand might be defined as the weekly sales of a product, the hourly material input requirements of the next work station on an assembly line, or the monthly number of crimes in a certain area of the city. Since PP&C involves making decisions now that concern *future* operations, the driving force in the decision process is future demand rather than the actual demand realized in past operations. That is, by its nature production planning involves making plans for future operations and thus requires predictions of the future demand for these operations.

Forecasting is the process of developing these predictions or estimates. The need for forecasting in PP&C goes beyond the need to predict future demand. We must also forecast the values of other constants or parameters that affect the PP&C decision, such as labor rates, raw material prices, and the cost of capital. In addition, forecasting has many obvious applications outside PP&C. For instance, the material flow requirements that drive the plant-layout decision (see Chapter 11) are really estimates of future flows since we are concerned with designing a layout to meet tomorrow's requirements rather than yesterday's flows. The data used in simulating a system (see Chapter 10) are actually forecasts in that we are trying to imitate how the system will work under various alternative conditions, not how it worked under a past set of conditions. In fact, most of the IE/OR tools discussed in this book are intended to be used with input data that are estimates of future realizations of those inputs and must be forecast by some means or other.

12.2.1 Forecasting Approaches: Subjective versus Quantitative Methods

A typical forecasting problem might be to obtain estimates of the monthly sales of a product during each of the next 12 months. These 12 estimates could be obtained through subjective means, such as asking salesmen and field representatives for their opinion and using the resulting answers as the forecasts, or through a quantitative analysis using mathematical models. Each approach has its relative advantages and disadvantages. On the one hand, subjective assessments can incorporate qualitative factors that are difficult to include in a mathematical model. On the other hand, a quantitative analysis is often more consistent than a subjective approach and can provide more information through manipulations and sensitivity analyses. Each approach has certain types of situations in which it is more accurate than the other—that is, produces estimates that turn out to be closer to the true, actual value. In general, quantitative approaches are more accurate when there is a substantial history available of the variable being forecast, and this history is indicative of how that variable will act on the future. In this chapter we concentrate on quantitative approaches.

12.2.2 Forecasting Approaches: Exogenous Variables versus Time Series

There are two basic categories of quantitative forecasting techniques: the use of exogenous variables and extrapolation of a time series. Both approaches use a mathematical model (see Chapter 2) to represent how the variable being forecast changes. In fact, they both use the following general model:

$$\hat{y} = b_0 + b_1x_1 + b_2x_2 + \cdots b_nx_n \tag{12.1}$$

where $\hat{y}$ is the forecast of the variable of interest y (e.g., y might be next week's sales and $\hat{y}$ the number used as its estimate or forecast); $x_1, x_2, \cdots, x_n$ are independent variables whose values influence the dependent variable of interest; and $b_0, b_1, \cdots, b_n$ are parameters or constants.

The primary analytical task in quantitative approaches to forecasting is to derive numerical values for these parameters $b_0, b_1, \cdots, b_n$. An example of the results of such analysis is the following forecasting model or equation.

$$\hat{y} = 80 + 2x_1 + 7x_2 - 3x_3 \tag{12.2}$$

In both the exogenous-variable and the time-series approach, statistical correlation is the basic method of deriving these parameter estimates. Once these estimates are derived, the forecast of the variable of interest can be directly calculated from the forecast equation. For example, if Eq. (12.2) were derived for a given forecasting problem, then the forecast that would result from x_1 taking on a value of, say, 10; x_2 taking on a value of, say, 2; and x_3 taking on a value of, say, 5, would be $\hat{y} = 99$.

The major difference between the two quantitative forecasting approaches is that in the exogenous-variable approach the independent variables [i.e., the x's in Eq. (12.1)] can be any external related factor, such as the U.S. gross national product or a stock market indicator. However, in the time-series approach, the independent variable is time itself. That is, the variable of interest is modeled as a function of the length of time or number of time periods from some base point, such as today, until the time period of interest. For example, the following model might represent the basic time series forecasting equation for weekly sales of a given product.

$$\hat{y}_t = b_0 + b_1 t + b_2 t^2 \tag{12.3}$$

A detailed illustration of the development of a forecast equation using the time-series approach is given in the next section.

12.2.3 Types of Time-Series Models

The time-series model given in Eq. (12.3) is called a quadratic model because it can represent a process that fluctuates at any increasing rate over time. Another type of time-series model is the trend model $\hat{y}_t = b_0 + b_1 t$, which represents a process that changes linearly over time. The simplest class of time-series models is the constant model $\hat{y}_t = b_0$, which represents a process that does not change in value from one time period to the next. As illustrated in Figure 12.2, regardless of the model used, the actual values realized by the process being forecast will not conform exactly to the model being used to represent that process. There will normally be a deviation between the actual value realized and the model's prediction of that value.

Once the type of time-series model is selected (through knowledge of the variable being forecast or direct observation of past observations of the variable), the major forecasting job that remains is to derive numerical estimates for the parameters in that model (b_0, b_1, etc.). There are several alternative methods for doing this, some of which give better results in the sense that the resulting forecasts turn out to be closer to actual values of the variable once they are realized in the future. These forecasting techniques or methods include exponential smoothing, regression analysis, and moving averages.

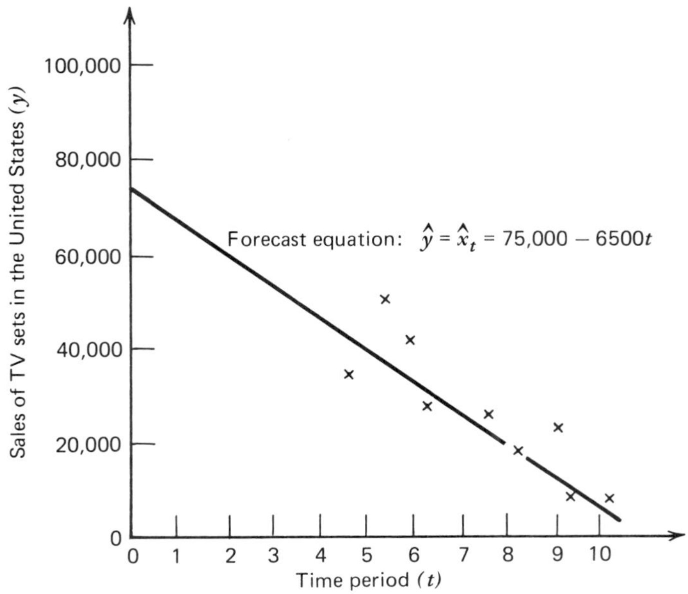

Figure 12.2 Example of a Forecasting Equation Using a Time-Series Approach

12.2.4 Simple Exponential Smoothing

Exponential smoothing is a mathematical method of using past observations of the variable being forecast to derive weighted "least-squares" estimators of the parameters in the time-series model being used. That is, exponential smoothing blends all historical data together to produce the parameter estimates in such a way as to lead to the minimization of the following quantity:

$$\sum_{t=1}^{T} W_t(y_t - \hat{y}_t)^2 \tag{12.4}$$

where T is the current time period.

This scalar is the weighted forecast error. In exponential smoothing, the weights W_t are assigned the value $(1 - \alpha)^{T-t}$ for $0 < \alpha < 1$, so that the weight given each piece of data decreases exponentially with the age of that data (α is a prespecified constant).

When a constant time-series model is being used, the exponential smoothing technique used is called *simple* (or single) *exponential smoothing,* as only one model parameter is involved. If a two-parameter trend model were used, a more complicated version of exponential smoothing, called *double exponential smoothing,* would be required. There are also higher-order versions of exponential smoothing that have been developed to estimate parameters in quadratic and higher time-series models. For purposes of illustration, we briefly discuss only simple exponential smoothing here. An excellent discussion of exponential smoothing in general can be found in Brown [2] or in Johnson and Montgomery [10].

In order to estimate the parameter in the constant time-series model, $y_t = b_0$, the forecast error sum given in Eq. (12.4) is minimized with respect to b_0. That is, the necessary and sufficient conditions for unconstrained optimization (see Chapter 7) are applied to the following function.

$$f(b_0) = \sum_{t=1}^{T} (1 - \alpha)^{T-t}(y_t - b_0)^2 \qquad \text{for } 0 < \alpha < 1 \tag{12.5}$$

Taking the derivative with respect to b_0 and setting it equal to zero leads to

$$b_0 \equiv S_T = \alpha y_t + (1 - \alpha)S_{T-1} \tag{12.6}$$

where S_T is the estimate used in period T for the parameter b_0, and α is the *smoothing constant,* which is assigned a value between 0 and 1 at the discretion of the analyst (a value between 0.2 and 0.5 is normally used for α). S_T is called the *smoothing statistic* and is actually a linear combination of all past observations of y_t. That is, through backward substitution it can be shown that

$$S_T = \alpha \sum_{t=0}^{T-1} (1 - \alpha)^t y_{T-t} + (1 - \alpha)^T y_0 \tag{12.7}$$

It should be noted that in exponential smoothing we are not attempting to estimate the variable of interest y_t *directly* but, rather, to estimate the parameters in the model chosen to represent y_t. In the case of the simple constant model this distinction is not critical, but in higher-order models it does become important.

Example 12.1

A major textbook publisher is interested in setting up a procedure to forecast his monthly sales of books. He wishes to use this procedure on a "rolling basis." That is, at the end of, say, May he wishes to predict sales for the entire month of June. Then, when the end of June comes, he will predict the sales to occur during July. He will continue moving from month to month, making an estimate of sales in month $(T + 1)$ given actual observations up through the end of month T.

A quantitative, time-series approach has been selected by the firm as the basis for its forecasting procedure. The first step in setting up the procedure is to select the type of time-series model that best represents the realization of monthly textbook sales as it changes over time. A plot of the last 8 months of sales activity is shown in Figure 12.3. On the basis of the assumption that future sales will act much like sales over the past 8 months, and after a visual analysis of the data in Figure 12.3, a constant time-series model was selected. In addition, a smoothing constant of $\alpha = 0.2$ was arbitrarily chosen.

The resulting forecast equation used by the firm to estimate sales in month $T + 1$,

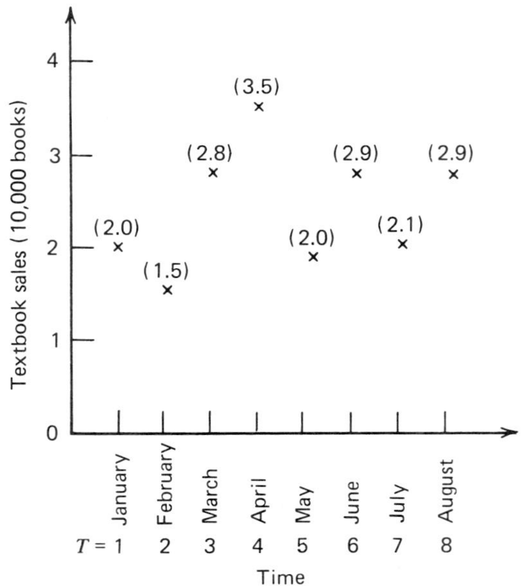

Figure 12.3 Historical Sales Activity for Example 12.1

given that sales data through the end of month T have been observed, is:

$$\hat{y}_{T+1} = S_T$$

where

$$S_T = (0.2)y_T + (0.8)S_{T-1}$$

Application of the forecast procedure to the company's sales process beginning at the end of August resulted in Table 12.1. By way of illustration, the forecast for September was found as follows (August is month "T"):

$$\hat{y}_{(\text{September})} = S_{(\text{August})}$$

and

$$S_{(\text{August})} = (0.2)y_{(\text{August})} + (0.8)S_{(\text{July})}$$

Using an initial starting "guesstimate" of $S_{(\text{July})} = 3.0$, then

$$S_{(\text{August})} = (0.2) \times (2.9) + (0.8) \times (3.0) = 2.98$$

so that the estimate of sales in September ($T + 1$) is

$$\hat{y}_{(\text{September})} = 2.98 \quad \text{or 29,800 books}$$

Similarly, as we move to the end of September, an *actual* sales of 25,800 books is

Table 12.1 Forecasts (in Thousands of Books) through November as Developed by Simple Exponential Smoothing

T Month	8 August	9 September	10 October	11 November	12 December	13 January
y_T	2.90	2.58	2.10	2.42	2.00	2.40
S_T	2.98	2.90	2.74	2.68		
$\hat{y}_T$	3.00	2.98	2.90	2.74		

recorded. The forecast error that is experienced for sales in September is (29,800 − 25,800) or 4000 books. The forecast we would make for sales in the next month October would be:

$$\hat{y}_{(\text{October})} = S_{(\text{September})} \quad \text{or } \hat{y}_{10} = S_9 = (0.2)y_9 + (0.8)S_8$$

or

$$\hat{y}_{10} = (0.2) \times (2.58) + (0.8) \times (2.98) = 2.90 \quad \text{or 29,000 books}$$ ■

It is important to note that we are forecasting one month at a time on a "rolling" basis. That is, if we are presently at the end of August, we know or have recorded the sales for the month of August and wish to use that actual y_T value to forecast next month's sales, y_{T+1}. After $\hat{y}_{T+1}$ is calculated at the end of August, we wait and do nothing more until time has expired and we reach the end of September. At that time, we repeat the forecast process to develop our next $\hat{y}_{T+1}$ (that is, the forecast for October). However, it should also be noted that a stationary forecast for more than one period in the future can be calculated at the end of the present month. That is, $\hat{y}_{T+n}$, where n is any positive number of time periods, can be calculated just as easily as $\hat{y}_{T+1}$. The same value of S_T is used. The associated estimate of b_0 is our best guess as to the true value of b_0 as of the present time (i.e., as of period T). That is, in the case of the constant model, $\hat{y}_{T+1} = \hat{y}_{T+2} = \cdots = \hat{y}_{T+n} = S_T = \hat{b}_0$.

Often the forecasting system used by a company will include a "tracking signal" to indicate when the forecast errors are out of control. In addition, some systems include an adaptive feature to automatically modify a portion of the forecast equation (such as the smoothing constant) when the tracking signal has flagged a problem. All of this is simply an attempt to have a forecasting system that produces more accurate forecasts. Details on features such as tracking signals can be found in Brown [2].

12.3 AGGREGATE PLANNING

There is a two-level or two-phase problem involved in determining the appropriate output or production rate in a manufacturing or service organization. First, we have to set an overall, aggregate output rate for each plant, such as 25,000 automobiles per month from the Atlanta plant and 50,000 automobiles per month from the Houston plant. Second, we then must decide how to schedule the machines and work stations *inside* these plants to achieve this overall plant output requirement. The problem of determining the best overall (first-level) plant output rate is often called the production smoothing or the *aggregate planning* problem. We discuss this problem and typical solution approaches in this section, and then treat the second-level scheduling problem in the next section.

Aggregate planning refers to the decision problem of determining the best production rate to use for a plant level facility from one time period to the next in view of conflicting cost considerations, such as personnel hiring and firing costs and inventory carrying costs. The end result of this decision is a *production plan,* which specifies the output to be achieved by the facility in each period during a prespecified planning horizon. This plan may follow very closely the changes in demand or sales as it fluctuates from one period to the next, or it may result in a constant, smooth production rate over the entire planning horizon. Figure 12.4 shows a typical graph of a plant's forecast of demand over time, and two alternative production plans for meeting this demand.

Notice in Figure 12.4 that if Plan 1 were used, the production output of the facility between the end of November and about the end of March would be greater than the cumulative sales or demand during that period. This would result in a buildup of finished goods inventory at the facility. Between August and November, however, there would be less production than sales so that the excess demand would have to be met by some other means, such as from inventories previously built up or from subcontracting or through the use of overtime. Each of these options is costly.

Like Plan 1, the second production plan has periods during which inventories are created and periods during which normal production is not adequate to meet demand. However, these periods are shorter in duration than in Plan 1. Plan 2 calls for *changes* in the regular production rate, increases or decreases to be achieved by varying the operating characteristics or availability of production equipment or by varying the size of the workforce through hiring and firing. Again, each option has a cost associated with it.

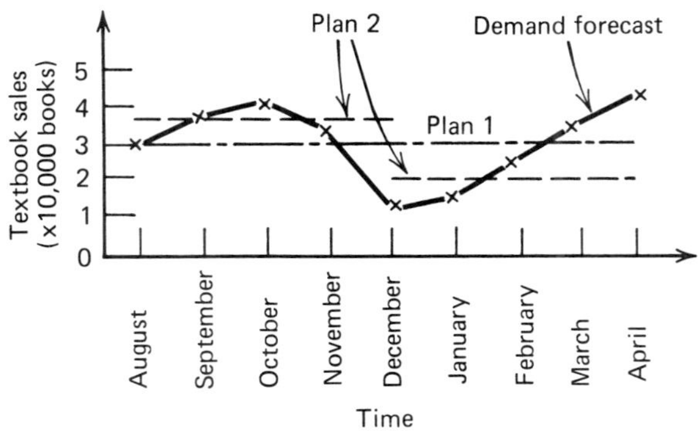

Figure 12.4 Typical Demand Forecast and Alternative Production Plans

The specific decision problem involved in aggregate planning is to determine the normal production plan to use over the next n periods along with any auxiliary plans (such as the use of subcontracting) that provides the lowest overall cost of operating the facility. Therefore, the decision variables involved in this problem typically include the regular time production rate each period, the level of subcontracting to use each period, the size of the workforce to have on hand each period, and the level of overtime to use in each period. An indirect decision that results from specifying the workforce size is the number of employees to hire and to lay off each period.

12.3.1 Alternative Solution Approaches

Aggregate production plans are often set by companies in a subjective manner based on top management's assessment of political factors, labor union considerations, and the influence of competition. For example, an international firm may wish to keep a foreign plant operating at the same production rate each month to avoid personnel layoffs and the associated governmental pressure that might result from such action. The indirect cost to the firm due to the foreign government's pressure tactics might far overshadow any direct cost penalty incurred in using a constant production plan as opposed to one that involves workforce changes.

On the other hand, there are many situations in which firms use a quantitative analysis of the aggregate planning problem to set their production plans. In such cases, the firm has several alternative approaches it can use to address the problem. These approaches are basically alternative ways to model the decision variables, the restrictions, and costs involved in the problem. However, they all involve the formulation of a mathematical model of the associated aggregate planning decision problem and a proposed method of solving that model. The difference is in the structure of the resulting model. For example, the model may turn out to be a linear programming problem, a nonlinear programming problem, or an unconstrained nonlinear optimization problem. In the next section we discuss one such approach.

12.3.2 A Linear Programming Approach

When the aggregate planning problem of a firm is modeled in such a way that all of the costs (inventory holding costs, workforce change costs, etc.) and all of the constraints involved are linear functions of the decision variables, then the resulting mathematical model can often be treated as a linear programming problem (see Chapter 8). As an illustration, consider the following approach first suggested by Bowman [1]. This particular formulation is applicable in situations in which the output requirements (demand) in each period can be met from any one of several alternative sources (regular time production in that period, overtime production in that period, or subcontracting). In addition, there is a single product or product group involved, inventory can be carried from one period to the next at a given unit cost, no shortages are allowed in any period, and the cost of obtaining units from any of the alternative sources is a linear cost that varies with the volume or the amount of that source used.

Let

x_{it} = the number of units of demand in period t to be met by alternative production source i ($i = 1, 2, \ldots, N$; $t = 1, 2, \ldots, T$).

D_t = the number of units demanded in period t.

P_i = the capacity of alternative source i to produce units.

c_i = the cost to produce one unit by source i.

h_t = the cost to carry a unit in inventory from period t to period $t + 1$.

I_t = the inventory on hand at the end of period t.

The objective of the problem is to find the level of use of alternative source i in period t (i.e., to specify values for all x_{it}) so as to minimize the combined cost of production and carrying inventory over the entire planning horizon (over periods $t = 1, 2, \ldots, T$). However, in specifying the production plan (the x_{it} values) the following restrictions must be satisfied:

1. The number of units produced in each period by a given alternative production source must not exceed the capacity of that source available in the period.
2. The inventory on hand at the end of a period must equal the inventory on hand at the end of the previous period plus all units produced in the period minus the demand in the period. (This "conservation of flow" restriction ensures that there are no uses or sources of production beyond those explicitly defined.)
3. The inventory on hand at the end of a period must not be negative. (This is to prevent shortages when demand occurs.)
4. The number of units produced by any given source in any period must not be negative. (Negative production has no physical interpretation here.)

These restrictions together with the objective stated earlier define the aggregate planning problem being dealt with. Using the terminology defined previously and the assumptions concerning linear costs, the problem can be represented by the following model.

$$\min_{\mathbf{x}} x_0 = \sum_{t=1}^{T} \left(\sum_{i=1}^{N} c_i x_{it} + h_t I_t \right)$$

$$\text{S.T.:} \quad (1) \quad \sum_{t=1}^{T} x_{it} \leq P_i \qquad \text{for all } i$$

$$(2) \quad I_t = I_{t-1} + \sum_{i=1}^{N} x_{it} - D_t \qquad \text{for all } t \qquad (12.8)$$

$$(3) \quad I_t \geq 0 \qquad \text{for all } t$$

$$(4) \quad x_{it} \geq 0 \qquad \text{for all } i \text{ and } t$$

Model (12.8) is a linear programming problem. Furthermore, it has a special structure that can be taken advantage of to invoke a highly efficient algorithm to use in its solution. Specifically, model (12.8) constitutes a *transportation problem,* as can be seen by representing the problem in a classical transportation problem tableau or matrix. This matrix is given in Table 12.2 for a six-source, three-period example.

Example 12.2

A certain chemical processing plant has a normal two-shift capacity to manufacture alcohol. A production plan is to be set for the next three months in which the least-cost method of meeting each month's sales requirements is to be established. The basic alternative production methods or sources available are the normal two-shift operation, and a higher cost, extra (third) shift in addition to the normal two shifts. In either case, units produced in one period can be used to satisfy sales in later periods.

Table 12.2 Transportation Matrix for an Aggregate Planning Problem[a]

	Month 1	Month 2	Month 3	Unused Capacity	Available Capacity
Initial inventory	0	h_1	h_2	0	I_0
Two-shift operation in month 1	c_1 $x_{1,1}$	$c_1 + h_1$ $x_{1,2}$	$c_1 + h_1 + h_2$ $x_{1,3}$	0 S_1	P_1
Third-shift operation in month 1	c_2 $x_{1,2}$	$c_2 + h_1$ $x_{2,2}$	$c_2 + h_1 + h_2$ $x_{2,3}$	0 S_2	P_2
Two-shift operation in month 2	NA $x_{3,1}$	c_3 $x_{3,2}$	$c_3 + h_2$ $x_{3,3}$	0 S_3	P_3
Third-shift operation in month 2	NA $x_{4,1}$	c_4 $x_{4,2}$	$c_4 + h_2$ $x_{4,3}$	0 S_4	P_4
Two-shift operation in month 3	NA $x_{5,1}$	NA $x_{5,2}$	c_5 $x_{5,3}$	0 S_5	P_5
Third-shift operation in month 3	NA $x_{6,1}$	NA $x_{6,2}$	c_6 $x_{6,3}$	0 S_6	P_6
Demand	D_1	D_2	D_3		Total capacity

[a] S_i is the unused capacity for source i and NA means not applicable because of the assumption that no "backorders" are allowed.

Production cost under the two-shift mode is \$20/ton, and \$24/ton under the third-shift mode, regardless of which month production takes place. Based on the cost of insurance, property tax and opportunity cost, the firm estimates that it costs them \$1 to hold 1 ton of product from one month to the next. There is 10 tons of product on hand initially, and 40 tons, 50 tons, and 15 tons of sales estimated in months 1, 2, and 3, respectively. In any month, 30 tons can be produced in the normal two-shift mode and an extra 20 tons can be produced using the third-shift mode. Using these estimates of required input data, the problem can be represented in matrix form as shown in Table 12.3.

The optimal solution for this transportation problem can be found by applying the techniques referenced in Chapter 8. The result is shown in Table 12.4 with the cell allocations representing the values assigned to the x_{it}'s. Specifically, the least-cost production plan is to use the initial inventory plus produce 30 tons in the normal two-shift operation to meet the demand in month 1, produce 30 tons during the normal two-shift operation in period 2 plus 20 tons on the third shift to meet demand in month 2, and produce 15 tons in the two-shift mode to meet demand in the last period.

Table 12.3 Transportation Matrix for Example 12.2

	Month 1	Month 2	Month 3	Unused Capacity	Available Capacity
Initial inventory	0	1	2	0	10
Two-shift operation in month 1	20	21	22	0	30
Third-shift operation in month 1	24	25	26	0	20
Two-shift operation in month 2	NA	20	21	0	30
Third-shift operation in month 2	NA	24	25	0	20
Two-shift operation in month 3	NA	NA	20	0	30
Third-shift operation in month 3	NA	NA	24	0	20
Demand	40	50	15	55	160

Table 12.4 Solution for Example 12.2

	Month 1	Month 2	Month 3	Unused Capacity	Available Capacity
Initial inventory	0 10	1	2	0	10
Two-shift operation in month 1	20 30	21	22	0	30
Third-shift operation in month 1	24	25 0	26	0 20	20
Two-shift operation in month 2	NA	20 30	21	0	30
Third-shift operation in month 2	NA	24 20	25	0 0	20
Two-shift operation in month 3	NA	NA	20 15	0 15	30
Third-shift operation in month 3	NA	NA	24	0 20	20
Demand	40	50	15	55	160

Note that there will be idle capacity in periods 1 and 3 with this plan. The model being used does not penalize this slack use of men and equipment. A penalty cost for this purpose could easily be added so that the optimal solution would then reflect the balance between idle cost and inventory-carrying cost. In addition, the use of *back ordering* could be included in the model to allow production in one period to be used to satisfy demand from previous periods. For a discussion of such modifications, see Reference 10. ■

Bowman's LP model is just one of several different approaches that have been suggested as viable ways to formulate and analyze the aggregate planning problem. Alternate LP models incorporate different assumptions regarding the underlying costs and plant operations, such as accounting for the cost of changing the size of the workforce from one period to the next. (Haussmann and Hess [8] describe one such approach.) Furthermore, nonlinear models have been proposed in which the various costs are assumed to change in a nonlinear fashion as decision variables such as workforce size and inventory level change from period to period. One such treatment is the HMMS model suggested by Holt, Modigliani, Muth, and Simon [9]. Their approach is to develop an unconstrained optimization model involving quadratic costs, which, through classical calculus, has an optimal solution in the form of linear decision rules, such as:

$$W_1 = \left(\sum_{t=1}^{T} a_t D_t \right) + bW_0 + cI_0 + d \tag{12.9}$$

Equation (12.9) gives the optimal workforce to have in the first period as a linear function of the demand D_t in each period of the planning horizon, as well as the initial workforce size W_0 and the initial inventory level I_0. The a_t's, b, c, and d are parameters to be estimated from cost data. A discussion of the details of these and other approaches to the aggregate planning problem can be found in Buffa and Taubert [4] or Johnson and Montgomery [10].

12.4 SCHEDULING

In the previous section we discussed the decision problem of determining the best overall production plan for a plant or firm. This production planning problem is an aggregate *scheduling* decision in that we are scheduling the use of one or more plants or other major facilities over a period of time. It is an *aggregate* problem because we are scheduling aggregated groups of products and the use of macrolevel facilities such as entire plants, rather than microlevel facilities such as individual machines. In this section, however, we are concerned with the microlevel problem of scheduling work operations and the use of their associated equipment. The essence of this operations scheduling problem is the selection of which operations are to be processed by which available equipment, and the determination of the timing for the start of these operations. (Note that this includes specification of the relative order of processing the operations in question.)

Many decision problems in government and industry involve scheduling work operations. For example, determining when each of a fleet of garbage trucks should depart its home base and which dumpsters it should visit to empty and in which order is a scheduling problem. Similar scheduling problems exist in planning the use of other delivery and transportation vehicles, such as oil tankers, milk trucks, and police patrol cars.

Another illustration of an operations scheduling problem is the determination of which buildings each person in a city's fire prevention bureau is to inspect during a given period of time, and in which order these inspections should take place. Similar manpower scheduling problems exist in planning the use of postmen, inspectors in heavily automated machining shops, and police patrolmen.

In this section we discuss operations scheduling in the context of the classical job shop. In such a situation, customer orders are received randomly over time and are processed as individual jobs on the appropriate machines in the shop. Processing a customer's job may require performing several different operations at one or more machines or work stations. Many of the techniques used to assist the shop scheduler in routing jobs through the shop are also applicable to other scheduling problem areas, such as those mentioned in the previous paragraph. In the remainder of this section we discuss traditional methods of analyzing the job shop scheduling problem, such as charts and graphs, as well as more recent quantitative methods. Although the treatment here will be brief, an in-depth treatment of scheduling can be found in Conway, Maxwell, and Miller [7] (see also References 4, 10, 13, 14, and 15).

12.4.1 Shop-Level Production Scheduling and Control

In a typical job shop or other manufacturing plant, there is often a position called the ''dispatcher'' or ''shop scheduler'' or, in some shops, the ''expeditor.'' He or she is the operations person who is responsible for the day-to-day routing of work through the shop. His or her job involves *scheduling* each work order that comes into the job by assigning work to be done to complete that work order to various machines or work centers in the shop. Since there is a limited amount of capacity and working time available at each work center, the dispatcher must often make a decision as to which of the work orders still outstanding will be assigned first to common work centers at which they all need to be processed. This allocation decision is part of his or her scheduling or production-planning function.

The dispatcher is also responsible for *monitoring* the progress of work orders through the shop. He or she must make sure that the *actual* work being done on each job is on or ahead of schedule with respect to its planned progression through the shop. When it is found that a job is behind schedule, the dispatcher must determine the reason and, if the priority of the work order involved warrants it, must expedite that job through the remainder of its production operations. But as the manner in which one particular job is to be processed through the shop is often changed, so must we change the plans for processing several other jobs that are competing for the same equipment. Essentially, this part of the dispatcher's function is the periodic *rescheduling* of jobs through the shop. (A good, pragmatic discussion of the scheduler's function can be found in Reference 13.)

12.4.2 Scheduling and Control Tools

The dispatcher usually relies on a variety of documents to determine what upcoming work needs to be processed through the shop and to understand the characteristics of each job. Such *input* information comes from items such as stores requisitions, scrap tickets, stores credit memos, flow process charts, and routing sheets (see Chapters 2, 3, and 4 for a discussion of such documents). For example, the route sheet for a job tells the dispatcher the specific operations necessary to complete that job, the equipment needed for each of these operations, the standard time to complete each operation, and precedence relationships as to the sequence in which the operations can be performed. Such information is the basic input to the scheduling decision.

In developing a specific schedule for the next several weeks or months, charts and diagrams as well as computerized techniques are often used. Traditionally, composite visual displays of the shop's equipment and the jobs being processed have been the primary tool for planning as well as monitoring shop schedules. A variety of such devices are commercially available, such as the "Schedngraph" and the "Boardmaster" [14]. Essentially, such devices are variations of a pegboard that uses colored strings and plastic slide inserts to display or model a snapshot view of the progress of jobs in the shop. Through trial and error, the dispatcher can use these visual charts to work up an acceptable schedule. The same device is used by the dispatcher to display actual shop conditions so that the current status of any job can quickly be observed either by the dispatcher or other interested parties such as shop foremen, purchasing agents, and salesmen.

Gantt Charts

One of the oldest but still most common visual displays used as an aid in scheduling is the Gantt chart. In fact, most commercial scheduling displays are just variants of the basic chart developed by Henry L. Gantt in 1917. (See Chapter 1 for a discussion of Gantt's work.)

An example of a typical Gantt chart is given in Figure 12.5. In this example, 12 jobs are scheduled on 6 different machines from January 10 through January 24. This Gantt chart not only shows the present and planned status of each job in the shop, but also indicates the loading on each machine as well as when the machines are scheduled to be down for maintenance. Also, the actual and planned utilization and idle time for each machine can readily be calculated from the chart.

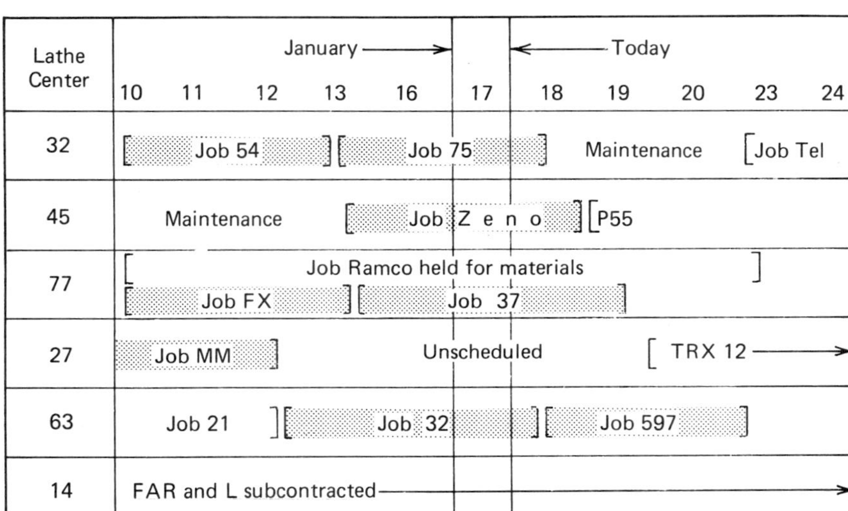

Figure 12.5 Illustration of a Gantt Chart (By permission of Prentice-Hall)

The assignment of jobs to machines shown in Figure 12.5 represents one specific schedule. Other, alternative schedules could be generated by simply rearranging the order of some of the jobs or by reassigning a job to a different machine. For example, a different schedule could be had by moving job 32 from machine or lathe center 63 to machine 27, starting it immediately after job MM is completed, and moving the start time of job 597 up to January 12 in order to take up the slack of machine 63.

But which of these two alternative schedules is better? There are several ways to measure the attractiveness of a given schedule. One of the more common criteria is the *mean* or average *flow time* of the jobs through the shop that is implied by the schedule. The total flow time is an overall schedule measure and is defined to be the sum of the completion times of all jobs—that is, the sum of the processing time for the jobs plus the length of time they have to wait in the shop before being completed. For instance, job 54 has a completion time of 3 working days, job 75 has a completion time (estimated) of 6 days, and so forth. The mean flow time for the schedule shown in Figure 12.5 is then (not counting job Tel, job P55, and job TRX 12):

$$\frac{3 + 6 + 6.5 + 3 + 7 + 2 + 2 + 6 + 9}{9} = 4.9 \text{ days}$$

Similarly, the mean flow time for the alternative schedule mentioned earlier is:

$$\frac{3 + 6 + 6.5 + 3 + 7 + 2 + 6 + 2 + 5}{9} = 4.5 \text{ days}$$

Therefore, the alternative schedule would be considered a better schedule if mean flow time were the relevant criterion for the dispatcher.

The Gantt chart has its advantages and disadvantages as a scheduling aid. It is easy to construct and to use, and it readily conveys necessary schedule information. However, its application is limited because of the physical limits a chart or board has in displaying a large number of machine centers or lengthy schedule horizon. Furthermore, although charts and boards are effective for portraying what a schedule looks like, they are not an efficient device to use to *generate* and analyze potential, alternative schedules.

Quantitative Techniques

Because of the limitations of charts and boards, much use has been made in the past few years of computerized, quantitative techniques both to portray a schedule and to generate a "good" schedule. Computer systems are now available in both a batch and a real-time interactive mode that allow the scheduler to maintain an up-to-date data base on the status of large numbers of machines and jobs, as well as provide him with recommended, optimal schedules to use in an upcoming period. Some of these computer systems even involve automated data entry either directly from the production equipment itself or from the workers on the shop floor.

While the *computer* technology used to assist the schedules is very advanced, the *optimization* technology used to generate recommended schedules is not. The primary reason for this is that the typical real-world job shop scheduling problem is extremely complex, so that it is difficult to formulate large problems in an accurate mathematical model that is still amenable to known optimization techniques. Specifically, we can readily derive optimal solutions to problems involving scheduling n jobs (that is, any number of jobs) on only one or two machines, or certain problems involving n jobs on three machines, but the general problem of scheduling n jobs on m machines is too formidable. Although researchers have developed fairly exact integer programming models of this general problem (see, for example, Reference 7), there are no known *efficient* solution techniques for these models. Nevertheless, it is often possible to make use of the smaller, simpler models in solving real-world scheduling problems by decomposing or suboptimizing the larger immediate problem. Similarly, the logic inherent in solution procedures for the smaller models is often useful in developing heuristic procedures for analyzing more realistic problems. In the next two sections we discuss the solution of some of these smaller models.

12.4.3 The *n*-Jobs, One-Machine Scheduling Problem

As with any decision problem, it is necessary to make certain assumptions about the scheduling problem and the underlying job shop before a model and solution can be developed. Common assumptions used in analyzing the n-jobs, one-machine scheduling problem are the following: (1) each job consists of one operation, (2) a *static* set of n jobs is available to be scheduled (as opposed to those jobs arriving in the shop dynamically over time), (3) the processing time (including setup) for each job is known with *certainty* (as opposed to being a random variable), and (4) the decision-making criterion is the *mean flow time* (as opposed to alternative measures of performance such as the schedule's *makespan*—the total amount of time required to completely process all the jobs—or the schedule's average *lateness* of jobs—the average deviation of the jobs' due dates from their actual completion dates).

As there is only one operation involved per job and only one machine, the scheduling decision for this problem is simply the determination of the *sequence* in which to process the n jobs on the one machine in order to minimize the mean flow time. The solution procedure used to generate the optimal sequence is called the SPT rule, or the shortest processing time rule. It states that the optimal sequence is the one found by ordering the n jobs by having the one with the smallest processing time required be the first job processed, the one with the next smallest processing time be the second job in the sequence, and so on until all n jobs have been ordered. It can be shown that this SPT rule does in fact minimize the mean flow time, as well as minimize the mean lateness of the jobs and the mean waiting time [10].

Example 12.3

The Centron Corporation manufactures recording–playback equipment to be used in automatic telephone message systems. Each device manufactured has a customer-specified circuit board as well as a common base assembly that is mass produced and inventoried. The company has one machine (driven by a minicomputer) that produces the custom-made circuit boards. Periodically, the shop collects customer orders and plans how these orders (jobs) are to be routed through the shop.

At the beginning of the current week, there are five customer orders to be scheduled with the following time requirements on the circuit board machine:

	Job				
	A	B	C	D	E
Setup time (hr)	0.5	0.2	1.3	0.5	0.8
Run time	1.0	0.8	1.0	0.6	1.2
Total	1.5	1.0	2.3	1.1	2.0

In order to minimize the average time the customer orders are in the shop (the mean flow time), the sequence in which the jobs should be scheduled is B, D, A, E, and C. This results in mean flow time of 7.9/5 = 1.58 hr (see Figure 12.6). ■

12.4.4 The *n*-Jobs, Two-Machine Scheduling Problem

The same set of assumptions listed earlier for the *n*-jobs, one-machine problem are also usually invoked for the two-machine problem. In addition, the following characteristics are often assumed in deriving optimal schedules for the two-machine case: (1) the job shop is a *flow shop* in which all jobs follow the same routing or technological order from one machine to the next (as opposed to a *randomly routed job shop*), and (2) no *passing* is allowed among the jobs—that is, the order in which jobs are processed on the first machine must be maintained on all subsequent machines.

A procedure was developed by S. M. Johnson [11] to generate the schedule that minimizes the overall makespan or completion time of the schedule. Johnson's procedure is as follows:

1. Select the job with the smallest processing time on either machine.

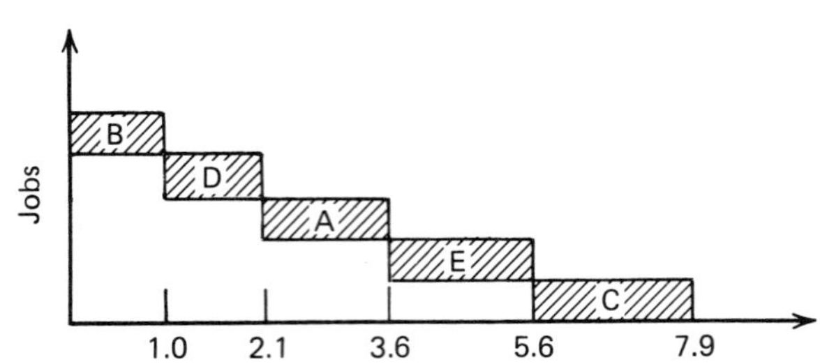

Figure 12.6 Flow Times for the Optimal Sequence for Example 12.3

2. If this time is associated with the first machine, put this job as the first job to be processed.
3. If the time is associated with the second machine, put the job as the last job to be processed.
4. Remove this job from the candidate list of jobs to be sequenced, and repeat steps 1 to 4 until all jobs have been scheduled. Ties are broken arbitrarily.

Example 12.4

Suppose that a particular job shop has five customer orders awaiting processing and that all five involve time on a metal-bending machine and at a subsequent heat-treating facility. The times are as follows.

	Job				
	A	B	C	D	E
Bending (hr)	2	5	1	2	5
Heat treating	3	4	2	4	6

Johnson's procedure would involve first choosing job C and placing it as the initial job in the optimal sequence since the 1-hr processing time occurred on the first machine. Next, job D (or alternatively, job A) would be selected and assigned as the second job in the sequence. Job A would then be selected and placed in the last position of the optimal schedule. Job B would be placed next to last; and, finally, job E would be placed as the third job behind job D. The optimal sequence is therefore C, D, E, B, and A with an overall completion time of 21 hr and a mean flow time of 4.2 hr.

■

12.4.5 Scheduling in Dynamic Job Shops

In large job shop situations involving many machines and jobs with randomly routed flows and dynamic job arrivals, Gantt charts are commonly used. The Gantt chart's main function in these cases is to serve as a snapshot model of the present and planned schedule. It can easily be updated as new jobs arrive and are worked into the existing schedule.

A Gantt chart functions only as a device to *display* the schedule. Additional tools and techniques are needed to assist in developing or *generating* the schedule itself. In a complex, dynamic scheduling situation, heuristic guidelines, called *dispatching rules*, are often used for this purpose. Dispatching rules function as criteria by which priorities are put on each job under consideration for processing at the moment. A commonly used dispatching rule is the SPT rule discussed earlier. If, for example, the scheduler had five jobs to be routed through the shop as of Monday morning, he or she could assign the top priority for access to machines to the job with the smallest overall processing time remaining.

There are numerous other dispatching rules that have been used and experimented with in various job shop situations. *First-come-first-serve* (FCFS) is a very common rule, along with *due-date*, which assigns jobs priorities equal to their promised completion date (jobs with lowest priority numbers are processed first). Others are *least-work-remaining*, in which the job priority equals the sum of the processing times for all operations not yet performed for that job, and *work-in-next-que*, in which the job priority is dependent on the backlog of work currently waiting at the next machine to be visited by that job based on the technological ordering of the job's operations. Even though it is one of the simplest dispatching rules, the SPT strategy is reported to be one of the most effective rules in terms of measure of performance such as mean flow time [6].

An alternative to the use of heuristic dispatching rules in dynamic job shop situations is the use of queueing theory. The flow of jobs through the shop is modeled with job arrivals, servers (machines) in series or parallel, and waiting lines between servers (see Chapter 9 for a discussion of queueing theory). However, only limited closed-form analytical results are available to assist in generating optimal schedules due to the mathematical complexities involved in having to model the job shop as a network of interrelated queues. A discussion of this approach can be found in Reference 7.

12.4.6 Other Operations Scheduling Problems

As mentioned earlier, there are many practical decision problems outside the realm of a job shop that are classified as scheduling problems. Some of these use the same analytical tools as used in job shop scheduling and sequencing problems, whereas others are more amenable to alternative approaches such as mathematical programming. An example of a problem that falls in this last category is that of scheduling ocean-going bulk chemical tankers.

An illustration of this ship scheduling problem can be found in the Ethyl Corporation, in which four bulk tankers are loaded with five or six gasoline additive products in Houston and are routed all over the world to make deliveries to bulk warehouses. A schedule is made up for deliveries 12 to 14 months into the future in anticipation of product sales out of each warehouse, and is revised monthly in view of updated information on sales forecasts and ship positions. The monthly scheduling decision involves determining the voyages for each ship (that is, the warehouses to be visited and the sequence in which they are to be called on), the type and amount of product to be loaded and delivered at each stop in each voyage, and the departure time for each stop. The criterion used in solving this scheduling problem is the combined cost of operating the ships, including fuel cost, port expenses, canal fees, and in-transit and warehouse inventory carrying costs.

An interactive, computerized system has been developed to assist the scheduling decision maker in generating optimal or satisfactory solutions. The basic components of the computer system are (1) a routine to evaluate the costs and feasibility of any given schedule, (2) an interactive module to guide the scheduler through user-suggested changes to the voyages, (3) a routine to generate a *feasible* schedule from scratch, and (4) a routine to generate the *optimal* schedule based on a given feasible starting schedule. This optimization routine is an adjacent extreme point search algorithm much like the Simplex method used to solve linear programming problems.

Complete development of the system would likely take Ethyl 4 to 5 man years. However, at the midway point of development, a working system had been installed with favorable results. Similar computerized systems are being used by several of the major oil refining companies, as well as by airlines in scheduling passenger flights.

12.5 SUMMARY

This chapter is concerned with one of the focal points of applications of IE/OR techniques, the topic of production planning and control. PP&C involves making decisions as to how work facilities are to be used to meet demand. PP&C problems arise in both the private and the public sector. They involve planning at the macro or plant level as well as at the micro or worker level. Analysis of PP&C problems may involve quantitative OR techniques such as linear programming and regression analysis, as well as more traditional, pragmatic tools such as a Gantt chart. In either case, PP&C relies heavily on fundamental IE time standards and documents such as route sheets for necessary input information before any analysis can take place.

Although the issues involved in PP&C range from inventory control to assembly-line balancing, only three major PP&C areas used are discussed in this chapter: forecasting, aggregate production planning, and scheduling. A more complete treatment of PP&C can be found in the various production control books referenced at the end of this chapter.

Forecasting involves predicting the value to be realized at some particular point in the future for some variable of interest. This variable could be almost anything, from a cost parameter such as raw material cost to the demand for a particular product. Subjective approaches to forecasting are very common, although there are situations in which quantitative techniques can lead to more accurate predictions. Exponential smoothing is an example of a type of quantitative approach often used in forecasting. Here, the variable being forecast is modeled as a time series, with its value being a function of the length of time from some base period such as today.

Aggregate production planning is concerned with the decision of the extent to which each available production option, such as overtime or subcontracting, should be used in each of several future time periods in order that the output requirements of a plant or other major facility be met. This decision results in a production plan that may call for a constant production rate from period to period, or for a rate that changes from one period to the next in an effort to map changing output requirements. A variety of quantitative approaches have been developed to address this aggregate planning problem. A mathematical programming approach suggested by Bowman is discussed in this chapter. It involves modeling the problem in such a way that a transportation linear programming model is formulated. Other approaches mentioned included the HMMS unconstrained optimization procedure, which results in a set of linear decision rules.

Scheduling was discussed in this chapter in the context of routing jobs through a job shop. However, scheduling is a decision problem issue that encompasses many situations and environments beyond job shops, such as routing or planning solid waste pickups or bulk ocean tanker deliveries. Many of the same models and techniques used to develop schedules in job shops are also utilized in analyzing non-job-shop scheduling problems. An example of this is the use of dispatching rules to set routing priorities for jobs or operations presently awaiting processing. Dispatching rules, such as FCFS or work-in-next-que, are primarily used in complex, dynamic scheduling situations. Gantt charts and other visual display models are also commonly used in schedule planning and monitoring in such situations. In those instances in which a static, less complex model is appropriate, optimal results can be obtained through an exact mathematical analysis. For example, under appropriate assumptions, the optimal sequence of jobs can be derived for the n-jobs, one-machine problem and the n-jobs, two-machine problem. The SPT rule and Johnson's procedure are the techniques used to develop these optimal sequences.

REFERENCES AND SUGGESTED READING

1. E. H. Bowman, "Production Scheduling by the Transportation Method of Linear Programming," *Operations Research*, **3,** No. 1, 100–103 (1956).
2. R. G. Brown, *Smoothing, Forecasting, and Predicting of Discrete Time Series*, Prentice-Hall, Englewood Cliffs, NJ, 1963.
3. E. S. Buffa, *Modern Production Management*, Wiley, New York, 1973.
4. E. S. Buffa and W. H. Taubert, *Production-Inventory Systems: Planning and Control*, Irwin, Homewood, IL, 1972.
5. R. B. Chase and N. J. Aquilano, *Production and Operations Management*, Irwin, Homewood, IL, 1973.
6. R. W. Conway, "An Experimental Investigation of Priority Assignment in a Job Shop," RM-3789-PR, The Rand Corp., Santa Monica, CA, 1964.
7. R. W. Conway, W. L. Maxwell, and L. W. Miller, *Theory of Scheduling*, Addison-Wesley, Reading, MA, 1967.
8. F. Haussman and S. W. Hess, "A Linear Programming Approach to Production and Employment Scheduling," *Management Technology*, **1,** No. 1, 46–51 (1960).
9. C. C. Holt, F. Modigliani, and H. A. Simon, "A Linear Decision Rule for Production and Employment Scheduling," *Management Science*, **2,** No. 1, 1–30 (1955) [see also *Mgt. Sci.*, **2,** No. 2, 159–177 (1956)].

10. L. A. Johnson and D. C. Montgomery, *Operations Research in Production Planning, Scheduling and Inventory Control*, Wiley, New York, 1974.

11. S. M. Johnson, "Optimal Two and Three Stage Production Schedules with Setup Times Included," *Naval Research Logistics Quarterly*, **1,** No. 1 (1954).

12. D. C. Montgomery, "An Introduction to Short-Term Forecasting," *Journal of Industrial Engineering*, **19,** No. 10, 500–503 (1968).

13. F. G. Moore and R. Jablonski, *Production Control*, McGraw-Hill, New York, 1969.

14. J. L. Riggs, *Production Systems: Planning, Analysis, and Control*, Wiley, New York, 1970.

15. M. K. Starr, *Systems Management of Operations*, Prentice-Hall, Englewood Cliffs, NJ, 1971.

16. Nicholas T. Thomopoulous, *Applied Forecasting Methods*, Prentice-Hall, Englewood Cliffs, NJ, 1980.

REVIEW QUESTIONS AND PROBLEMS

12.1 Define or explain each of the following, giving examples where appropriate:

(a) PP&C
(b) Forecasting
(c) Exogenous-variables approach to forecasting
(d) Time-series approach to forecasting
(e) Exponential smoothing
(f) Double exponential smoothing
(g) Smoothing statistic
(h) Aggregate planning
(i) A production plan
(j) Bowman's approach to aggregate planning
(k) HMMS model
(l) Operations scheduling
(m) Gantt chart
(n) Mean flow time
(o) Makespan
(p) Flow shop
(q) Johnson's procedure
(r) SPT
(s) Dispatching rules

12.2 Discuss how and where OR techniques and traditional IE tools overlap in PP&C.

12.3 Derive the estimate of the single parameter in the constant time-series model as is used in simple exponential smoothing. That is, give the detailed steps involved in going from Eq. (12.5) to Eq. (12.6).

12.4 Explain the logic behind the statement that "single exponential smoothing is a procedure that updates the smoothed statistic S_T proportionally to the difference between the smoothed statistic and the data or observation."

12.5 A common time-series forecasting procedure used for constant models is the "N-period moving average." In this approach, the estimate of the constant model's single parameter is

$$b_0 = \left(\frac{1}{N}\right) \sum_{t=T-N+1}^{T} y_t$$

where y_t is the tth actual observation for $t = 1, 2, \ldots, T$, and N is the number of the most recent periods of data we wish to include in the average. In this approach, the average age of the data used to estimate b_0 is:

$$\left(\frac{1}{N}\right) \sum_{K=0}^{N-1} K = \frac{(N - 1)}{2}$$

Derive the equation that gives the value of smoothing constant α in simple exponential smoothing that leads to an estimate of b_0 in which the average age of the data used is the same as with the moving average procedure. (*Hint:* the weight given to data K periods ago is $\alpha(1 - \alpha)^K$ in simple exponential smoothing whereas it is 1 in moving average.)

12.6 A local Ford dealer is interested in forecasting his sales of Thunderbird cars in the next three months. His records indicate sales in the last eight months have been: 5, 8, 2, 6, 10, 9, 9, and 12 (12 is the number of cars sold in the current month, July)

(a) Develop forecasts for the next three months—August, September and October—with a 4-period moving average developed on a stationary basis at the end of July.

(b) Develop the same forecasts but using simple exponential smoothing with a smoothing constant of 0.4 and a smoothing statistic value of 10 for June.

(c) Develop the same three forecasts using a two-period moving average on a *rolling* basis, rather than stationary at the end of July, if actual sales in August turn out to be 9, and 14 in September and 13 in October. (That is, August sales are forecast at the end of July, September sales at the end of August, and October sales at the end of September.)

(d) Develop the same forecast as in c using the exponential smoothing procedure defined in b.

(e) Discuss the relative merits of the above forecasting methods as opposed to other possible approaches for this particular situation.

12.7 Complete Table 12.1 for December and January.

12.8 Develop simple exponential smoothing forecasts for the problem in Example 12.1 using a smoothing constant of 0.6 rather than 0.2 as was used in the example. Discuss the sensitivity of forecast errors and of reaction time to process changes to the magnitude of the smoothing constant.

12.9 Discuss the relative merits of (1) subjective forecasts, (2) an exogenous-variables approach, and (3) a time-series approach in forecasting each of the following:

(a) Prediction of fires throughout a city.

(b) The number of fatalities each year in automobile accidents.
(c) The need for cash on hand at a bank each week.
(d) The output for a plant producing wire rope.
(e) The monthly production at a gasoline refinery.
(f) The attendance each game for a professional baseball team.

12.10 Rewrite model (12.8) so that it has the appearance of a transportation problem model.

12.11 Modify model (12.8) to allow back orders.

12.12 Give the numerical version of model (12.8) for the aggregate planning problem addressed in Example 12.2.

12.13 An airline maintenance shop has a capacity to overhaul 50 airplane engines per month. An extra 10 engines per month can be overhauled by an outside subcontractor. Current forecasts call for the shop to turn out 35 overhauled engines next month, 50 engines the following month, and 40 each month thereafter. Shop management wishes to plan manpower requirements as well as cash-flow needs during the next three months and therefore needs a monthly production plan for internal shop output as well as subcontracting usage. The standard cost to overhaul an engine in the shop is \$3500, whereas the cost to have the subcontractor handle the overhaul is \$4000. The shop estimates that it costs them about 10 percent of the engine's standard overhaul cost to carry an overhauled engine in inventory for one month. If there are presently 5 overhauled engines in inventory, and if shop management desires to have 5 engines in inventory at the end of *each* month, determine the least-cost production plan for the next three months.

12.14 What sign (of magnitude) would you expect on the parameters b and c in Eq. (12.9)? Why?

12.15 Discuss the strong points and weaknesses of Bowman's LP approach to the aggregate planning problem.

12.16 By investigating the open literature (i.e., technical journals, books, etc.) find and describe two approaches to the aggregate planning problem that were not mentioned in this chapter.

12.17 How are Gantt charts used in job shop scheduling situations?

12.18 Identify at least three traditional IE tools (including charts and diagrams) that can provide necessary input information for the scheduling activity in a job shop. Describe how this information is used in the scheduling activity.

12.19 The International Electronics Company (IEC) manufactures custom-made electric motors in essentially a job shop environment. A common component in all motors made by IEC is a flexible shaft coupling. The route sheet or operations sheet for this part is shown in Figure 12.7. The only difference in this part from one customer's order to the next is the specification on dimensions, which in turn affects the standard machining times.

The shop scheduler for IEC presently (beginning of the first week in June) has 4 customer orders that have to be produced. Because there are currently no shafts in inventory, the dispatcher must immediately schedule the processing of 4 shafts through the shop. As can be seen from the route sheet, this requires scheduling the work at 5 work centers: lathe (Operations 1 and 2), grinding (Operations 3 and 7), milling (Operation 4),

OPERATION SHEET

PART NO.	**ISSUE**	**PART NAME**
A-12963	A	FLEXIBLE SHAFT COUPLING

MODEL Standard Part

MATERIAL C-1018 Steel $1\frac{1}{2}$-in. diameter

			Time	(hr.)
OPER.	**DEPT.**	**OPERATION**	**SETUP**	**MACHINE**
1	Engine lathe	Hold in three jaw chuck	0.3	0.8
		Centerdrill		
		Drill 15/32 × 7/8 in. deep		
		Bore 0.500 $^{-0.000}_{-0.010}$ diam. × 5/8 in. deep		
		Face end of part square		
		Turn 0.625 $^{-0.000}_{-0.003}$ diam. × 1/2 in. long		
		Cut off 29/32 long (7/8 1/32 extra)		
2	Engine lathe	Hold in three jaw chuck on 0.625 diam.	0.3	0.6
		Face—Hold 7/8 in. dimension		
		Turn 1.000 $^{-0.000}_{-0.001}$ diam. × 9/32 in. long		
		Bore 0.8665 diam. to 0.855 $^{-0.003}_{-0.003}$ × 9/32 deep		
		Bore 3/4 diam. × 1/32 deep		
3	Internal grinder	Hold in collet on 0.625 diam.	0.5	0.1
		Grind 0.8665 $^{-0.0000}_{-0.0004}$ diam. × 9/32 deep		
4	Milling machine	Hold in dividing head on 0.625 diam.	0.5	0.2
		Mill 1 1/4 in. square (4 cuts)		
5	Burring department	Deburr (milling burrs)	—	0.03
6	Jig boring	Hold in vise on 1 1/4 square	0.5	0.2
		Centerdrill and drill 4 #31 holes		
7	Thread grinding	Hold on 0.8665 mandrel	1.5	0.2
		Grind 5/8 -27 R.H. -3 thread × 3/8 long		
8	Burring	Deburr (drilled holes and threads)	—	0.02
9	Outside processing	Cadmium plate costs $0.15 each	—	—
10	Inspection	Inspect	—	0.01

Figure 12.7 Operations Sheet for a Flexible Shaft Coupling

burring (Operations 5 and 8), and boring (Operation 6). The standard times for the 4 different orders are as follows:

	Customer Order No.			
	A	B	C	D
Machine time, Operation 1 (hr)	0.8	1.5	2.0	0.8
Machine time, Operation 2 (hr)	0.6	1.0	0.6	0.8
Machine time, Operation 4 (hr)	0.2	2.0	1.0	1.5

All other times are the same for all 4 jobs as is shown in the route sheet in Figure 12.7. The technological order of processing the required work operations is also the same for all jobs.

(a) Develop a Gantt chart that reflects the processing of the 4 shafts at the 5 work centers if the dispatcher uses the following schedule sequence: B, C, A, D. (Use $\frac{1}{2}$-hour intervals on the horizontal axis of the chart; schedule all operations for job B before scheduling operations for job C, etc.)

(b) Determine the makespan and the mean flow time of the schedule used in part (a).

(c) Find, if possible, an improved schedule, and give its associated Gantt chart.

12.20 For the situation described in Problem 12.19, suppose that the dispatcher is interested only in loading the milling machine. Under the criterion of average flow time, what would be the optimal sequence for processing the four shafts?

12.21 For the situation described in Problem 12.19, suppose the dispatcher is interested in loading each of two engine lathes in the lathe department with the four shaft jobs. His strategy is to assign Operation 1 to lathe I for all four jobs and to assign Operation 2 to lathe II for all jobs. All jobs must maintain the same processing sequence, with Operation 1 preceding Operation 2, and the same sequence of jobs used at lathe I must also be used at lathe II. Find the schedule for processing the four jobs that gets them all out of the lathe department as soon as possible.

12.22 Five R&D requests are to be handled by a certain specialist in the research group of a particular chemical manufacturer. The estimated times to complete each of the 5 projects along with their due dates are given below. Determine the priority that should be given to each job in order that the average request is delayed as little as possible. Calculate the schedule's makespan, mean flow time, and mean lateness.

	Project				
	A	B	C	D	E
Time required (wk)	10	6	8	15	8
Due date (weeks from today)	20	30	28	25	25

12.23 Four jobs are to be processed on two machines. Processing times (hr) are given below. Apply Johnson's rule to determine the optimal schedule. Are there alternative schedules just as good?

	Job			
	A	B	C	D
Machine I	2	3	3	5
Machine II	1	4	3	2

Chapter 13

Inventory Control

13.1 INTRODUCTION

Inventories are idle goods or assets. However, its "idleness" does not mean that an inventoried item is not serving a vital purpose. For instance, in production planning, inventories provide a source of goods that can be used to satisfy customer demands rather than having to use plant capacity in that time period to produce those same goods. Inventories can also be used as a buffer to meet unexpected demands while additional stock is on order, as well as a mechanism to take advantage of periods of low raw material prices, or to provide a wide range of items for display to customers.

Most businesses have one or more inventory situations that they must manage on a day-to-day basis. A manufacturing organization typically has an inventory of raw material, an inventory of finished goods, and several different inventory points of in-process units. An oil refinery usually has an inventory of crude oil, as well as storage of finished gasoline and other products, and inventories in the pipeline system throughout the refinery. Management of any of these different types of inventories involves monitoring the input and withdrawals of inventoried items, as well as making decisions as to the best means of replenishing the inventory.

13.1.1 Operation of an Inventory System

The day-to-day operation of a typical inventory situation involves two basic types of transactions: withdrawals of stock, and resupply or restocking of a batch or lot of stock. As an example, consider the operation of the inventory system for stocking a particular-size fan belt in a retail auto parts store. On Monday morning there might be, say, 16 belts on hand as stock. During the course of the week, the level of belts on hand will fluctuate as customers purchase ("demand") one or more belts, and one or more belts are obtained from a resupply vendor and put into stock as replenishment. A typical week of operation is shown graphically in Figure 13.1. Notice that there are several customers that purchase a single belt during Monday, no demand Tuesday,

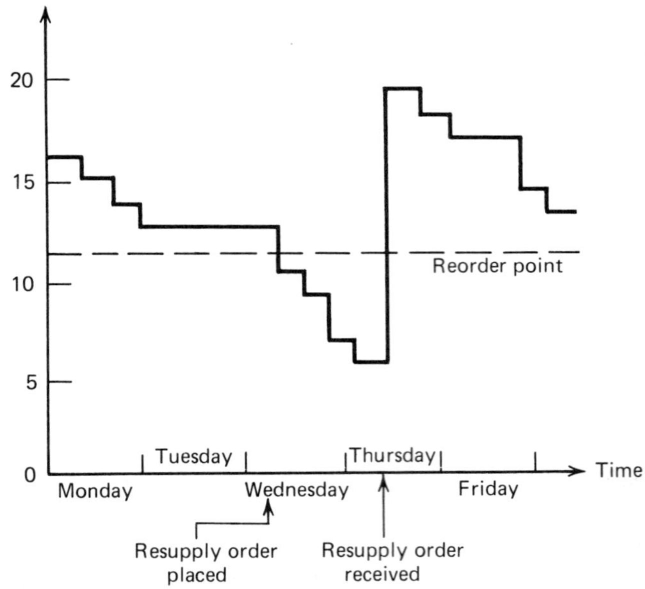

Figure 13.1 Illustration of the Fluctuation of the Level of an Inventory System

and a two-belt purchase by a customer during Wednesday. In addition, a dozen belts are put into stock on Thursday after having been ordered from the resupply vendor on Wednesday. The time interval between the placement of the resupply order and its arrival at the store is called the resupply "lead time."

13.1.2 The Meaning of Inventory Control

Inventory control is the activity of managing a given inventory situation. As indicated earlier, this normally involves maintaining an accounting system to monitor the transactions that affect the level of the inventory, and making inventory replenishment decisions. The accounting or record-keeping system might be manually maintained, or it may be computerized. In any event, the purpose is the same—to monitor inventory transactions. A typical record-keeping system is the "Kardex file" in which a 5×8 index card is maintained for each inventoried item. All inventory transactions concerning an item are recorded on the item's card. An example of such a card is shown in Figure 13.2.

There are usually two basic replenishment decisions that are made periodically: when to reorder stock, and how many units to reorder when that order is placed. Therefore, inventory control not only involves bookkeeping or accounting, but also making periodic decisions as to reorder times or points and reorder quantities. In the remainder of this chapter, we will concentrate on reorder decisions as opposed to inventory bookkeeping. Specifically, we will discuss the derivation of optimal reorder points and reorder quantities for any given inventory situation.

Date	Received No.	Received Qty	Unit Cost	Issued No.	Issued Qty	Bal OH	$ Amt
INVENTORY			3.16			8	
8-5			3.16	Scrap	1	7	
8-8			3.16	E1804	5	2	
8-10			3.16	Scrap	1	1	
8-27	52329	26	3.50			27	
9-5			3.50	E1902	12	15	
9-12			3.50	E1940	26	(11)	
9-13			3.50	E1941	4	(15)	
9-24	52371	79	3.25			64	
	avg	Cost	3.25			64	208.00
10-2			3.25	E1860	20	44	
10-8			3.25	E1871	10	34	
10-15			3.25	E2000	10	24	
10-16			3.25	E2011	12	12	
10-16			3.25	E2012	10	2	
10-18			3.25	E2013	2	0	
11-5			3.25	E2015	8	(8)	
11-6	52470	80	3.30			72	
11-15			3.30	Scrap	2	70	

Date	Received No.	Received Qty	Unit Cost	Issued No.	Issued Qty	Bal OH	$ Amt
10-21	52469	50	3.40			120	
11-21			3.40	E2110	5	115	
11-29			3.40	Maint.	2	113	
12-2			3.40	E2112	30	83	
12-5			3.40	E2115	42	41	
12-12			3.40	E2200	10	31	
12-12			3.40	Scrap	6	25	
12-21			3.40	E2221	12	13	
12-22	52475	50	3.27			63	

Notes

Part No.	Description	Dwg No.	Acct No	Unit of meas	Location	Lead Time	Min Level
3502-01	TRANSFORMER CT1	A51		EACH	STOCK RM	3 days	25

Figure 13.2 Illustration of a Kardex File

As is the case with quantitative analysis of any decision problem, the first step in deriving optimal inventory reorder points and quantities is to formulate a mathematical model to represent the decision at hand. Such models typically include an objective function of minimizing the costs that are directly affected by changes in the inventory reorder policy, and any constraints on this policy due to limited resources such as floor space for storage. The resulting model is therefore either an unconstrained optimization problem or a mathematical programming problem, depending on whether or not any constraints are explicitly included in the model.

An optimal inventory policy can then be derived from this model by using an appropriate solution procedure or algorithm. For instance, if the model for selecting a reorder point and reoder quantity for the inventory of fan belts illustrated by Figure 13.1 turned out to be a linear programming problem, then the Simplex procedure could be used to find the optimal reorder point and quantity. However, inventory models usually are unconstrained so that classical optimization procedures are normally used in their solution.

13.2 CLASSIFICATION OF INVENTORY SITUATIONS

Although every inventory situation is somewhat unique, it is possible to identify various aspects or characteristics that many situations have in common. This allows a single model and analytical procedure to be designed to derive the framework of the optimal reorder strategy that is applicable to all inventory situations in the same class (that is, those with similar characteristics). The primary characteristics used to classify inventory situations are briefly described below.

13.2.1 Costs Considered

The types of costs explicitly modeled and included in the analysis constitute one way to classify inventory situations. The basic costs normally considered for inclusion in the analysis are the cost of placing an order, the cost of carrying inventory from one day to the next, and the cost of any shortages incurred (that is, units demanded but unavailable). For one particular inventory situation, the relevant cost of a shortage may be the lost profit from a sale; in another situation, shortages may not be allowed so that no shortages cost need be included in the analysis.

13.2.2 Inventory Replenishment Process

The rate of the replenishment or input of the reorder quantity has a major effect on the analysis of an inventory situation. The replenishment rate will be either infinite or some finite value, such as 10 units per hour. If the rate is infinite, the units constituting the reorder quantity are put into inventory all together as a batch or lot "instantaneously" upon arrival. This is characteristic of the replenishment of a retail sales item, or of raw material at a plant. In contrast, some inventory situations involve placing the units constituting the reorder quantity into inventory one at a time. In this case, replenishment takes place at a finite rate over time. A typical example of such situations is an in-process inventory in a plant, such as the buildup of partially assembled radios between two stations on an assembly line. The mathematical models and solution procedures used to derive the optimal reorder strategy for inventory situations with a finite replenishment rate are different from those used for situations with an infinite rate.

13.2.3 Stochastic Nature of the Inventory Situation

Many of the attributes of a typical inventory situation are stochastic in nature rather than occurring with certainty. For example, the demand on the inventory is usually a random variable even though we may have an excellent forecast of the value the demand will take on. The techniques used to model and analyze inventory systems in which one or more factors are explicitly assumed to be random variables are more complicated than those used for entirely deterministic situations. When a stochastic approach is used, the two factors most commonly modeled as random variables are demand and resupply lead time.

13.2.4 Review Policy

There are two basic policies used in monitoring or reviewing the actual inventory level so as to determine if and when a resupply order should be placed. The first is the *continuous review policy*, in which the status of the inventory level is constantly monitored. As soon as the level reaches the predetermined reorder point, the reordering process is initiated. The other policy is the *periodic review policy*, in which the status of the inventory system is monitored only at specific, prearranged points in time, such as the end of each day or week. If the inventory level is below the reorder point at the time it is reviewed, a resupply order would be placed. The major difference between the two review policies is that in a continuous review system the decision as to whether or not to place an order is made after each and every demand transaction, whereas in a periodic review situation, this decision is made only at certain predetermined points.

13.2.5 Item Interdependence

Often the items being stored as inventory interact with one another in a variety of ways. For instance, they may compete for limited floor space, or they may be linked to one another as interrelated subassemblies. In such cases, the decision made in replenishing the inventory of one item affects the cost of carrying or resupplying another item. To determine least-cost, feasible inventory policies for these items means that they must all be analyzed together—as a system. On the other hand, when items are independent of one another they can be analyzed by themselves. Most "lot sizing" practices in use in industry treat items as if they were independent.

When the analysis of an inventory problem is reported in a journal or in internal company documentation, the inventory situation is usually described in terms of the classifications given above. For example, a company's chief IE might be documenting his efforts in improving the reordering system for raw materials in the firm's packaging department, and report that the inventory situation in question was appropriately modeled as "involving instantaneous input, probabilistic demand and lead time, with ordering and carrying costs being the only relevant cost factors." In the next two sections, we investigate two relatively common inventory situations that can be described in terms of the above classifications. Both involve a single item having no interaction with other parts being stocked.

13.3 WILSON LOT SIZE MODEL

Perhaps the most commonly used inventory model is the Wilson lot size model, or the simple economic lot size model (the "EOQ" model). This mathematical model assumes that the underlying inventory situation has the following characteristics.

1. The relevant costs are carrying cost and ordering cost.
2. No shortages are permitted.
3. The replenishment rate is infinite.
4. All aspects of the inventory system are deterministic.
5. A continuous review system is used for reorder purposes.

In this section we develop a model to represent the decision problem implied in deriving an optimal reorder strategy for inventory situations with the above characteristics. As mentioned earlier, this model (called the Wilson lot size model) is the most commonly used inventory model. This is not to say that the inventory situation represented by this model is the most commonly occurring situation in practice; rather, whatever the actual inventory situation being analyzed, it is very often *modeled* as if it has the same characteristics as the Wilson lot size model situation. Hence, the actual situation being addressed might involve, say, a probabilistic lead time, but the analyst might treat lead time as deterministic by using its expected value as *the* (constant) value of lead time. The frequent use of the Wilson lot size model assumptions is usually a result of an implicit trade-off between the model's accuracy and its utility. This is the same trade-off found in any problem solving–modeling situation.

13.3.1 Graphical Model

The first part of the analysis of an inventory problem is usually the construction of a graphical illustration of how the inventory level typically fluctuates over time. Such a graph is given in Figure 13.3 for the Wilson lot size model situation. Notice that a cycle in the graph begins with the receipt of the previous order and ends right before the receipt of the next order. The events occurring during a cycle are exactly the same from one cycle to the next. Although this is normally not the case in actual practice, we are dealing here with a model or representation, one that is often very accurate. Based on this graph, we can readily calculate the average inventory level that occurs during a cycle and from that we can find the average inventory level during the year; this is the basis for determining the inventory carrying cost to include in our model.

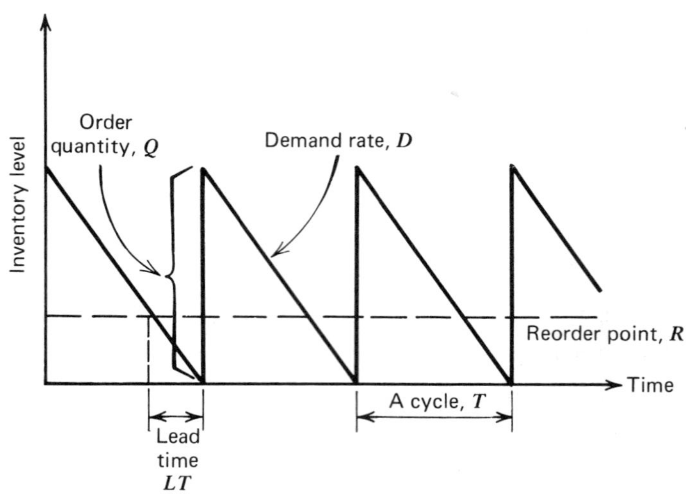

Figure 13.3 Graphical Model of the Inventory Situation Underlying the Wilson Lot Size Model

13.3.2 Math Model

The next step in the analysis is to formulate a mathematical model of the decision problem of finding the best values to use for the reorder point and reorder quantity in view of the actual costs and demand involved. In the Wilson lot size model situation, this normative model is the minimization of the average annual cost of carrying inventory plus the average annual cost of placing orders, without any restrictions or constraints on the feasible values for the reorder point and reorder quantity other than the fact that they both must be nonnegative.

The decision as to the best value for the reorder point R is independent of the decision on the reorder quantity Q. This is true in the Wilson lot size model situation because of the deterministic nature of the demand. In fact, the reorder point can be calculated directly from Figure 13.3. The reorder point is simply the inventory level required at the beginning of a lead time so that the inventory would be reduced to exactly zero at the end of the lead time. Therefore, if the lead time is LT years and the demand rate is D units per year, the reorder point is

$$R^* = (D) \times (LT) \tag{13.1}$$

This allows the inventory model to be expressed in terms of only one decision variable Q.

The carrying-cost component of the model is formulated as the average annual inventory level times the cost of carrying one unit of inventory for a complete year. The average level can be defined as the average number of inventory unit-years carried during a year period, and can be derived by using the graph in Figure 13.3.

To do this, first find the number of unit-years of inventory carried during a cycle. This is simply the area of the triangle that represents the inventory carried during the cycle, having a height of Q, a base of T, and a slope of D. Therefore, the area for a cycle is $\frac{1}{2}(Q) \times (T)$, or since slope $D = Q/T$, then the cycle's area can be expressed as $(Q^2/2D)$. Next, find the average number of cycles that will occur in a year. Since a cycle is T yr long, then there is an average of $1/T$ cycles in a year. Hence, the total area for a year, or the average number of unit-years carried, is $(Q^2/2D)(1/T)$, or $(Q^2/2D)(1/Q/D)$, or $(Q/2)$. Letting the cost of carrying one unit in inventory for a year be C_h, a model of the annual carrying cost is $(C_hQ/2)$ dollars per year.

The cost parameter C_h is expressed in dollars per unit of inventory. It is usually defined as the dollar value of an inventoried item times a percentage carrying charge. This percentage factor comprises components such as property-tax and insurance cost on inventoried goods, warehousing charges if any, and a cost for the money or capital tied up in the inventoried item. This last component, the cost of capital, is usually the predominant component. For example, the total carrying charge in a particular situation might be 25 percent, with the cost of capital being as much as 20 percent.

There is no universal agreement as to the exact definition or makeup of the cost of capital component. Usually, it is interpreted as an opportunity cost incurred for tying up capital in "idle" inventory as opposed to using that capital in an alternative use. Under this "alternative use" interpretation, the appropriate value of this cost factor would normally be the average rate of return achieved on investments in the organization. However, in assigning a value to this factor, the key question to ask is "if the inventory were significantly reduced (or, alternatively, increased), how would the funds be used and what would that use net the company?"

The second cost component of the decision problem model is the cost of placing orders. In the Wilson lot size model situation, ordering cost is defined as the average number of orders placed during a year times the cost of placing a single order. Referring to Figure 13.3, note that there is a cycle for every order placed. Therefore, the number of orders placed during a year is equal to the number of cycles occurring in a year, which is $1/T$, or from the geometry of the graph, D/Q. Letting C_0 be the cost of placing one order, the ordering cost component can be expressed as C_0D/Q.

The decision problem model or inventory model can now be written as:

$$\min_{Q}(\text{Total average annual cost}) = \text{Carrying cost} + \text{Ordering cost}$$

or

$$\min_{Q} TC = \frac{C_hQ}{2} + \frac{C_0D}{Q} \tag{13.2}$$

This normative model represents the decision problem of selecting the best inventory reorder policy, given that the characteristics of the inventory situation at hand match the assumptions of the Wilson lot size case as outlined earlier. When these assumptions are not met, a different (but similar) model is formulated. However, the same approach would be used in modeling the various cost components. For example, in developing the average annual carrying cost, we would use the inventory graph to find the area of inventory carried in a cycle and multiply it by the number of cycles occurring per year. The difference is that the resulting formulas would not be exactly the same as found for the Wilson lot size case, because the graph would likely be different.

13.3.3 Solution

Note that model (13.2) is an unconstrained optimization problem in which the decision variable is Q. The objective function is nonlinear in Q due to the second term in model (13.2). The solution can be found through classical calculus by finding the first derivative and setting it equal to zero:

$$0 = dTC/dQ = \frac{C_h}{2} - \frac{C_0D}{Q^2} \tag{13.3}$$

The optimal solution is then

$$Q^* = \sqrt{2C_0D/C_h} \tag{13.4}$$

This formula, (13.4), is the Wilson lot size (descriptive) model or Wilson lot size formula. It gives the optimal reorder quantity for a given set of values for the parameters C_0, C_h, and D. The optimal value of the reorder point R is found from Eq. (13.1).

Example 13.1

As an illustration of the use of the Wilson lot size model, consider the auto-parts-store example mentioned in Section 13.1.1. The store is interested in setting up a formal recording system for the fan belt in question. Based on records over the past 8 months, the store concluded that they can anticipate a monthly demand of around 50 belts. In addition, their records show that when a resupply order is placed with their vendor, it usually takes about 3 days to receive that order and put the new belts into stock.

If the belts cost the store \$3.50 each, and if the store uses an annual carrying cost rate of 30 percent and incurs \$5 expense whenever an order is placed and processed, then the optimal policy for it to follow is:

$$R^* = (50 \text{ belts/month}) \times (3 \text{ days}) \times (1 \text{ month/30 days}) = 5 \text{ belts}$$

and

$$Q^* = \sqrt{\frac{2(5)(50/\text{month})(12 \text{ month/yr})}{(0.30)(3.50)}} = 76 \text{ belts}$$

and the average annual cost of this policy is:

$$TC^* = (0.30) \times (3.50) \times (76/2) + (5) \times (50) \times (12/76) = \$79.37$$

Therefore, the least-cost policy for the store to follow is constantly to monitor the stock of belts and to place a resupply order for 76 belts when there are 5 or fewer on hand. If this policy is followed, the average annual cost of carrying these belts will equal the average annual cost of placing orders. That is, $(0.30)(3.50)(76/2) \cong (5)(50)(12)/76$.

Suppose that the store's vendor only sells in lots of, say, 20 belts. In this case, the optimal reorder quantity of 76 is infeasible. If $Q = 80$ were used instead, notice that the annual cost would be \$79.50 as opposed to \$79.37, an increase of only 0.16 percent. This lack of sensitivity in the total cost of the inventory system to changes in the value of Q actually used is true in general. That is, the convex curve resulting from graphing TC as a function of Q is very flat at the minimum point Q^*. In fact, for the Wilson lot size model, it can be shown that a 200 percent increase or decrease in the Q actually used as compared to Q^* causes the average annual cost to increase by only 25 percent [2]. ■

13.4 PRODUCTION LOT SIZE MODEL

In the Wilson lot size model, once a resupply order is initiated and received, it is assumed to be placed into stock as a complete batch, or at an "infinite input rate." This is characteristic of retail situations or inventorying raw material. However, as mentioned earlier, many situations, such as in-process inventory, involve putting the order quantity or lot into stock at a finite rate. For instance, if the second work station on an assembly line has an output rate of 5 units per hour, then the replenishment of the inventory in front of the *third* work station will take place at a finite rate of 5 per hour, rather than as an entire day's supply all at once.

The model to be developed for this inventory situation is based on the same characteristics as listed earlier for the Wilson lot size model, except for the use of a finite input rate. In addition, the same approach will be used in developing and solving an inventory model for this case as was done for the previous model.

13.4.1 Graphical Model

A graphical model of this production lot size situation is given in Figure 13.4. Notice that it is similar to Figure 13.3, except that the cycle is broken into two parts, a production cycle and a demand cycle. During the production phase, units are generated and placed into stock at a rate of P units/year and, at the same time, are consumed or demanded at a rate of D units/year. Production ends after T_p years, and only demand takes place in the remaining T_d years of the cycle. Notice that if P is very large, this production lot size model degenerates into the Wilson lot size case.

13.4.2 Math Model

To derive the average annual carrying cost, we must first compute the unit-years of stock carried in any one cycle. This is simply the area under the inventory graph for the production cycle plus the area for the demand cycle. That is:

$$\text{Area} = \left(\frac{1}{2}\right)(T_p)(h) + \left(\frac{1}{2}\right)(T_d)(h) = \left(\frac{1}{2}\right)(h)(T_p + T_d)$$

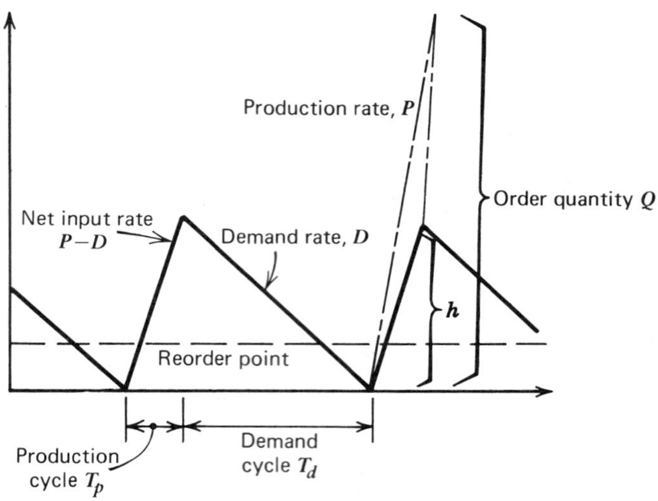

Figure 13.4 Graphical Model of the Production Lot Size Inventory Situation

The maximum on-hand inventory h can be found as follows: From the triangle formed in the demand cycle, note that $T_d = h/D$. From the triangle formed in the production cycle, $T_p = h/(P - D)$. During the overall cycle, $T = Q/D$. Therefore:

$$T = T_p + T_d$$

or

$$\frac{Q}{D} = \frac{h}{(P - D)} + \frac{h}{D}$$

or

$$\frac{Q}{D} = \left(\frac{P}{D(P - D)}\right)h$$

or

$$h = \left(\frac{D(P - D)}{P}\right)\left(\frac{Q}{D}\right) = \left(1 - \frac{D}{P}\right)Q$$

This implies that:

$$T_p = \frac{h}{(P - D)} = \frac{(1 - D/P)Q}{(P - D)} = \frac{Q}{P}$$

and

$$T_d = \frac{h}{D} = \frac{(1 - D/P)Q}{D}$$

and

$$T = T_p + T_d = Q\left(1 - \frac{D}{P}\right)\left[\frac{1}{(P - D)} + \frac{1}{D}\right]$$
$$= Q\left(1 - \frac{D}{P}\right)\left[\frac{P}{D(P - D)}\right] = \frac{Q}{D}$$

so that the area per cycle is:

$$\text{Area} = (1/2)(1 - D/P)(Q)(Q/D)$$

The average number of overall cycles per year is $1/T$ or D/Q. Therefore, the average annual carrying cost is:

$$\left(\frac{C_hQ}{2}\right)\left(1 - \frac{D}{P}\right)\left(\frac{Q}{D}\right)\left(\frac{D}{Q}\right)$$

or

$$\left(\frac{C_hQ}{2}\right)\left(1 - \frac{D}{P}\right)$$

Since the average number of orders placed per year is D/Q, the ordering cost is C_0D/Q. The total model can then be written as:

$$\min_{Q} TC = \left(\frac{C_hQ}{2}\right)\left(1 - \frac{D}{P}\right) + \frac{C_0D}{Q} \tag{13.5}$$

To compute the reorder point, we must recognize that R might fall in either the production cycle or the demand cycle. That is[1]:

$$\begin{aligned} R^* &= (LT) \times (D) && \text{if } LT \le T_d \\ &= (T - LT)(P - D) && \text{if } LT > T_d \end{aligned} \tag{13.6}$$

13.4.3 Solution

Model (13.5) can be solved using the necessary conditions from classical calculus. Specifically:

$$0 = \frac{dTC}{dQ} = \frac{C_h}{2(1 - D/P)} - \frac{C_0D}{Q^2}$$

and solving for Q:

$$Q^* = \sqrt{2C_0D/C_h(1 - D/P)} \tag{13.7}$$

This result holds as long as $(1 - D/P) > 0$, that is, as long as the production rate is greater than the demand rate.

Example 13.2

Suppose that the vendor in Example 13.1 delivered fan belts to his various outlets in a succession of deliveries so that a given outlet's order might "dribble in" in two or three batches rather than as a single lot. For the store in question, assume that typically they receive 20 belts a day from the time they place an order with the vendor until all units are in stock.

Since model (13.5) is appropriate for this case, the least-cost policy for the store is now

$$Q^* = \sqrt{\frac{2(5)(50/\text{month})(12\ \text{month/yr})}{(0.30)(3.50)[(1 - (50/\text{month})/(20/\text{day})(30\ \text{day/month})]}}$$

or

$$Q^* = 79 \text{ belts}$$

[1]In calculating reorder points, it has been assumed that $LT \le T$. If this is not the case, adjustments must be made (see Ref. 2).

and since $LT = 3$ days is less than $T_d = 43$ days:

$$R^* = (3) \times \left(\frac{50}{30}\right) = 5 \text{ belts}$$ ■

13.5 OTHER MODELS

As can be seen from the two models developed previously, the characteristics of the inventory situation dictate the appearance of the model and hence its solution. If we change one or more of the assumed characteristics underlying a specific inventory situation, we have to reformulate the model and obtain a different expression for Q^* and R^*. The characteristics involved in some situations cause the cost model to be so complex that it is very difficult to derive a closed-form expression for Q^* and R^*. In such cases we often turn to simulation as an analytical tool.

Several of the situations for which closed-form results have been derived are described below. The details of these models along with formulations of several other cases can be found in Hadley and Whitin [2].

13.5.1 Wilson Lot Size with Back Orders

When back orders of shortages are allowed (i.e., units demanded when the on-hand stock is zero are filled whenever the next resupply quantity is received), an additional decision variable is present—S, the desired number of back orders on the books when a procurement of Q units arrives. If there is a back-order cost of C_b dollars per unit-year of back orders that are incurred, a cost model for this extension of the basic Wilson lot size model is:

$$\min_{Q,S} TC = \frac{C_h(Q - S)^2}{2Q} + \frac{C_0 D}{Q} + \frac{C_b S^2}{2Q} \tag{13.8}$$

The solution to this model is:

$$Q^* = (\sqrt{(C_b + C_h)/C_b})(\sqrt{2C_0 D/C_h}) \tag{13.9}$$

and

$$S^* = \sqrt{2DC_0 C_h / C_b(C_b + C_h)} \tag{13.10}$$

and

$$R^* = (LT)(D) - S^* \tag{13.11}$$

13.5.2 Production Lot Size with Back Orders

Consider the finite replenishment rate model given earlier. In addition to the time-dependent C_b cost of back orders, assume there is also a fixed cost C_s incurred for each back order (for example, an expediting expense). The solution to the resulting model is:

$$Q^* = \sqrt{\left(\frac{C_b + C_0}{C_b}\right)} \sqrt{\left(\frac{P}{P - D}\right)\left(\frac{2DC_0}{C_h}\right) - \frac{C_s^2 D^2}{C_h(C_b + C_h)}} \tag{13.12}$$

and

$$S^* = \frac{(1 - D/P)}{(C_b + C_h)} \left(\sqrt{\left(\frac{P}{P - D}\right)(2DC_0C_h)\left(1 + \frac{C_h}{C_b}\right) - \frac{C_h C_s^2 D^2}{C_b}} - C_s D\right) \tag{13.13}$$

and, R^* is the reorder point given in Eq. (13.6) minus S^*.

13.6 MATERIAL REQUIREMENTS PLANNING

The models developed in the previous section are applicable in situations where the item being analyzed is independent of other items contained in inventory. That is, use of reorder quantities such as the Wilson lot size formula carries with it the implicit assumption that factors such as limited floor space or interrelated demands among items are not significant enough to require being incorporated into the inventory cost model. Although this is often true, there are numerous situations in which such sub-optimization leads to serious oversights and poor decision making. Therefore, several techniques and approaches have been developed to include explicit consideration of item interdependences in the model and analysis. One such methodology is material requirements planning (MRP).

MRP is useful in situations in which the demand for one item affects or dictates the demand for other items. This is the case with many manufacturing operations in which products are made up of various subassemblies and subcomponents; the number of end or parent items needed in a period dictates the number of each subcomponent needed. Another situation in which demands are interrelated and MRP is applicable is a multiechelon system of warehouses or other supply/distribution facilities. Here, the demand at the retail level for an item dictates the demand for that item (and its subassemblies) at storage points back through the distribution chain.

MRP begins with a *master schedule* for how much and when each end item is needed in the future, and translates or "explodes" this into a schedule of the requirements for subassemblies, component parts, and raw material. The result is a *time-bucketed* demand series for each inventoried item. In the process of converting the end item usage into time-bucketed requirements for lower-level parts, MRP takes into account the amount of inventory on hand as well as the amount already on order. In other words, demands are the *net* requirements projected for each future time period. An illustration of this conversion process is provided in the following example.

Example 13.3

A firm that makes fasteners manufactures a U-bolt to serve as a clamp for connecting two pieces of wire rope together. As shown in the bill of materials in Figure 13.5, there are three components for this fastener: a threaded bolt in the shape of U, a top clamp that slides over the bolt, and two springs that attach to the clamp, which are used to fasten the assembly together. The firm has received two orders for these fasteners from its major customer: 6000 units due four days from now and 8500 due four days after that. It takes two days to manufacture the bolts, one day to produce the clamp, and the springs can be ordered and delivered from a local vender on a one-day turnaround basis. Initially, there are 1000 bolts on hand, no clamps, and 2000 springs available. No orders are presently outstanding.

The result of using an MRP system to plan the future requirements for the three components is shown in Figure 13.6. As implied in this illustration, the calculation of a component's net requirements in a period begins with carrying down its needs in supplying planned order releases for all higher-level assemblies in the period in which those releases are scheduled (as dictated by their required purchase or manufacturing lead times). This constitutes a schedule of gross requirements for the component. Then, the amount of inventory *projected* to be on hand at the beginning of each period is netted against these gross requirements, producing a time series of discrete "lumps" of demand for the component. A predetermined lot sizing strategy such as the Wilson lot size model is then employed to calculate the number of units of that item to order so that the net demand requirements just calculated will be met. This calculation results in a schedule of planned order *receipts*. Finally, the purchasing or manufacturing of

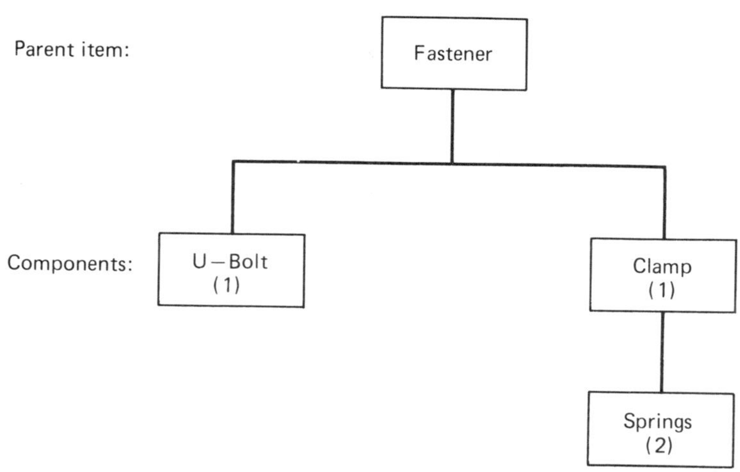

Figure 13.5 Bill of Materials for a U-Bolt Fastener

Master Schedule	Period								
	0	1	2	3	4	5	6	7	8
	0				6000				8500

Bolt:	Period								
	0	1	2	3	4	5	6	7	8
Gross requirements					6000				8500
Scheduled receipts									
Projected on hand	1000	1000	1000	1000	1000	0	0	0	0
Net requirements					5000				8500
Planned order receipts					5000				8500
Planned order releases			5000				8500		

Clamp:

Gross requirements					6000				8500
Scheduled receipts									
Projected on hand	0	0	0	0	0	0	0	0	0
Net requirements					6000				8500
Planned order receipts					6000				8500
Planned order releases				6000				8500	

Spring:

Gross requirements				12,000				17,000	
Scheduled receipts									
Projected on hand	2000	2000	2000	2000	0	0	0	0	0
Net requirements				10,000				17,000	
Planned order receipts				10,000				17,000	
Planned order releases			10,000				17,000		

Figure 13.6 MRP Calculations for the Wire Fastener Example

these orders is scheduled by backing up the number of time periods or buckets necessary so that the orders can be initiated and produced within their associated lead times. ■

The processing and calculations that take place in MRP were described in Example 13.3. Note that one of the steps in this processing was to use a predetermined lot size strategy to convert an item's time-bucketed net requirements into a schedule of planned order receipts. There are several options that can be used here. For instance, the calculations shown in Figure 13.6 were based on a lot-for-lot strategy. In this case, the demand for an item in a given period is satisfied by production in that same period—no inventories ever build up, but ordering cost can be high. An alternative is temporarily to ignore the interdependences among the items and calculate the lot sizes as if they were completely *independent* items, as was done in Sections 13.3 to 13.5. For example, if the production lot size formula of Eq. (13.7) had been used for the U-bolts in Figure 13.6 instead of a lot-for-lot strategy, and if the order quantity Q had turned out to be 7000 units, then the MRP planned would have turned out to be as shown in Figure 13.7.

Finally, it is important to recognize exactly what MRP is and is not. MRP is as much a philosophy as it is a technique, and as much an approach to scheduling as it is to inventory control. It is a production planning system. It takes as *input* a master schedule of future demand for end items and automately backs this down into a schedule of what to order and when to make that order. The amount of each order planned by MRP is an off-line, input decision. MRP is not a lot sizing model or strategy in the sense of the models of Sections 13.3 to 13.5. Rather, it is a material requirements planning system that automates much of the record keeping and inventory accounting that should normally take place anyway in scheduling the production of demand-dependent items.

13.7 SUMMARY

This chapter presents an overview of the procedures used to plan and control inventories. These procedures include pragmatic analogue models such as a Kardex card system, as well as mathematical models such as the Wilson lot size model. Their purpose is to enhance our ability to monitor inventory costs and to keep these costs as low as possible.

There are several cost factors whose values change as some aspects of the inventory replenishment process changes. These costs include ordering or setup cost, carrying cost, and shortage costs. They can be affected by varying the inventory replenishment decisions, such as the reorder quantity and the reorder point. Therefore, the primary analytical activity in inventory control is deriving the best values for these decisions. The specific models and techniques used to optimize these decisions depends on the characteristics of the underlying inventory system. Factors such as how the actual replenishment takes place, whether or not demand and lead time are deterministic or

Master Schedule

Period								
0	1	2	3	4	5	6	7	8
0				6000				8500

Bolt:

Period	0	1	2	3	4	5	6	7	8
Gross requirements					6000				8500
Scheduled receipts									
Projected on hand	1000	1000	1000	1000	1000	2000	2000	2000	2000
Net requirements					5000				6500
Planned order receipts					7000				7000
Planned order releases			7000				7000		

Clamp:

	0	1	2	3	4	5	6	7	8
Gross requirements					6000				8500
Scheduled receipts									
Projected on hand	0	0	0	0	0	0	0	0	0
Net requirements					6000				8500
Planned order receipts					6000				8500
Planned order releases				6000				8500	

Spring:

	0	1	2	3	4	5	6	7	8
Gross requirements				12,000			17,000		
Scheduled receipts									
Projected on hand	2000	2000	2000	2000	0	0	0	0	0
Net requirements				10,000			17,000		
Planned order receipts				10,000			17,000		
Planned order releases			10,000			17,000			

Figure 13.7 MRP Calculations Using a Predetermined Lot Size

stochastic, and whether the status of the inventory is reviewed periodically or continuously, all affect the structure of the models used in the optimization process.

Two common inventory situations are analyzed in detail in this chapter. The first has characteristics found in many retail inventory systems. The mathematical cost model used to derive the optimal reorder quantity and reorder point represents an unconstrained classical optimization model. It can be solved by applying the necessary and sufficient conditions described in Chapter 7. The resulting formula for the optimal value of the reorder quantity is often called the Wilson lot size model.

The second situation discussed in this chapter is typical of in-process inventories in a manufacturing environment. The main difference between this case and the retail situation underlying the Wilson lot size model is the rate of inventory replenishment. In the retail case it is instantaneous, whereas in the production situation it has some finite value. The optimization models are very similar.

Several other inventory situations are mentioned at the end of this chapter. These include variants of the two basic models in which back orders are allowed. This permits demand to be satisfied at a date after it initially occurs. Such models are more difficult to analyze but are often more accurate representations of the underlying inventory situation. These complex models and the simpler ones developed in the earlier sections of the chapter all are applicable when the item being inventoried has no interaction with other items. One procedure often used when this independence does not hold is called MRP, which is as much a production scheduling procedure as it is a specific inventory-control device. MRP should not be confused with the models presented in Sections 13.3 to 13.5; it is not a lot size calculation. It takes up where equations such as the Wilson lot size formula leave off.

REFERENCES AND SUGGESTED READING

1. E. S. Buffa and W. H. Taubert, *Production-Inventory Systems: Planning and Control*, Irwin, Homewood, IL, 1972, Chapters 3 and 4.
2. G. Hadley and T. M. Whitin, *Analysis of Inventory Systems*, Prentice-Hall, Englewood Cliffs, NJ, 1963.
3. L. A. Johnson and D. C. Montgomery, *Operations Research in Production Planning, Scheduling, and Inventory Control*, Wiley, New York, 1974, Chapter 2.
4. S. F. Love, *Inventory Control*, McGraw-Hill, New York, 1979. Contains a good dsecription of inventory-related costs.
5. J. Orlicky, *Materials Requirement Planning: The New Way of Life in Production and Inventory Management*, McGraw-Hill, New York, 1975.
6. G. W. Plossl and O. W. Wright, *Production and Inventory Control*, Prentice-Hall, Englewood Cliffs, NJ, 1967. Provides a good description of pragmatic approaches to inventory control.
7. N. V. Reinfold, *Production and Inventory Control*, Reston Publishing, Reston, VA, 1982.

8. A. F. Veinott, "The Status of Mathematical Inventory Theory," *Management Sciences*, **12,** No. 11 (July 1966). Contains a rigorous review of grass-roots inventory control research.

9. O. W. Wright, *The Executive's Guide to Successful MRP II*, Prentice-Hall, Englewood Cliffs, NJ, 1982.

REVIEW QUESTIONS AND PROBLEMS

13.1 Define or explain the following terms:

(a) Inventories
(b) Inventory control
(c) Continuous review policy
(d) Periodic review policy
(e) Wilson lot size model
(f) Kardex file
(g) Production lot size model
(h) MRP
(i) Master schedule

13.2 Classify each of the following inventory situations in terms of their primary characteristics or components:

(a) Stocking gasoline in storage tanks at a service station
(b) Automobiles at a new car dealer
(c) Finished nails coming out of a wire-nail-drawing machine
(d) Water in a city's water reservoir

13.3 Prove that the value of the reorder quantity for the Wilson lot size model as given by Eq. (13.4) is a gobal minimum [as opposed to (a) a local minimum, or (b) a maximum. *Hint:* see Chapter 7].

13.4 Explain how the ordering cost and carrying cost values used in Example 13.1 could arise in that auto parts store. That is, describe the possible makeup of these cost items.

13.5 Calculate the optimal reorder strategy for the situation in Example 13.1 if:

(a) The carrying cost is increased to 50 percent.
(b) The carrying cost is only 10 percent.
(c) The ordering cost is increased by 100 percent.

13.6 Develop a possible Kardex file card for the situation in Example 13.1 and simulate the operations of this card by listing several usage and restocking transactions.

13.7 Calculate the optimal reorder strategy for the situation in Example 13.2 if:

(a) The lead time is 45 days.
(b) The input replenishment rate is 51 units per month.

13.8 The ABC job shop company has a large number of orders each month for assemblies that require a $\frac{1}{2}$-in. valve as a main component. The shop can make the valve itself, or purchase the valve from an outside vendor. If the valve is made in-house, the estimated machine setup time to produce a batch of valves is 2 hr, and the standard time is 30 min to produce an individual valve. Labor rates for the machine operator in question are \$15/hr, and allocated overhead is another \$5/hr. Direct material costs \$50 per valve. Storing finished valves for any length of time requires tying up the company's capital, which they have to borrow at an annual rate of interest of 12 percent. If the valves are purchased from outside vendors, it normally takes 2 weeks for the valves to be received after the order is placed. Placement of the order takes about 4 hr of clerical time, costing \$10/hr. The vendor charges \$85 for each valve purchased from him.

(a) If the ABC Company has an annual requirement of 1000 valves, should they purchase or manufacture them, assuming they will operate using optimum inventory policies?
(b) What is the breakeven volume at which it would be equally desirable to purchase or to manufacture the valves? (Set up the solution; do not solve.)

13.9 Derive the cost minimization model for the inventory situation involving:

(a) Continuous review policy
(b) Deterministic lead time and demand
(c) Only relevant cost components are ordering costs, carrying cost, and a back-order cost for each unit back ordered
(d) A finite replenishment rate

Show all steps in the derivation.

13.10 A tire retailer inventories a certain brand of inner tube that costs him \$1.90 per tube if he orders in lots of 1000 or more, and \$2.00 per tube if he orders less than 1000. He has tracked his sales of this tube over the last 3 years and feels that they follow a normal probability distribution with a mean of 2000 tubes per year and a standard deviation of 50 tubes. It costs \$20.00 to place an order, and, figuring taxes, insurance, and capital costs, about \$0.16 to carry one tube in inventory for one year. What is the optimal inventory policy for this retailer? What is his total average annual inventory cost under the optimal strategy? The total number of orders per year? If he has a third purchase discount option of \$1.80 per tube for orders greater than 2000 tubes, what is his optimal inventory policy?

13.11 How sensitive is the lot size decision associated with the Wilson lot size model to changes in the various model parameters? Support your answer with an analysis using the model.

13.12 Develop a graph of reorder quantity versus total average annual cost for the Wilson lot size model of Example 13.1. On the same graph, show the plot of carrying cost versus reorder quantity, as well as the plot of ordering cost versus reorder quantity.

13.13 Write a computer program that will (1) find the optimal reorder strategy for the production lot size model in view of a number of alternative price discounts (see Problem 13.10), and (2) produce a sensitivity analysis report in which all parameters in the model are varied + and − 20 percent.

13.14 Discuss the role that each of the following might (or does) play in inventory control:

(a) A bill of material
(b) A production route sheet
(c) A flow process chart
(d) Classical optimization theory
(e) Description mathematical models

(f) Analogue models
(g) Math programming

13.15 Explain the statement ''MRP takes up where equations like the Wilson lot size formula leave off.''

13.16 Rework Example 13.3 using a master schedule that calls for 4500 fasteners due two days from now, another 4500 due four days later, and 5000 fasteners due one day after that.

13.17 Rework Example 13.3 using a version of the fastener that requires one *additional* spring for each U-bolt used (i.e., 3 springs are now needed: 2 to support each clamp and 1 to support each U-bolt).

13.18 Rework Example 13.3 using the following lot sizes for the individual components: U-bolts, $Q = 10{,}000$; clamp, $Q = 3000$; and spring, $Q = 3000$.

13.19 Rework Example 13.3 assuming that 20,000 springs are currently on order and are due to be received two weeks from now.

Chapter 14

Quality Control

14.1 NATURE OF QUALITY CONTROL

The quality of manufactured products is a concern of management that can be traced as far into history as manufacturing itself. Until recently the focus of attention of quality control was almost exclusively on the product itself. That is, the function of quality control was, and still is in large measure, an assessment of the ability of a manufactured product to perform its intended function for a prescribed period of time. Today, however, the horizons of quality control have been broadened to include not only the surveillance of the quality of the primary products of manufacturing but also many of its unintended by-products. The Environmental Protection Agency has established standards of quality for the vehicular emission of pollutants into the atmosphere and the disposal of waste production in the nation's waterways. Similar standards have been established for noise control near airports and within factories. Thus the manufacturing organization must exercise concern for both the quality of its saleable products and many of their unintended by-products.

Before elaborating on the techniques that may be applied to the control of product quality, a discussion of the nature of the problem and its setting is in order. In general, quality control is concerned with detecting the presence or absence of some desirable or undesirable characteristics in the product or with the degree to which a characteristic is present. For example, in the manufacture of glassware, quality control would be concerned with the detection of cracks in the glass, an obviously undesirable characteristic. A furniture manufacturer would be concerned with the detection of scars on the finish of tables. A dentifrice manufacturer would certainly exercise the controls necessary to ensure that its customers do not receive an unintended bonus in a tube of toothpaste, such as a fly or mosquito. From the point of view of quality control, these characteristics represent properties of the product that are either present or absent without reference to the degree of their presence or absence. Such characteristics are referred to as *attributes*.

In many instances quality control is interested not only in the simple presence or absence of a characteristic, but also in the degree to which it is present. For the purposes of this chapter, we refer to measurable characteristics as *variables*. As an illustration, consider the problem faced in the production of an appliance that is guaranteed for one year. To avoid significant replacement or repair costs, the organization would have to exercise those controls necessary to ensure that a significant number of units were not produced that would last for a period of time less than that guaranteed. In this instance the characteristic to be controlled is the life of the product up to the first failure, which is a measurable quantity and therefore a variable. In a similar manner, a soft-drink manufacturer must control the volume of soft drink in each bottle, whereas the dentrifrice manufacturer would attempt to control the amount of toothpaste in each tube as it comes off the filling line.

The primary emphasis of this chapter is on statistical techniques for detecting adherence to or departures from quality standards. This aspect of quality control is an after-the-fact function in that it attempts to detect whether the output of a process adheres to prescribed standards of quality. The quality of the product, as revealed through the detection phase of quality control, is dependent upon the design standards for the product and the ability of the manufacturing process to maintain those standards. The standards of quality for a product are prescribed in the design specifications. Poor product design will ultimately result in failure of the product to perform its intended function. Poor design thus leads to poor quality.

The purpose of a manufacturing process is to produce items of product that conform to design specifications. On the one hand, a manufacturing process may perform this function, but still turn out products of substandard quality because of poor design standards. On the other hand, the process may turn out poor-quality products simply because it is not capable of consistently meeting the design specifications.

Two process problems are noteworthy in this regard. First, the process may be such that it produces a high proportion of defective ware under the best of conditions. That is, the inherent variability of the process is so high that it is unable to meet design standards consistently even when it is under control. The second problem is related to process malfunction. Every manufacturing process will fail to operate in its intended manner occasionally. In some cases these failures are immediately obvious. However, a manufacturing process frequently continues to operate after failure but at a level that yields product of substandard quality. In such cases, the failure can often be detected only by careful inspection of the units of product coming off the line. The quality problem here is one of process control. Hence, a process that produces satisfactory product when under control may be unsatisfactory because it goes out of control too frequently.

As the above discussion suggests, the quality of manufactured product may be improved by raising design standards or by refinement of the manufacturing process. In some cases the design requirements are so high and the manufacturing control so tight that the quality of the final product is assured, thereby eliminating a need for monitoring the output of the process. More often the monitoring effort is necessary to identify those cases where product quality has fallen below standard. This effort is usually carried out by selecting a random sample of product from the output of the manufacturing process, inspecting each unit in the sample, and drawing a conclusion about the overall quality of the population of items from which the sample was drawn.

The areas of application of quality control can be broadly classified as *process control* and *acceptance sampling*. As discussed above, process control is carried out by monitoring the output of a production process to determine when it is in and when it is out of control. When an out-of-control condition is identified, steps are taken to identify and correct the condition, bringing the process back into control. Acceptance sampling is used to determine whether specific manufactured lots meet quality standards. The result of acceptance sampling is a decision to accept or reject the lot, based upon the result of sampling inspection of units selected from the lot. Acceptance sampling is frequently applied to lots of materials that are purchased, to internally produced lots before they are delivered to the customer, or to in-process lots before they are used in subsequent phases of manufacture of the final product. Thus the focus of attention of acceptance sampling is the disposition of a specific lot of product, whereas that of process control is on the process that produced the lot. Obviously, a sampling inspection system could be used for the dual purpose of acceptance sampling and process control.

14.2 STATISTICAL BASIS FOR QUALITY CONTROL

Very few manufacturing systems are capable of producing defect-free product 100 percent of the time. Our concern in this chapter is with the majority of production processes that occasionally go out of control at random points in time and that produce some defective units at random even when they are in control. To place the problem

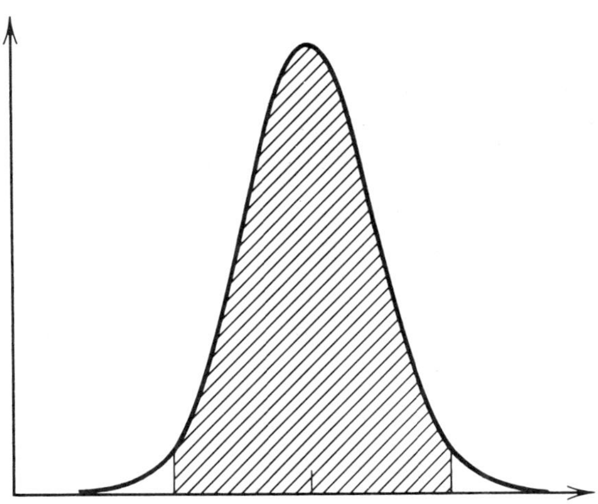

Figure 14.1 Proportion of Product within the Specification Limits

in perspective, consider an extrusion process that is to produce metal tubes 8-in. in length. The tubes are useable if they are between 7.96 in. and 8.04 in.; that is, the *specification limits* are 7.96 in. (S_L) and 8.04 in. (S_U). The extrusion process is such that when it is under control the length of the tubes produced is normally distributed with mean 8.00 in (μ) and standard deviation 0.02 in. (σ). Therefore, if x is the length of a tube, the probability that the tube will be within the specification limits is the shaded area under the normal distribution in Figure 14.1 or is given by Eq. (14.1).

$$
\begin{aligned}
P[\text{length within specification limits}] \\
&= P[S_L \leq x \leq S_U] \\
&= P[7.96 \leq x \leq 8.04] \\
&= P\left[\frac{S_L - \mu}{\sigma} \leq z \leq \frac{S_U - \mu}{\sigma}\right] \\
&= P\left[\frac{7.96 - 8.00}{0.02} \leq z \leq \frac{8.04 - 8.00}{0.02}\right] \\
&= P\left[z < \frac{8.04 - 8.00}{0.02}\right] - P\left[z \leq \frac{7.96 - 8.00}{0.02}\right] \\
&= P[z < 2.0] - P[z < -2.0] \\
&= 0.9772 - 0.0228 \\
&= 0.9544 \qquad (14.1)
\end{aligned}
$$

In Eq. (14.1) $P(z \leq k)$ is the probability that z is less than or equal to k and may be read from the distribution function of the standard normal random variable in the Appendix. Thus the probability that any unit will be outside the specification limits is 0.0456; we would expect about 4.56 percent of the units produced to be defective with respect to length. Now let us suppose that the mean of the process shifts from the in-control state of 8.00 in. to an out-of-control state of 8.02 in. This shift is shown graphically in Figure 14.2, where the proportion of defective units produced is represented by the shaded area. The probability that a unit produced in the out-of-control state will fall within the specification limits is given by

$$
\begin{aligned}
P[\text{length within specification limits}] \\
&= P[7.96 \leq x \leq 8.04] \\
&= P\left[\frac{7.96 - 8.02}{0.02} \leq z \leq \frac{8.04 - 8.02}{0.02}\right] \\
&= P[-3.00 \leq z \leq 1.00] \\
&= P[z \leq 1.00] - P[z \leq -3.00] \\
&= 0.8413 - 0.0014 \\
&= 0.8399 \qquad (14.2)
\end{aligned}
$$

or we would expect about 16.01 percent of the units produced to be defective with respect to length when the process is out of control with mean 8.02 in.

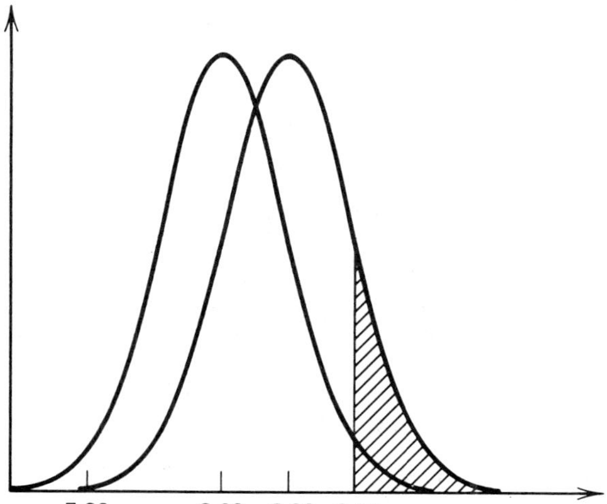

Figure 14.2 Distribution of Tube Length in the In-Control and Out-of-Control States

The example we have just presented illustrates the case of a system that will produce some defective units even when it is in control. However, when the mean length of tubes produced increases to a value greater than the desired dimension, the proportion of defective ware resulting is greater as we would expect. If 5 percent defective ware is deemed acceptable and 16 percent is deemed unacceptable, then the quality-control system used should tell us when the mean dimension of tube length has shifted from 8.00 in. to 8.02 in. or above or to 7.98 in. or below.

The preceding discussion indicates that a change in the value of the mean of the distribution of the dimension of tubes causes a change in the proportion of defective units produced. A change in the standard deviation of the length will also alter the proportion of defective units produced. For example, suppose the standard deviation of length were to increase from 0.02 in. to 0.04 in. as shown in Figure 14.3. The shaded area represents the probability that the length of a given tube is within the specification limits and is calculated by:

$$
\begin{aligned}
P[\text{length within specification limits}] \\
&= P[7.96 \leq x \leq 8.04] \\
&= P\left[\frac{7.96 - 8.00}{0.04} \leq z \leq \frac{8.04 - 8.00}{0.04}\right] \\
&= P[-1.00 \leq z \leq 1.00] \\
&= P[z \leq 1.00] - P[z \leq -1.00] \\
&= 0.8413 - 0.1587 \\
&= 0.6826 \qquad (14.3)
\end{aligned}
$$

Thus an increase or decrease in the process mean or an increase in the process standard

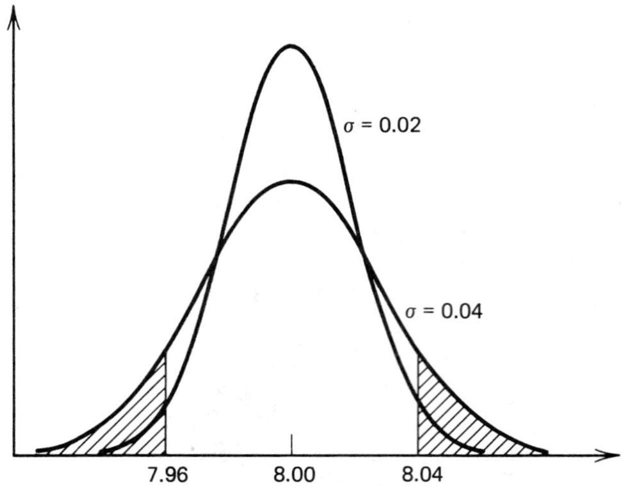

Figure 14.3 Effect of the Change in the Standard Deviation of Tube Length on the Probability That a Tube Will Be Defective

deviation will increase the probability that a given tube will have a length outside the specification limits.

As the above examples illustrate, if the quality characteristic to be controlled is a random variable, the distribution of the random variable must be known to describe the probabilistic behavior of the characteristic. To completely specify the distribution of a random variable, one must determine a functional relationship for the associated probability mass or probability density function and the numerical values of the parameters of the function. When a manufacturing process goes out of control from a quaility point of view, the implication is that something about the probability distribution of the characteristic to be controlled has changed. When a quality-control investigation indicates that the process has gone out of control, it indicates that the probability distribution of the characteristic has changed. However, it does not indicate the source of the change but merely that an *assignable course of variation* should be sought and corrected if detected. In the case of acceptance sampling, detection of an unacceptable lot of product would also indicate that the distribution of the characteristic to be controlled has changed in an undesirable manner. In this case, however, the undesirable condition leads to rejection of the lot rather than to an investigation of the process to determine why the overall quality of the lot is below standard.

14.3 PROCESS CONTROL

The purpose of a process quality-control system is to indicate when the manufacturing process has reached an out-of-control state. This is achieved by selecting a sample of items from the production process and, based upon the results of inspection of the sample, drawing a conclusion about the overall quality of the product being manufactured by the process. For example, if 100 items were drawn from one day's production and 90 of the units were found to be defective, we would probably conclude that the process was turning out substandard product—a condition that should be rectified before the start of production on the next day. In contrast, if all of the items in the sample were of acceptable quality, we would have little basis for concluding that the process is out of control.

Before attempting to develop a process quality-control system, one should first determine whether or not a quality-control system would be cost effective. For example, if the process rarely turns out a defective unit, implementation of a reliable quality-control system may not be worthwhile since it may only tell management what it already knows—that the process is quite reliable. At the other extreme, the manufacturing process may be so unreliable that it turns out a high proportion of defective product under the best of circumstances. Again, an effective quality-control system would only serve to tell us what we already know—that the process is usually out of control.

Example 14.1

A company makes watches that are designed to be accurate to ± 1 min/day. Management has decided that 98 percent of the watches produced should maintain these specifications. The first 2000 watches produced were inspected and the distribution of error in minutes per day was found to be normal with mean 0.00 and standard deviation 0.90 min/day. During this period of production, the assembly process was very carefully controlled and is believed to have been operated at maximum effectiveness. Is the process capable of maintaining the specifications prescribed?

The probability that the process will produce a watch with error x less than or equal to ± 1 min/day is given by:

$$\begin{aligned} P[-1.0 \le x \le 1.0] &= P\left[\frac{-1.0 - 0.0}{0.90} \le z \le \frac{1.0 - 0.0}{0.90}\right] \\ &= P[z \le 1.11] - P[z \le -1.11] \\ &= 0.8666 - 0.1335 \\ &= 0.7330 \end{aligned} \tag{14.4}$$

Hence, under the most strictly controlled conditions, the process will produce watches with an error in accuracy outside the specification limits about 27 percent of the time, whereas management requires that this figure be no greater than 2 percent. Thus the production process is not capable of meaningful control without modification. ■

Example 14.2

A particular line of calculators is guaranteed for a period of 1.0 year and 95 percent of the calculators are to operate without malfunction for the guarantee period. A pilot run of the production process indicates the life of this line is exponentially distributed with a design mean life of 5.0 years. Will this production process meet the design specifications? If a design mean life of 5.0 yr is not adequate, what should the design mean life be?

The density function of calculator life x is given by:

$$f(x) = \lambda e^{-\lambda x} \qquad 0 < x < \infty \tag{14.5}$$

where the mean of x is given by $1/\lambda$. Thus

$$f(x) = 0.20e^{-0.20x} \qquad 0 < x < \infty \tag{14.6}$$

The probability that a calculator will operate without malfunction for at least 1.0 yr is given by:

$$\begin{aligned} P[x \geq 1.0] &= \int_{1.0}^{\infty} 0.20e^{-0.20x}\, dx \\ &= e^{-0.20} \\ &= 0.8187 \end{aligned} \tag{14.7}$$

Thus about 18 percent of the calculators can be expected to fail within the guaranteed period, and the design mean life is inadequate. To determine the required design mean life we must find the value of λ such that

$$P[x \geq 1.0] = 0.95 \tag{14.8}$$

or

$$\int_{1.0}^{\infty} \lambda e^{-\lambda x}\, dx = 0.95 \tag{14.9}$$

Thus

$$e^{-\lambda} = 0.95$$

and

$$\begin{aligned} \lambda &= -\ln(0.95) \\ &= 0.0513 \end{aligned} \tag{14.10}$$

Hence, the design mean life must be 19.50 yr. ■

Examples 14.1 and 14.2 illustrate cases where product of satisfactory quality cannot be manufactured under current operating conditions. Example 14.1 dealt with a manufacturing process that was so inherently variable that quality standards could not be met. In Example 14.2 the cause of poor product performance could be attributed to either the product design specifications or variability in the manufacturing process. In both cases, a quality-control system would be of little value unless the design specifications, the manufacturing process, or the quality standards were modified, since it would only reflect what is already known. Let us now examine a situation where design specifications and process capabilities are compatible with established quality standards.

Example 14.3

Beer bottles are filled on an automated line. Each bottle is to contain 12 fluid ounces of beer. The filling line can be adjusted such that the average contents per bottle can be set at any desired level. The distribution of actual contents per bottle is normal with mean setting on the filling line μ and standard deviation 0.040 oz. If the filling line is operating properly, 97 percent of the bottles filled should contain between 11.90 and 12.10 fluid ounces of beer. Determine whether or not the filling line is capable of maintaining the specified level of quality if the line is set to a mean of 12.00.

According to the quality standards given, the probability that a randomly selected bottle will contain between 11.90 and 12.10 oz of beer should be 0.97. Now if x is the contents of a bottle and μ is the mean set for the filling line:

$$\begin{aligned} P[11.90 \le x \le 12.10 \mid \mu &= 12.00] \\ &= P\left[\frac{11.90 - 12.00}{0.040} \le z \le \frac{12.10 - 12.00}{0.040}\right] \\ &= P[-2.50 \le z \le 2.50] \\ &= 0.9876 \end{aligned} \tag{14.11}$$

Therefore, the filling line will meet quality specifications when the mean contents per bottle is set at 12.00. ■

Although the filling line in Example 14.3 is capable of meeting established quality standards when the mean is set to 12.00 fluid ounces, if the process slips out of adjustment we would expect some deviation from those standards. For example, if the mean slipped from 12.00 to 12.04, the proportion of bottles having contents outside the specification limits 11.90, 12.10 is given by

$$\begin{aligned} P[11.90 \le x \le 12.10 \mid \mu &= 12.04] \\ &= P\left[\frac{11.90 - 12.04}{0.040} \le z \le \frac{12.10 - 12.04}{0.040}\right] \\ &= P[-3.50 \le z \le 1.50] \\ &= 0.9330 \end{aligned} \tag{14.12}$$

Thus the manufacturer would be interested in determining when the filling line has slipped out of adjustment such that $\mu \neq 12.00$. This determination is the function of quality control.

In the case just cited, we are interested in developing a quality-control technique that will allow us to determine when the process mean has shifted from its desired value. Let μ_0 be the desired value of the process mean and μ the actual value of the process mean, which may or may not be equal to μ_0. Since we have no way of knowing the precise value of μ we must estimate its value. The best estimate of a population mean is the *sample mean* $\bar{x}$. Let $x_1, x_2, \ldots, x_n$ be observed contents of a bottle of beer. Then $\bar{x}$ is given by

$$\bar{x} = \frac{1}{n} \sum_{i=1}^{n} x_i \tag{14.13}$$

That is, in the case of the filling line discussed in Example 14.3 we would select n bottles at random from the line and measure the contents of each yielding the values $x_1, x_2, \ldots, x_n$.

Since $\bar{x}$ is an estimate of u, it should provide us with some information as to whether $\mu = \mu_0$ or not. That is, if $\mu = \mu_0$ we would expect the value of $\bar{x}$ to be fairly close to μ_0. However, since $\bar{x}$ is only an estimate of μ, we would not expect it to be equal to μ_0 even when $\mu = \mu_0$. Since the values $x_1, x_2, \ldots, x_n$ that make up $\bar{x}$ are random variables, $\bar{x}$ is also a random variable and any inferences as to the value of μ based upon the value of $\bar{x}$ must be probabilistically based.

To draw meaningful inferences about the value of μ based upon an observed value of $\bar{x}$, we must know the distribution $\bar{x}$. For sufficiently large n the distribution of $\bar{x}$ is approximately normal. A useful rule of thumb is that $\bar{x}$ may be considered to be normally distributed for $n \geq 30$. If σ is the standard deviation of x_i, $i = 1, 2, \ldots, n$ and:

$$\mu = E(x_i) \tag{14.14}$$

then $\bar{x}$ has mean μ and standard deviation $\sigma/\sqrt{n}$. Hence, for $n \geq 30$ the probability density function of $\bar{x}$ is given approximately by

$$f(\bar{x}) = \frac{\sqrt{n}}{\sigma\sqrt{2\pi}} e^{-n(\bar{x}-u)^2/\sigma^2} \qquad -\infty < \bar{x} < \infty \tag{14.15}$$

If the observed value of $\bar{x}$ deviates only slightly from μ_0, we may have little reason to believe that $\mu \neq \mu_0$. For large deviations, however, the conclusion that $\mu \neq \mu_0$ may be justified. Thus, we require limits for $\bar{x}$, C_L and C_U—called *control limits*, such that if $C_L \leq \bar{x} \leq C_U$ we conclude that $\mu = \mu_0$ and that $\mu \neq \mu_0$ if $\bar{x} < C_L$ or $\bar{x} > C_U$. To assign values to the control limits we must recognize that $\bar{x}$ is a random variable. As such, even if $\mu = \mu_0$ we may find that $\bar{x} < C_L$ or $\bar{x} > C_U$ and erroneously conclude that $\mu \neq \mu_0$. This error is referred to as a *Type I* error. Furthermore, the difference between μ and μ_0 may be substantial and yet $C_L \leq \bar{x} \leq C_U$, leading us to the erroneous conclusion that $\mu = \mu_0$, a *Type II* error. Let

$$\alpha = P[\text{Type I error}] \tag{14.16}$$

$$\beta = P[\text{Type II error}] \tag{14.17}$$

If we assign a value to α (usually between 0.00 and 0.10), the control limits C_L and C_U must satisfy the relationship:

$$\begin{aligned}\alpha &= P[\text{Type I error}] \\ &= P[\bar{x} < C_L \text{ or } \bar{x} > C_U | \mu = \mu_0] \\ &= 1 - P[C_L \leq \bar{x} \leq C_U | \mu = \mu_0]\end{aligned} \tag{14.18}$$

It should be noted that the Type I error can occur only when $\mu = \mu_0$. From Eq. (14.18) we have:

$$
\begin{aligned}
1 - \alpha = P[C_L \le \bar{x} \le C_U | \mu = \mu_0] \\
= P\left[\frac{C_L - \mu_0}{\sigma/\sqrt{n}} \le z \le \frac{C_U - \mu_0}{\sigma/\sqrt{n}}\right] \\
= P\left[z \le \frac{C_U - \mu_0}{\sigma/\sqrt{n}}\right] - P\left[z \le \frac{C_L - \mu_0}{\sigma/\sqrt{n}}\right] \qquad (14.19)
\end{aligned}
$$

since $\bar{x}$ is assumed to be normally distributed with mean $\mu = \mu_0$ and standard deviation $\sigma/\sqrt{n}$. If we choose C_L and C_U such that $C_U - \mu_0 = \mu_0 - C_L$, then the criterion used to test the hypothesis that $\mu = \mu_0$ is said to be *uniformly most powerful*. Since the normal distribution is symmetric, we have:

$$P\left[z \le \frac{C_U - \mu_0}{\sigma/\sqrt{n}}\right] = 1 - \frac{\alpha}{2} \qquad (14.20)$$

$$P\left[z \le \frac{C_L - \mu_0}{\sigma/\sqrt{n}}\right] = \frac{\alpha}{2} \qquad (14.21)$$

That is, the area under the density function of z below $(C_L - \mu_0)/(\sigma/\sqrt{n})$ and above $(C_U - \mu_0)/(\sigma/\sqrt{n})$ is $\alpha/2$ as shown in Figure 14.4. $(C_U - \mu_0)/(\sigma/\sqrt{n})$ is then the value of the cumulative standard normal random variable corresponding to a cumulative

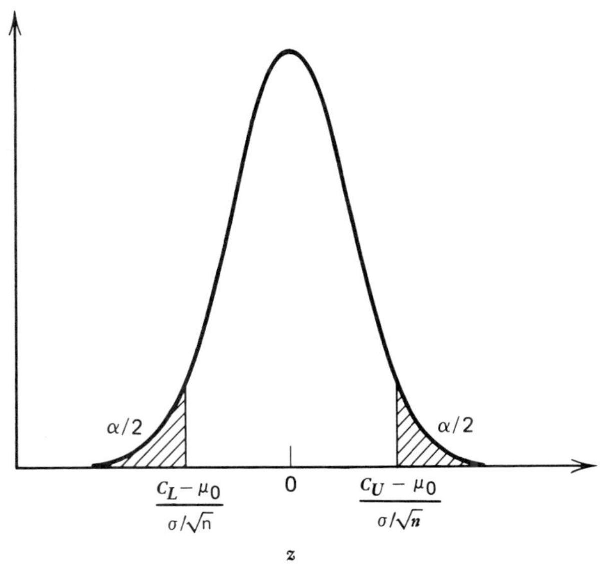

Figure 14.4 Tail Areas Under the Standard Normal Distribution

probability of $(1 - \alpha/2)$. Hence

$$\frac{C_U - \mu_0}{\sigma/\sqrt{n}} = z_{1-\alpha/2} \tag{14.22}$$

By a similar argument

$$\frac{C_L - \mu_0}{\sigma/\sqrt{n}} = z_{\alpha/2} \tag{14.23}$$

Since the probability density function of the standard normal random variable z is symmetric we have:

$$z_{\alpha/2} = -z_{1-\alpha/2} \tag{14.24}$$

Solving Eq. (14.22) and (14.23) for C_L and C_U we have:

$$C_L = \mu_0 - \frac{\sigma}{\sqrt{n}} z_{1-\alpha/2} \tag{14.25}$$

$$C_U = \mu_0 + \frac{\sigma}{\sqrt{n}} z_{1-\alpha/2} \tag{14.26}$$

Using these control limits if $C_L \leq \bar{x} \leq C_U$ we would conclude that $\mu = \mu_0$ and would have no reason to search for an assignable cause of variation in the process. On the other hand, if $\bar{x} < C_L$ or $\bar{x} > C_U$ we would conclude that $\mu \neq \mu_0$ and an assignable cause of variation in the process should be sought and corrected if detected.

Example 14.4

A process quality-control system is to be developed to control the diameter of glass tubes. The desired diameter is 3.00 in. The standard deviation of tube diameter is 0.01 in. From each day's production, 100 tubes are to be selected at random, the diameter of each measured, and the sample mean $\bar{x}$ calculated. The limits for the quality control system, C_L and C_U, are to be specified such that the probability of concluding that $\mu \neq 3.00$ when in fact $\mu = 3.00$ is to be 0.01. Find the limits C_L and C_U.

Because the probability of a Type I error is to be 0.01:

$$\begin{aligned} z_{1-\alpha/2} &= z_{0.995} \\ &= 2.575 \end{aligned} \tag{14.27}$$

From Eq. (14.25) and (14.26) we have:

$$C_L = 3.00 - \frac{0.01}{\sqrt{100}}(2.575) \tag{14.28}$$

$$C_U = 3.00 + \frac{0.01}{\sqrt{100}}(2.575) \tag{14.29}$$

or

$$C_L, C_U = 2.9974, 3.0026 \tag{14.30}$$

If $2.9974 \leq \bar{x} \leq 3.0026$ we conclude that $\mu = 3.00$ and that the process is functioning as intended. If $\bar{x} < 2.9974$ or $\bar{x} > 3.0026$ we conclude that $\mu \neq 3.00$ and that an adjustment of the manufacturing process is required to bring it under control. ■

Implementation and maintenance of a process quality-control system can usually be simplified through the use of a *control chart* for $\bar{x}$ such as that shown in Figure 14.5. Each time a sample is drawn the value of $\bar{x}$ is plotted on the control chart to highlight out-of-control points such as those circled in Figure 14.5.

Example 14.5

Samples of size 100 were selected from each of 20 days' production for the manufacturing process described in Example 14.4 with the following results.

Day	$\bar{x}$	Day	$\bar{x}$
1	3.0014	11	3.0036
2	3.0003	12	3.0005
3	2.9991	13	2.9986
4	3.0016	14	2.9980
5	2.9962	15	3.0041
6	2.9958	16	3.0020
7	2.9981	17	3.0001
8	3.0020	18	3.0011
9	2.9998	19	3.0017
10	3.0004	20	2.9978

Record the values of $\bar{x}$ on a control chart for $\bar{x}$, and determine those days for which the process was likely to have been out of control.

The control limits for $\bar{x}$ were developed in Example 14.4, and the control chart for $\bar{x}$ is shown in Figure 14.6. The out-of-control points are circled on the control chart and occurred on the fifth, sixth, eleventh, and fifteenth days. ■

An out-of-control point on a process control chart indicates that the process probably requires adjustment to reestablish a state of control. However, the search for the source of trouble may occasionally prove fruitless since there is always a possibility that an out-of-control point is the result of a Type I error. Of course, the frequency of incidence of the Type I error can be controlled by appropriate selection of α. Frequently 3 sigma limits are used for the control limits C_L and C_U, which yield a value of α of 0.0028. Although these limits are appealing in that they yield a small value of α, they may prove to be inappropriate since the probability of occurrence of the Type II error may prove to be unacceptably large.

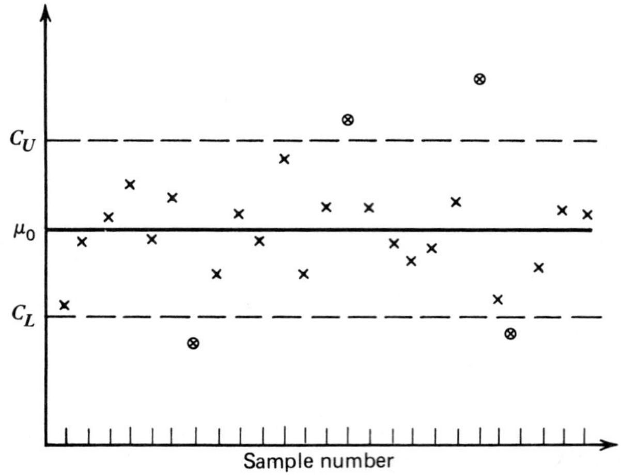

Figure 14.5 Control Chart for $\bar{x}$

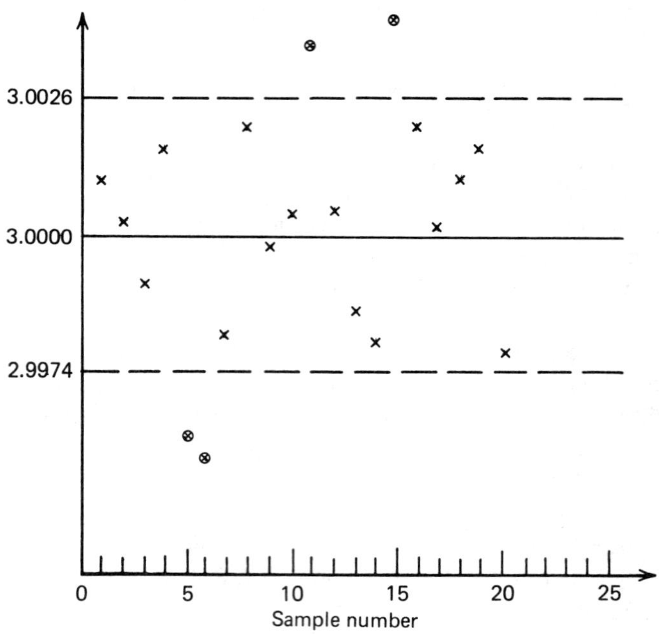

Figure 14.6 Control Chart for $\bar{x}$ for Example 14.5

As the process mean μ deviates further from the desired value μ_0 we would expect the probability of concluding that $\mu = \mu_0$ to diminish. That is, if

$$d = |\mu - \mu_0| \tag{14.31}$$

and

$$P_d = P[C_L \leq \bar{x} \leq C_U \| \mu - \mu_0| = d] \tag{14.32}$$

then the quality-control procedure will possess the property that P_d will decrease as d increases. Since $\bar{x}$ is assumed to be normally distributed:

$$\begin{aligned} P[C_L \leq \bar{x} \leq C_U \| \mu - \mu_0| = d] &= P[C_L \leq \bar{x} \leq C_U | \mu - \mu_0 = d] \\ &= P[C_1 \leq \bar{x} \leq C_U | \mu - \mu_0 = -d] \end{aligned} \tag{14.33}$$

where d is positive. If $\mu - \mu_0 = d$, then $\mu = \mu_0 + d$, and $\bar{x}$ is normally distributed with mean $\mu_0 + d$ and standard deviation $\sigma/\sqrt{n}$. Thus

$$\begin{aligned} P_d &= P[C_L \leq \bar{x} \leq C_U | \mu - \mu_0 = d] \\ &= P\left[\frac{C_L - \mu_0 - d}{\sigma/\sqrt{n}} \leq \bar{x} \leq \frac{C_U - \mu_0 - d}{\sigma/\sqrt{n}}\right] \end{aligned} \tag{14.34}$$

If we plot P_d versus d we obtain an *operating characteristic* curve (O.C. curve) such as that shown in Figure 14.7. The O.C. curve in Figure 14.7 shows that P_d decreases with increasing values of d.

The sensitivity of a quality-control system to deviations of μ from μ_0 can be increased by increasing n. As n is increased P_d decreases for all positive values of d. This property is illustrated in Figure 14.8. Thus, to reduce the probability of a Type II error, or increase the *power* $(1 - \beta)$, we can simply increase the sample size n for the quality-control system.

Example 14.6

The wall thickness of a plastic container is 0.10 ± 0.01 in. The distribution of wall thickness is normal with mean 0.10 in. and standard deviation 0.003 in. when the process is in control. A quality-control system for mean wall thickness is to be developed using a sample size of 100 and $\alpha = 0.90$. Determine the control limits for $\bar{x}$ and sketch the O.C. curve.

From Eq. (14.25) and (14.26) we have:

$$\begin{aligned} C_L &= \mu_0 - \frac{\sigma}{\sqrt{n}} z_{0.95} \\ &= 0.10 - \frac{0.003}{\sqrt{100}}(1.645) \\ &= 0.09951 \end{aligned} \tag{14.35}$$

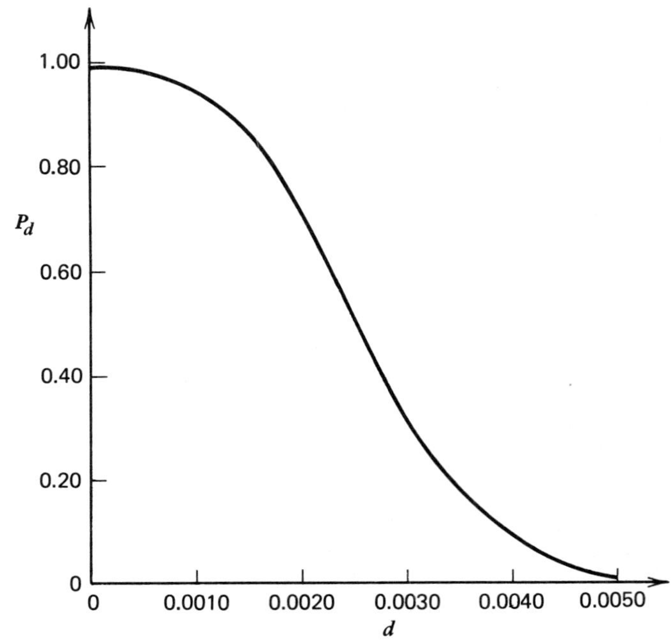

Figure 14.7 Operating Characteristic Curve for the Process Quality-Control System in Example 14.4

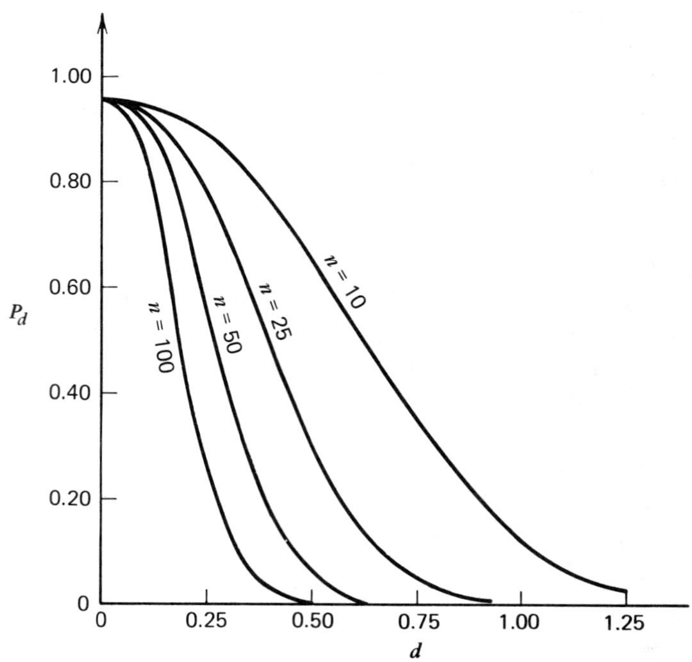

Figure 14.8 Operating Characteristic Curves for Various Sample Sizes Where $\alpha = 0.045$

and

$$
\begin{aligned}
C_U &= \mu_0 + \frac{\sigma}{\sqrt{n}} z_{0.95} \\
&= 0.10 + \frac{0.003}{\sqrt{100}} (1.645) \\
&= 0.10049 \qquad (14.36)
\end{aligned}
$$

To calculate the probability that the quality-control system will lead us to the conclusion that $\mu = \mu_0$ for various values of $d = |\mu - \mu_0|$ we use Eq. (14.34). Therefore, for any value of d we have:

$$
\begin{aligned}
P_d &= P\left[\frac{0.09951 - 0.10 - d}{0.003/\sqrt{100}} \leq z \leq \frac{0.10049 - 0.10 - d}{0.003/\sqrt{100}}\right] \\
&= P[-1.633 - 3333.333d \leq z \leq 1.633 - 3333.333d] \\
&= P[z \leq 1.633 - 3333.333d] - P[z < -1.633 - 3333.333d] \qquad (14.37)
\end{aligned}
$$

The calculations for P_d for various values of d are summarized in Table 14.1. The O.C. curve for this process quality-control system is shown graphically in Figure 14.9. ■

O.C. curves are often used to determine whether or not a given quality-control plan will provide satisfactory discrimination between in- and out-of-control conditions. To illustrate, consider the O.C. curve in Figure 14.9. If $\mu = \mu_0$ ($d = 0$), the probability that the quality-control system will indicate an in-control condition is 0.90. Suppose that we would like the quality-control system to indicate an out-of-control condition if $\mu = 0.1010$ or $\mu = 0.990$ with probability at least 0.70. Either condition yields $d = 0.001$.

Table 14.1 Calculation of P_d for Various Values of d for Example 14.6.

d	z_1 ($1.633 - 3333.333d$)	z_2 ($-1.633 - 3333.333d$)	$P[z \leq z_1]$	$P[z \leq z_2]$	P_d
0.0000	1.633	−1.633	0.9488	0.0512	0.8976
0.0001	1.300	−1.966	0.9032	0.0246	0.8786
0.0002	0.966	−2.300	0.8330	0.0107	0.8223
0.0003	0.633	−2.633	0.7366	0.0042	0.7324
0.0004	0.300	−2.966	0.6179	0.0015	0.6164
0.0005	−0.034	−3.300	0.4865	0.0005	0.4860
0.0006	−0.367	−3.633	0.3568	0.0001	0.3567
0.0007	−0.700	−3.966	0.2420	0.0000	0.2420
0.0008	−1.034	−4.300	0.1506	0.0000	0.1506
0.0009	−1.367	−4.633	0.0859	0.0000	0.0859
0.0010	−1.700	−4.966	0.0446	0.0000	0.0446
0.0011	−2.034	−5.300	0.0210	0.0000	0.0210
0.0012	−2.367	−5.633	0.0090	0.0000	0.0090
0.0013	−2.700	−5.966	0.0035	0.0000	0.0035

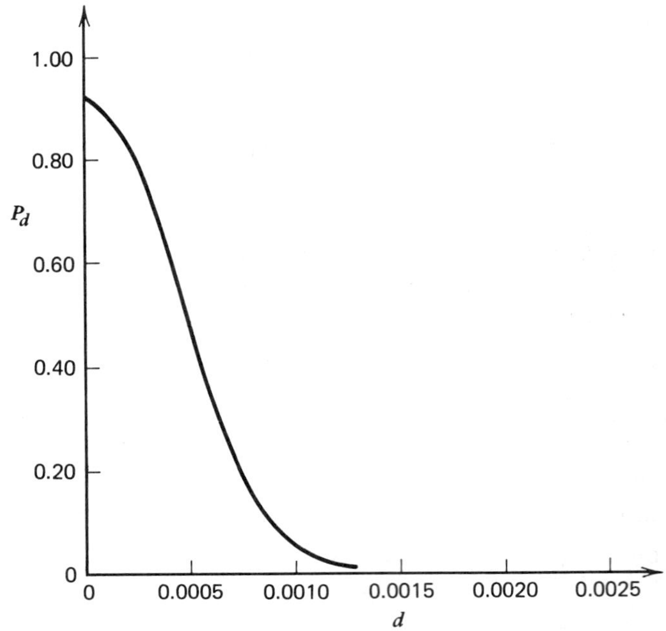

Figure 14.9 O.C. Curve for the Quality-Control System in Example 14.6

Using the O.C. curve in Figure 14.9, we find that $P_d = 0.0446$ when $d = 0.001$. Thus the probability that an out-of-control condition will be indicated when $\mu = 0.1010$ or $\mu = 0.9990$ is 0.9554, which satisfies the required level of protection. However, if the probability of an indication of an out-of-control condition were to be at least 0.70 when $\mu = 0.1004$ or $\mu = 0.9996$, the quality-control plan would be unacceptable since at $d = 0.0004$, $P_d = 0.6164$. Hence, the probability that the quality-control plan in Example 14.6 will indicate an out-of-control condition when $\mu = 0.1004$ or $\mu = 0.9996$ is only 0.3836. In such cases we can improve the quality-control system by increasing the sample size.

In this section we have discussed quality-control systems for the process mean. We have already indicated that a change in the standard deviation of the process can alter the quality of the product produced. Quality-control systems for the process standard deviation can be developed in a manner similar to that used for the process mean. However, discussion of the development of such systems is beyond the scope of this book. (For a discussion of control of the process standard deviation and other topics relating to process quality control, see References 4, 6, or 9.

14.4 ACCEPTANCE SAMPLING

As we have already mentioned, the applications of quality control are not restricted to the control of variation of the manufacturing process. In many instances we are simply interested in whether a given lot of product is of acceptable quality or not without regard to the status of the manufacturing process when the lot was produced. For example, suppose that a company purchases a certain product in lots of 10,000 units. Upon receipt of such a lot, the company may wish to assess the overall quality of the lot before it accepts and pays for the lot. In such a situation the organization purchasing the lot is not interested in or capable of controlling the process that produced the lot.

Acceptance sampling has as its purpose the segregation of lots of acceptable or unacceptable quality on the basis of the results of inspection of a sample of units drawn at random from the lot. Where the characteristic to be controlled is a measurable quantity, we might apply a variables quality-control system for the mean or standard deviation such as those discussed in the previous section on process control. The only difference in the application is that acceptance or rejection of the hypothesis that the mean or standard deviation is at its desired value results in acceptance or rejection of the lot, rather than a decision to seek or not to seek out an assignable cause of variation in the production process. Of course, a given quality system might be used for the dual purpose of acceptance sampling and process control where a dual application is appropriate.

Example 14.7

An army medical facility purchases thermometers in large quantity. The thermometers contained in each shipment are to have a mean error of 0.0°F. The standard deviation of error per shipment has been found to be 0.10°F and has proved to be relatively constant from one shipment to another. However, mean error per shipment does tend to vary. The medical facility has developed an acceptance sampling plan for the mean with control limits $C_L = -0.0438$ and $C_U = 0.0438$, where the sample size is 20 and $\alpha = 0.05$. A shipment of thermometers has been received and a random sample of 20 units has been inspected with the results shown in the table below. Should the lot be accepted or rejected?

Unit Number	Error	Unit Number	Error
1	−0.0480	11	0.1680
2	0.0070	12	−0.2210
3	0.0350	13	0.0400
4	0.1330	14	−0.0030
5	0.0330	15	−0.0280
6	−0.2130	16	−0.1670
7	0.0560	17	−0.0710
8	0.0590	18	0.1480
9	−0.0270	19	0.0840
10	−0.1680	20	−0.0130

The sample mean error $\bar{x}$ for the 20 samples is given by

$$\bar{x} = -0.0098$$

Since $C_L \leq \bar{x} \leq C_U$ we would conclude that the lot is of acceptable quality. ■

A frequently chosen alternative to acceptance sampling by variables is acceptance sampling by attributes. Sampling by attributes is also used as an alternative to variables sampling in the case of process control, but to a lesser extent. In sampling by attributes we simply record the number of defects found in the sample. The number of defects in the sample is treated as an indication of the number or proportion of defects in the lot from which the sample was drawn.

Let L be the number of items in the lot, N the number of defective items in the lot, n the sample size, and x the number of defective items in the sample. Then x has a hypergeometric distribution with probability mass function given by

$$p(x) = \frac{\binom{N}{x}\binom{L-N}{n-x}}{\binom{n}{x}} \qquad x = 0, 1, 2, \ldots, n \tag{14.38}$$

The proportion defective for the lot, q, is given by

$$q = \frac{N}{L} \tag{14.39}$$

The probability mass function of x is then given by

$$p(x) = \frac{\binom{qL}{x}\binom{L-qL}{n-x}}{\binom{n}{x}} \qquad x = 0, 1, 2, \ldots, n \tag{14.40}$$

When the lot size is large, as is often the case, the hypergeometric distribution is computationally tedious to use. If the sample size is small relative to the lot size, the hypergeometric distribution may be approximated by the binomial distribution, and $p(x)$ is given by

$$p(x) = \binom{n}{x} q^x (1-q)^{n-x} \qquad x = 0, 1, 2, \ldots, n \tag{14.41}$$

Finally, for most quality-control applications, the binomial distribution may be approximated by the Poisson, where

$$p(x) = \frac{(nq)^x}{x!} e^{-nq} \qquad x = 0, 1, 2, \ldots \tag{14.42}$$

The number of defective units found in the sample is used to determine whether a submitted lot should be accepted or rejected. The criterion used for acceptance or rejection of a lot is the *acceptance number* c. Specifically, if $x \leq c$ the lot is accepted and if $x > c$ the lot is rejected.

If q_0 is the proportion of defective items that is deemed acceptable, then the probability of a Type I error α is the probability that the lot is rejected if the actual lot proportion defective q equals q_0, or

$$\alpha = P[\text{lot rejected} | q = q_0] \tag{14.43}$$

or

$$\begin{aligned} \alpha &= P[x > c | q = q_0] \\ &= 1 - P[x \leq c | q = q_0] \end{aligned} \tag{14.44}$$

To specify the required acceptance sampling plan, we must determine the value of c that satisfies Eq. (14.44) for a specified value of α. If x is assumed to be Poisson distributed we have:

$$P[x \leq c | q = q_0] = \sum_{x=0}^{c} \frac{(nq_0)^x}{x!} e^{-nq_0} \tag{14.45}$$

and c must satisfy the relationship:

$$\sum_{x=0}^{c} \frac{(nq_0)^x}{x!} e^{-nq_0} = 1 - \alpha \tag{14.46}$$

Since x and c are discrete variables, there may be no value of c that satisfies Eq. (14.46) precisely. Hence, we attempt to find the value of c that satisfies Eq. (14.46) as nearly as possible.

Example 14.8

A motel chain purchases glasses in lots of size 10,000. From each lot, a sample of size 110 is to be drawn and inspected. A glass containing a crack or break is considered defective. A lot that contains a proportion defective of 1 percent or less is considered acceptable. The probability of rejecting a lot that is 1 percent defective is to be 0.02. What should the acceptance number c be?

From the statement of the problem we have:

$$n = 110$$
$$q_0 = 0.01$$
$$\alpha = 0.02$$

Application of Eq. (14.46) yields:

$$0.98 = \sum_{x=0}^{c} \frac{(1.1)^x}{x!} e^{-1.1} \tag{14.47}$$

The calculations required for the determination of c are summarized in Table 14.2. The analysis in Table 14.2 shows that $c = 3$ most nearly satisfies Eq. (14.47)

The operating characteristic curve describes the probability of lot acceptance for a continuum of values of q. As we would expect, the probability of lot acceptance is greater than $(1 - \alpha)$ for $q < q_0$ and is less than $(1 - \alpha)$ for $q > q_0$. ■

Table 14.2 Trial and Error for the Value of c for Example 14.8

c	x	$\frac{(1.1)^x}{x!} e^{-1.1}$	$\sum_{x=0}^{c} \frac{(1.1)^x}{x!} e^{-1.1}$
0	0	0.3329	0.3329
1	1	0.3662	0.6991
2	2	0.2014	0.9005
3*	3	0.0738	0.9743
4	4	0.0203	0.9946

Example 14.9

Plot the O.C. curve for the acceptance sampling plan in Example 14.8. What is the effect of decreasing the sample size to 50 or increasing it to 200 on the O.C. curve?

Let

$$n' = 50 \qquad n'' = 200$$

If the value of α is to be 0.02 when $q_0 = 0.01$ for all three sampling plans, the corresponding acceptance numbers c' and c'' are defined as shown in Table 14.3, where c' and c'' must satisfy the relationships:

$$0.98 = \sum_{x=0}^{c'} \frac{(0.50)^x}{x!} e^{-0.50} \tag{14.48}$$

$$0.98 = \sum_{x=0}^{c'} \frac{(2.00)^x}{x!} e^{-2.00} \tag{14.49}$$

The results shown in Table 14.3 yield the sampling plans:

$$n' = 50 \qquad c' = 2$$
$$n'' = 200 \qquad c'' = 5$$

■

Table 14.3 Trial-and-Error Solution for c' and c'' Corresponding to $n' = 50$ and $n'' = 200$ in Example 14.9

Determination of c'				Determination of c''			
c'	x	$\frac{(0.50)^x}{x!} e^{-0.50}$	$\sum_{x=0}^{c'} \frac{(0.50)^x}{x!} e^{-0.50}$	c''	x	$\frac{(2.00)^x}{x!} e^{-2.00}$	$\sum_{x=0}^{c''} \frac{(2.00)^x}{x!} e^{-2.00}$
0	0	0.6065	0.6065	0	0	0.1353	0.1353
1	1	0.3033	0.9098	1	1	0.2707	0.4060
2*	2	0.0758	0.9856	2	2	0.2707	0.6767
				3	3	0.1805	0.8572
				4	4	0.0902	0.9474
				5*	5	0.0361	0.9835

The O.C. curves for these and the acceptance sampling plan in Example 14.8 are shown in Figure 14.10, where

$$P_q = P[\text{lot acceptance} \mid q] \tag{14.50}$$

Figure 14.10 illustrates the type of information provided by the operating characteristic curve. As in the case of process control, as the parameter to be controlled, in this case proportion defective, deviates in an undesirable manner from the desired value of that parameter, the probability of accepting the lot decreases. It should be noted that an undesirable shift from the desired value of q, q_0, occurs only when $q > q_0$, while any shift from the desired mean μ_0 in the case of process control or acceptance sample is undesirable when this is the parameter controlled. Figure 14.10 also demonstrates that the sensitivity of an acceptance sampling plan to undesirable shifts in proportion defective can be increased by increasing the sample size for a constant value of α. That is, for a constant value of α, P_q decreases for a given value of q as n increases.

Acceptance sampling plans of the type just discussed are referred to as single sampling plans since they require a single sample to determine whether a lot should be accepted or rejected. Where the cost of inspection is high, it may be desirable to implement a *double sampling plan*. A double sampling plan specifies two sample sizes, n_1 and n_2, and two acceptance numbers, c_1 and c_2. An initial sample n_1 is drawn and

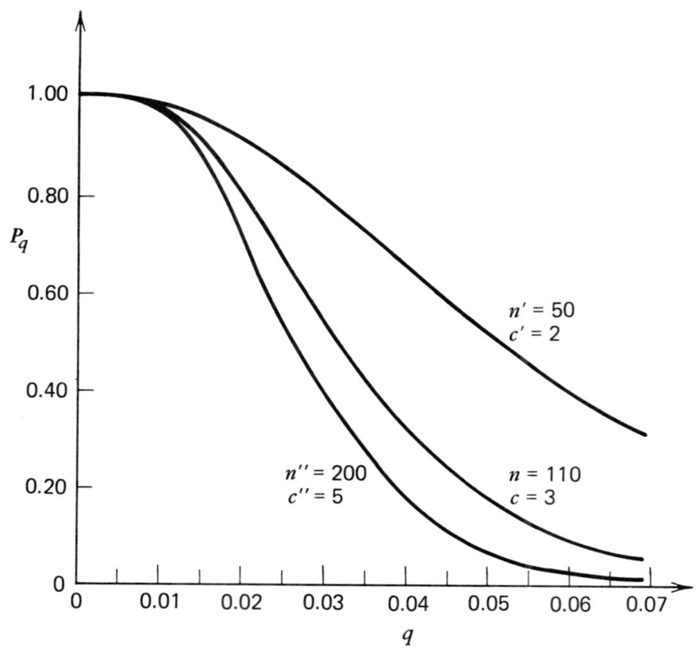

Figure 14.10 Operating Characteristic Curves for the Problem in Example 14.9

the number of defective units x_1 is recorded. If $x_1 \leq c_1$ the lot is accepted, and if $x_1 > c_2$, it is rejected. If $c_1 < x_1 \leq c_2$ a second sample of n_2 units is selected. If x_2 is the number of defective units in the second sample, the lot is accepted if $x_1 + x_2 \leq c_2$ and is rejected otherwise. Double sampling may be extended to *multiple sampling* where three or more samples are drawn. Let m be the number of samples that may be selected. Associated with each sample we define a sample size n_i and an acceptance and *rejection number*, c_i and r_i, respectively, $i = 1, 2, \ldots, m$, where $c_1 < c_2 < \cdots < c_m$, $r_1 < r_2 < \cdots < r_m$. If $x_1 \leq c_1$ the lot is accepted and if $x_1 \geq r_1$ the lot is rejected. If $c_1 < x_1 < r_1$ a second sample of n_2 units is selected, and if $x_1 + x_2 \leq c_2$ the lot is accepted and is rejected if $x_1 + x_2 \geq r_2$. If $c_2 < x_1 + x_2 < r_2$ a third sample of n_3 units is drawn, and $x_1 + x_2 + x_3$ is compared with c_3 and r_3. This process is repeated until the lot is accepted or rejected. For the mth sampling plan, $r_m = c_m + 1$ so that a decision must be reached by the time the mth sample is inspected. Finally, multiple sampling may be extended to sequential sampling where there is no limit on the number of items that may be inspected.

14.5 SUMMARY

In this chapter we present the fundamental principles of quality control, emphasizing the types of problems to which quality control may be effectively applied, process control and acceptance sampling, and the statistical basis for the development of quality-control plans. For a detailed discussion of statistical quality control, see References 4, 6, or 9.

REFERENCES AND SUGGESTED READING

1. K. V. Bury, *Statistical Models in Applied Science,* John Wiley, New York, 1975.
2. K. E. Case, G. K. Bennett, and J. W. Schmidt, "Cost-Based Acceptance Sampling," *Industrial Engineering,* **4,** No. 11 (November 1972).
3. D. J. Cowden, *Statistical Methods in Quality Control,* Prentice-Hall, Englewood Cliffs, NJ, 1957.
4. A. J. Duncan, *Quality Control and Industrial Statistics,* Irwin, Homewood, IL, 1959.
5. I. N. Gibra, *Probability and Statistical Inference for Scientists and Engineers,* Prentice-Hall, Englewood Cliffs, NJ, 1973.
6. E. L. Grant and R. Leavenworth, *Statistical Quality Control,* 4th ed., McGraw-Hill, New York, 1972.
7. A. Hald, *Statistical Theory with Engineering Applications,* John Wiley, New York, 1952.
8. B. L. Hansen, *Quality Control: Theory and Applications,* Prentice-Hall, Englewood Cliffs, NJ, 1963.
9. J. M. Juran (Ed.), *Quality Control Handbook,* 2nd ed. McGraw-Hill, New York, 1962.

10. W. A. Shewhart, *Economic Control of Quality of Manufactured Product,* Van Nostrand, New York, 1931.

11. R. C. Vaughn, *Quality Control,* Iowa State University, Ames, Iowa, 1974.

REVIEW QUESTIONS AND PROBLEMS

14.1 In Example 14.1, how much would the standard deviation of error have to be reduced for 98 percent of the watches to maintain an accuracy of ±1 min/day?

14.2 In Example 14.3, how much can the mean vary from 12.00 fluid ounces such that 97 percent of the bottles filled will contain between 11.90 and 12.10 fluid ounces of beer?

14.3 In Example 14.4, suppose that the process is to produce tubes with diameter 3.00 ± 0.02 in. 95 percent of the time if the mean is 3.00. Is the process capable of meeting this requirement?

14.4 Find the control limits C_L and C_U for the problem in Example 14.4 if the probability of a Type I error is 0.10.

14.5 The control limits for a quality control system are $C_L = 2.95$ and $C_U = 3.05$. If $\mu_0 = 3.00$ $\sigma = 0.20$, and $n = 100$, what is the probability of a Type I error?

14.6 A production line produces parts that are to be 3.000 in. in length. The standard deviation of length has been found to be 0.020 in. The following data have been collected from the production line over a 1-hr period.

Part Length (in.)	Part Length (in.)	Part Length (in.)
2.990	2.993	2.998
3.029	3.004	3.010
2.975	2.971	2.978
3.014	3.010	3.003
3.021	2.979	3.018
2.992	2.971	3.011
3.003	2.988	3.026
3.024	2.993	2.995
2.995	3.034	3.028
3.011	2.994	3.018
2.986	3.001	3.014
3.027	3.033	2.999
3.037	2.945	3.004
2.998	3.016	3.030
3.030	2.994	2.982
3.022	2.991	2.985
3.011	2.984	2.981

If $\alpha = 0.05$ would you conclude that the line is producing parts with mean length 3.000 in.?

14.7 Given the scenario in Problem 14.6, would you conclude that the process is in control, based upon the following data?

Part Length (in.)	Part Length (in.)	Part Length (in.)
2.983	3.020	3.027
3.020	2.996	3.019
3.016	3.028	2.999
3.018	3.002	3.008
3.053	3.016	3.004
2.993	3.030	3.010
2.997	3.012	3.000
3.010	3.027	3.024
3.023	3.011	3.014
3.013	2.984	3.005

14.8 A variables quality-control system is used to determine whether or not a production process is in control for the day during which the sample was drawn. If the control limits are established such that $\alpha = 0.10$, what is the probability that the quality-control system will indicate that it is out of control?

(a) 1 day out of 3 when it is in control all three days
(b) 2 days out of 3 when it is in control all three days
(c) 3 days out of 3 when it is in control all three days

14.9 In Example 14.4, if the true mean diameter μ is 3.0010 for a given day's production, what is the probability that the quality-control system will indicate that the process is out of control?

14.10 Plot the operating characteristic curve for the quality-control system in Problem 14.5.

14.11 Plot the operating characteristic curve for the quality-control system in Problem 14.6, where $n = 51$.

14.12 Given the quality-control system in Problem 14.6, what is the probability that the process will be correctly identified as in and out of control on three consecutive days if μ_i is the true process mean on the ith day and $\mu_1 = 3.001$, $\mu_2 = 3.000$, $\mu_3 = 2.998$?

14.13 For the situation discussed in Example 14.4, the quality-control system is to indicate an out-of-control condition with probability 0.90 if $\mu = 2.997$ or $\mu = 3.003$. What sample size would be necessary to meet this condition?

14.14 In Example 14.6 the quality-control system is to indicate an out-of-control condition with probability 0.80 if $\mu = 0.0995$ or $\mu = 0.1005$. What should the sample size be?

14.15 Plot the operating characteristic curve for the quality-control system in Example 14.6 using the sample size in Problem 14.14.

14.16 In Problem 14.6, what sample size would be necessary if the process is to be defined as out of control 95 percent of the time if the mean is less than 2.99 or greater than 3.01?

14.17 Solve the problem in Example 14.8 if the sample size is 75 instead of 110.

14.18 Solve the problem in Example 14.8 if $\alpha = 0.10$ instead of 0.02.

14.19 In Example 14.8, if $q = 0.008$, what is the probability that the lot will be rejected?

14.20 Plot the operating characteristic curve for the quality-control system in

(a) Problem 14.17
(b) Problem 14.18

14.21 For the problem in Example 14.8, find the acceptance number and sample size such that

$$P[\text{lot rejection} | q = 0.01] = 0.01$$

$$P[\text{lot rejection} | q = 0.04] = 0.20$$

14.22 In Problem 14.6, what proportion of defective units would be expected from the production process when the mean is 3.000 in.?

14.23 In Problem 14.6, a part is to be considered defective if its length is less than 2.97 or greater than 3.03. If $n = 51$, what should the acceptance number c be if 1 hr's production is to be accepted 90 percent of the time when $\mu = 3.000$?

14.24 Given the attributes sampling plan in Problem 14.23 and the data in problem 14.6, would you accept the production lot from which the sample was drawn?

14.25 Given the scenario in Problem 14.23, what should the acceptance number c be if $n = 30$?

14.26 Given the attributes sampling plan in Problem 14.25 and the data in Problem 14.7, would you accept the production lot from which the sample was drawn?

14.27 Let x be a Poisson random variable with probability mass function given by

$$p(x) = \frac{\lambda^x}{x!} e^{-\lambda} \qquad x = 0, 1, 2, \ldots$$

Show that

$$p(x + 1) = \frac{\lambda}{x + 1} p(x)$$

14.28 Let x be a binomial random variable with probability mass function given by

$$p(x) = \binom{n}{x} q^x (1 - q)^{n-x} \qquad x = 0, 1, 2, \ldots, n$$

Show that

$$p(x + 1) = \left(\frac{n - x}{x + 1}\right)\left(\frac{q}{1 - q}\right) p(x)$$

14.29 Let $q = 0.005$ and $n = 100$. Find the value of x such that the absolute error in approximating the binomial probability mass function by the Poisson probability mass function is maximized. What is the maximum absolute error?

SECTION FIVE

TYING IT ALL TOGETHER THROUGH PRODUCTIVITY

As explained in the next and final chapter, productivity is the notion of how much mileage we are getting out of the various resources we put into a given job. Getting good productivity performance from the labor force, raw materials, energy, and equipment is crucial to the long-term profitability and competitiveness of a firm. A productive company is a cost-competitive company. It prospers in good times and can weather recessions.

Improving productivity can involve increasing the output of a job or operation while keeping the amount of inputs required at a fixed level. It can also involve holding the output fixed while reducing the input requirements. This second approach to increasing productivity (and, to a degree, the first approach as well) is precisely the primary target of industrial engineering and operations research. In fact, this book could have just as appropriately been titled "IE/OR: The Productivity Improvement Discipline."

Each of the previous chapters has a direct or an indirect bearing on productivity. Some of the topics presented are useful in understanding and analyzing productivity problems. Other topics describe various tools and techniques that are used in industry to improve the productivity performance of an individual worker, a particular operation, or an entire system of workers and facilities.

To fully appreciate this connection between productivity and both traditional IE and operations research, it is necessary to discuss productivity as a topic in itself. Therefore, we devote our final chapter to the subject of productivity—its definition, how it is measured, and how it can be improved.

Chapter 15

Productivity

15.1 THE MEANING OF PRODUCTIVITY

Productivity is a concept that concerns how well we use our resources. Essentially, productivity is the relationship between the level of the output of a process or system and the level of input resources required to achieve that output. In other words, productivity is "how much bang we get for the buck"—for example, 120 parts produced per worker, or perhaps 89 gallons of gasoline refined per barrel of crude oil used. A productive plant is one that has a large production volume of good units (that is, acceptable or nondefective products) relative to low amounts of material, energy, labor, capital, and other resources consumed.

Productivity is closely related to the profitability of an organization. Like productivity, profitability is a way to look at the relationship between the output of a process and the associated inputs. The primary difference is in how this relationship is defined or modeled. Profits are the net results of subtracting the costs of inputs from the revenues generated by the sale of production output. Productivity, on the other hand, is normally expressed as the *ratio* of the production output divided by the inputs. And productivity measures are usually developed on a *deflated* basis; that is, they remove the effects of price inflation. Profitability accounting does not. Another difference is that whereas profitability is dollar-oriented, productivity is physically oriented—production units, manpower, and so forth.

Normally, a productive firm is also a profitable firm. This is because the more productive we are in using resources, the more mileage we get out of those resources. It takes less time, effort, and material to produce the same amount of parts or units. However, because of the sales orientation of profits, an organization may be very efficient and productive in the use of its resources, but market conditions might not allow it to sell its products at a high enough price to earn a net profit.

As an example, consider an independent commercial trucker who transports goods from one town to another. Through engine tune-ups, proper tire maintenance, and investment in a wind-deflecting airfoil, the trucker increased the gasoline mileage capability of his truck by 35 percent. His "energy productivity"—that is, the efficiency with which he used the major energy-related resource in his trucking operation—increased significantly. However, his pocketbook may never see the "bottom-line" effect of this productivity gain if the price he has to pay for gasoline escalates or inflates to the extent that his trips still cost more than they did before he made the improvements in gasoline productivity.

15.1.1 Components of Productivity Management

Almost every business has an accounting system that tracks and reports for management use information on the profitability of the business. This accounting system is based on specific measures or models of costs and profits, such as the cost effects of depreciation of equipment. In addition to this measurement and reporting, most firms are actively engaged in efforts to improve profits, either through formal "profit improvement programs" or through informal, independent projects aimed at reducing costs or increasing revenues.

Likewise, with respect to productivity, many organizations also have efforts aimed at the same three areas or components: *measurement, reporting,* and *improvement.* The ultimate purpose of focusing attention on the concept of productivity is improvement. However, we have to know what areas of our operations *need* to have their productivity improved. Where are the problems? In order to answer this question, we need to be able to *measure* and analyze the level of productivity of our operations. And in order to ensure that adequate management attention is focused on the problem areas, a *reporting* system needs to be maintained. This system provides a necessary monitoring of productivity performance.

15.1.2 Productivity Programs

Although productivity improvement has been an informal concern of most firms since Fredrick Taylor's era, widespread use of formal company programs to measure, report, and improve productivity is a relatively recent development.

The present-day extent of these programs is reflected by a survey of the 1300 largest U.S. corporations conducted by Sumanth and Einspruch [11]. Seventy-five percent of the industrial companies responding to the survey reported having a formal productivity *measurement* program. Twenty-three percent of these companies had departments at the corporate level specifically responsible for productivity programs. Eighteen percent had such departments at the plant level. The orientation of these departments is often toward the industrial engineering function, as evidenced by the fact that more than one third of the departments had heads whose title is IE-related, such as manager of industrial engineering. Nine percent of these departments used titles exclusively oriented to productivity (for example, productivity coordinator).

Sixty-two percent of the responding industrial companies have formal productivity *improvement* programs. Forty-one percent of these programs were located at the corporate level, while 48 percent were located at the plant or division level.

15.1.3 Productivity and IE/OR

As indicated by the relatively large percentage of IE-related productivity departments in the Sumanth and Einspruch survey, there is a very fundamental relationship between productivity and the industrial engineering/operations research discipline. Essentially, IE/OR *is* the practice of productivity analysis and improvement. Through the techniques involved in areas such as engineering economy (see Chapter 6) and work measurement (Chapter 4), the IE/OR analyst has developed and utilized methods to measure and analyze productivity, primarily at the workplace level (rather than at the firm level). This has included looking at measures such as the length of time required to produce one part on a machine, that is, the productivity of the machine. "Standards" have also been developed to assist in evaluating the reported productivity of an operation.

Improving the productivity of plant operations is a primary objective of most IE/OR methods and studies. Even though words such as "improving efficiency" or "reducing unit cost" are usually used rather than "improving productivity," the result is the same. This is true of all of the traditional IE activities discussed in Section Two, as well as all of the OR techniques described in Section Three and the application areas mentioned in Section Four. For instance, queueing theory (Chapter 9) is often used to find ways to increase the throughput of a waiting-line situation having a fixed number of servers. Such a study is essentially an effort to increase productivity as defined as output divided by input or, in this case, the number of customers served (i.e., throughput) divided by the number of servers used. Likewise, mathematical programming (Chapter 8) is frequently used to determine how certain resources, such as men or materials, can be most efficiently employed and at the same time used to maximize the output of those resources. Clearly, this is an attempt to increase the output of a process while at the same time reducing the required input, that is, increasing the productivity of the process.

In the remainder of this chapter, we shall discuss two topics in detail. First, we describe some of the approaches often used to *measure* productivity. In this discussion we present several models commonly used that capture a partial view of the productivity of a process or plant, as well as those that capture the total productivity picture. The second major topic presented will be methods to *improve* productivity, including one outside the IE/OR domain—quality circles.

15.2 PRODUCTIVITY MEASUREMENT

The discussion here focuses on quantitative measures of the level of productivity of a particular plant or business, rather than the productivity of a work station or, at the other extreme, national productivity. That is, of concern here is productivity measurement at the *firm's* level. In particular, the discussion includes a description of the models used in the two basic classes of measures—partial measures and total or multiple factor measures. (Discussion of *national* productivity measurement can be found in several articles referenced in References 7 or 10.)

15.2.1 Partial Measures

There are no formal, universally accepted definitions of productivity measures. Companies define their own measures to suit the nature of their business. However, these measures are usually expressed as a ratio of an indicator of output or accomplishment relative to (divided by) an indicator of the inputs consumed in producing that output.

These ratios are defined by an organization in such a way as to shed light on a particular function or resource within the organization rather than the composite productivity of *all* resources used by the firm. Such ratios are called *partial* productivity measures since they only reflect part of the total productivity picture of the organization. For example, the measure of (tons of product produced)/(pounds of raw material used) is often employed to monitor the efficiency of the raw material conversion in a given period of time for a production process. However, this ratio reflects the efficient utilization of only one resource—raw material. It does not capture the effects of labor, capital, and other resources used to produce the output of the firm.

In general, the types of partial productivity ratios used by an organization can be grouped into two basic areas according to their scope:

1. *Overall ratios*. These are used to capture the productivity of the entire organization rather than one specific department or function. Typical of this type of measure are ratios like (sales)/(employee) or (profit)/(capital dollar employed), or (university graduates produced)/(number of faculty members).
2. *Objective ratios*. These are ratios that reflect the productivity of a specific activity or area in the organization. Such a measure might be set up to monitor the performance of an ongoing activity in the organization, such as to keep track of the weekly labor efficiency of the casting department in a foundry. Alternatively, such a measure might be used to assess the performance of a one-time project in terms of predefined goals or objectives of the project. Typical objectives ratios are (number of capital projects completed)/(number of projects started), (tons of product casted)/(number of man hours worked in the casting department) or (actual progress in a labor negotiation)/(scheduled progress to date).

Partial productivity ratios such as those cited above refer to outputs and inputs in a specific period of time. For example, (sales)/(employee) refers to the sales revenue booked by a company in a particular time period, such as a month, and the number of employees on roll in that same month. Note, however, that the time period of the input need not be the same as the period for the output.

In analyzing the productivity performance of a resource or operation *indices* are usually formed from the basic productivity ratio as reported for different periods of time. For example, to examine the performance of an operation over time, the following index could be defined:

$$\text{Productivity index in period } t = \frac{\text{Productivity of current period}}{\text{Productivity in a base period}}$$

or

$$P_t = \frac{\sum_i O_{it} \Big/ \sum_j I_{jt}}{\sum_i O_{iT} \Big/ \sum_j I_{jT}} \tag{15.1}$$

where

O_{it} = output level in period t as measured in terms of output factor i (e.g., output i = 1 might be sales revenue).

I_{jt} = input required in period t as measured in terms of input factor j (e.g., input j = 1 might be number of employees).

O_{iT}, I_{jT} = output and input levels, respectively, as measured in the base period, $t = T$.

For example, if the productivity of the workers involved in the painting department of a firm is to be analyzed, an index might be used that is based on the ratio of square feet of painted parts in a week to the number of department man-hours worked in the week. Because of its reflection of normal working conditions, the fifth week in the year might be selected as the base period. Assume that 220 man-hours were worked in this base week, and that 25,200 ft^2 of parts were painted. If in the 30th week 28,500 ft^2 were painted and 260 man-hours were worked, then the productivity index for week t = 30 would be

$$P_{30} = \frac{28{,}500/260}{25{,}200/220} = 95.7 \text{ percent}$$

In other words, the paint department experienced a 4.3-percent decline in its labor productivity (as measured by the *particular* indicator chosen), even though its output increased by 13 percent.

When this type of index is routinely developed over time, the *trend* of the productivity performance of a resource or operation can easily be established. This information is essential to a complete analysis of the resource's productivity. A simple graph of these indices can give management a quick feel for areas that need attention. As an illustration, the plot of two series of indicators for an aluminum plant is shown in Figure 15.1. One series depicts the company's labor productivity as measured by the ratio of net sales revenue for the year to the average number of employees on roll in the year. The other series measures the productivity of the capital invested in buildings, equipment, and inventory. The ratio used here was net sales revenue for the year to total dollars invested in capital items (equipment, facilities, etc.) Note that Figure 15.1 readily points out that management attention needs to be given to capital useage rather than labor efficiency.

15.2.2 Use and Interpretation of Partial Measures

There are several general guidelines that are normally followed in constructing and using partial productivity measures. These include the following.

Figure 15.1 Illustration of Productivity Trends

1. When dollars are involved in either the output or input factor in a ratio, a *deflator* should be used to convert the given dollar figure into an equivalent amount expressed in "base period or constant dollars." That is, the effect of inflation should be removed when developing an index that compares dollars in a base period with dollars in a future period. The proper deflator to use depends on the factor involved. For example, the percentage change in the Consumer Price Index (CPI) from the base period to the period of interest might be used to convert current wageroll costs into base period wageroll cost. Other available deflators are the Wholesale Price Index (WPI) and the GNP Implicit Price Deflator. The use of an index to deflate a current dollar figure is really a shortcut for calculating the deflated amount using base-period prices and current period volumes. For instance, the 1978 dollar value of a store's sales of tennis rackets in 1981 is the 1978 price per racket multiplied by the number of rackets sold in 1981.
2. The productivity measure should be defined with its end use in mind. For instance, measures can be set up to reflect the "mileage" we are getting out of our resources, whereas other measures can be defined to capture how "efficiently" we are working. Measures such as (constant sales dollars)/(number of employees) reflect the amount of output we get from a unit of input

resource, that is, the mileage we get from a unit of input. These types of measures do not necessarily provide information on how *efficiently* we use a resource since there are many factors other than efficiency that affect the measure. For example, if we go from a level of $200/man of sales to a level of $175/man, the reason may be that our marketing people did a poor job in securing sales orders even though our production employees worked much more efficiently. We must develop ratios such as (standard manhours required)/(actual manhours used) to reflect efficiency properly.

3. The reasons for changes in productivity measurements are not always what they seem to be. In other words, the cause of the change in a productivity ratio's value from one period to another may not be a true "productivity" change but may be due to extraneous factors such as shifts in the economy, quality of raw material fluctuations, or the introduction of new products. For example, if the ratio of tons produced per man-hour worked is used as the measure of labor productivity in a chemical plant, there are several factors that might cause this ratio to change other than how hard or how cleverly the employees work. Specifically, the ratio will increase if the chemical reactions involved become more effective owing to, say, a new catalysis being used. Many other extraneous factors influence this ratio, such as the production schedule itself, raw material availability, and equipment breakdowns due to poor maintenance. Note, however, that although these extraneous causes do not reflect changes in labor "efficiency," they nevertheless contribute to the mileage we are getting out of the labor workforce.
4. Because of complications such as mentioned in 3 above, partial measures are not stand-alone indicators. They should be used in conjunction with other partial productivity measures as well as with other data.
5. Productivity measures are descriptive mathematical *models*. They are representations of the relationship between an organization's output and its associated resources consumption or inputs. Like any other model, when we define a productivity measure we are making an implicit trade-off between the accuracy of the model and its utility. That is, a measure such as (pounds casted)/(BTUs of electricity consumed) may accurately portray our energy productivity picture, but it would have little utility if our current data system does not capture the BTUs of electricity used in the casting process. Another modeling property that is applicable to productivity measures is GIGO (Garbage In, Garbage Out). Specifically, the measure may be properly formulated for our end purpose, but if the data used in the measure are in error then our conclusions about productivity will be worthless.

15.2.3 Total Factor Measures

By far the most common approach to tracking productivity in an organization is through the use of one or more partial productivity measures like those described above. And usually these concentrate on labor productivity rather than on other resources such as raw materials or fuels. However, some firms employ models or measures that incorporate the effects of several resources or inputs simultaneously.

These models are called total factor or *multiple factor models*. Such measures offer the advantage of depicting the impact on output of the effect of changes in several diverse factors rather than of just one factor such as labor. For example, a 1981 study of Fortune 500 companies revealed that the most profitable and most successful 16 firms *lagged* behind the average Fortune 500 company in labor productivity. The average (sales)/(employee) value of the 16 most profitable firms was 24 percent less than that of the average company's $84,000/employee [8]. Because these were 16 highly successful companies, they must have been productively using other resources rather than labor. Use of a total factor model rather than the single labor measure would indeed have shown this fact and allowed a composite, overall picture of the firms' productivity to be obtained.

Perhaps the first multiple-factor productivity model to be widely advocated was that suggested by Craig and Harris [4]. Their model for measuring productivity in period t is:

$$P_t = \frac{O_t}{I_{Lt} + I_{Ct} + I_{Rt} + I_{Qt}} \tag{15.2}$$

where

O_t = the total output of good units produced in period t, measured in deflated or base-year dollars.

I_{Lt} = the base-year dollar value of all labor consumed in period t.

I_{Ct} = the base-year dollar value of capital consumed in period t.

I_{Rt} = the base-year dollar value of raw material consumed in period t.

I_{Qt} = the base-year dollar value of miscellaneous goods and services consumed in period t.

The output quantity used in Eq. (15.2) is the sum of the deflated dollar value of units produced in each product line, not units sold. This quantity also includes revenue generated by the firm from nonproduction sources, such as rents and income from outside investments. The four input factors represented in Eq. (15.2) are common to most organizations. However, not every organization would define or calculate them in the same fashion. Craig and Harris suggest defining the labor input factor I_{Lt} as the sum of the deflated labor cost of man-hours worked in all job categories in the firm. This cost includes base wage rates or salaries as well as fringe benefits. Note that since the labor cost is based on man-hours worked, the labor factor captures the effects of overtime requirements. An alternative is to use only nonovertime hours.

The raw material factor is intended to cover outside purchased materials and parts. Miscellaneous inputs would include "all resources except labor, capital, and raw material/purchased parts." This would cover items such as maintenance supplies, utilities, income and property taxes, and office supplies.

The capital input factor is defined by Craig and Harris to be the amount of money, in deflated dollars, that would be spent by the firm if it leased its capital requirements from an outside leasing company. Essentially, this quantity is the sum of the annual *economic* depreciation over all fixed assets operated by the firm. This economic depreciation is not the accounting or "book" depreciation value of an asset; rather, it is the asset's net current replacement value spread over its useful or economic life. In Craig and Harris' model, this economic depreciation also includes the annual opportunity or interest cost the leasor would charge the firm. The capital factor normally includes nonfixed asset consumptions, such as inventory carrying costs and the annual cost of having cash on hand and accounts receivables outstanding.

A second example of an approach to measuring the firm's overall productivity is to analyze its period-to-period change in profit in terms of two components: productivity changes and inflation effects. That is, the change in profit is dissected into the amount that resulted from changes in the firm's overall productivity and the amount due to the net increase in sales prices over increases in the prices of labor, materials, and other inputs—that is, net "price recovery." For example, if a firm had a growth in gross profit from one year to the next of, say, \$8 million, price recovery might be responsible for \$3 million, whereas productivity improvement contributed \$5 million. If the firm were *less* productive in the second year, then the \$8 million profit might be made up of a \$10 million price recovery contribution and a −\$2 million productivity contribution (loss).

The basic product of this analysis is a graph showing the performance in profitability, productivity, and price recovery over time. Figure 15.2 gives examples of such a graph for two different plastic parts manufacturers. In case (a), the profit center has a steady growth in profitability compared to its base year. This growth is from positive price recovery rather than from productivity performance. The reverse situation is shown in case (b). This second situation is usually more preferred, since in periods of high inflation and recessionary markets, sustained positive price recovery is very difficult.

This "profitability = productivity + price recovery" analysis can be structured in different ways. That is, alternative definitions of profitability can be used—such as gross profit, net income, or operating profit—and different schemes can be employed in splitting the profitability changes between productivity and price recovery. One specific approach that is advocated by the American Productivity Center involves first obtaining ratios of productivity and price recovery changes from one period to the next, and converting these into dollar values. Another approach is to use a "constant margin" analysis (see Miller [14] for details).

In this analysis, the gross profit in a period is compared with the gross profit that would have been realized had the firm achieved the same profit *margin* (that is, gross profit divided by net sales revenue) as they realized in a base period. This anticipated profit at the base period's margin constitutes a standard that is used to judge how well profits grew since the base period. This comparison can be represented as follows.

$$\Delta G_t^A = (R_t^A - C_t^A) - \left(\frac{R_{t-1}^A - C_{t-1}^A}{R_{t-1}^A}\right)(R_t^A) \qquad (15.3)$$

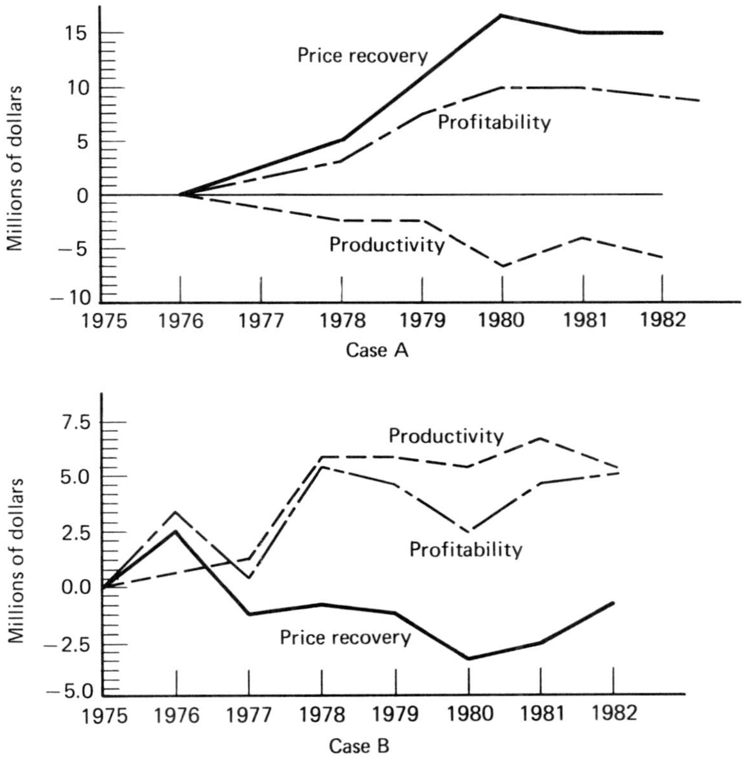

Figure 15.2 Actual Illustrations of the Results of a "Profitability = Productivity + Price Recovery" Analysis

where

R^A_{t-1}, R^A_t = the actual or reported net sales revenue in period $t - 1$ (the base period) and period t, respectively.

C^A_{t-1}, C^A_t = the actual cost of sales in periods $(t - 1)$ and t.

ΔG^A_t = the gain (or shortfall if $\Delta G^A_t < 0$) of gross profit in period t beyond its standard at constant, base period margin.

A positive value of ΔG^A_t indicates a gain in profitability of the firm. However, a portion of this gain may result from revenue addition owing to net price recovery, and the remainder owing to cost reductions due to productivity gains. In order to determine the appropriate allocation of profit gain between these two factors, we can make a comparison similar to Eq. (15.3), but use only *real* or deflated revenues and costs. That is, find:

$$\Delta G^D_t = (R^D_t - C^D_t) - \left(\frac{R^D_{t-1} - C^D_{t-1}}{R^D_{t-1}}\right)(R^D_t) \qquad (15.4)$$

where

R_{t-1}^D, R_t^D = the deflated dollar value of net sales revenue in periods $t - 1$ and t, respectively.

C_{t-1}^D, C_t^D = the deflated dollar value of cost of sales in periods $t - 1$ and t.

ΔG_t^D = the gain (or shortfall) of gross profit owing to productivity effects beyond those of the base period.

The comparison (i.e., subtraction) implied in Eq. (15.4) is equivalent to comparing the overall or multiple-factor productivity in period t with that of period $(t - 1)$. That is, we are comparing directly the ratios of deflated output R_t^D divided by deflated overall input C_t^D in the two periods. Therefore, changes in the deflated portion of profits ΔG_{t-1}^D are due directly to changes in productivity.

The deflated output and inputs in Eq. (15.3) are analogous to those used in the previously mentioned multiple-factor productivity model of Craig and Harris, Eq. (15.2). For instance, the deflated cost of sales C_t^D is the sum of the deflated value of labor consumed I_{Lt}, capital consumed I_{Ct}, raw material used I_{Rt}, and miscellaneous cost inputs I_{Qt}. Obviously, there are different ways of defining these (and other) cost components, and different ways of deflating them. There is really no one "right" way of structuring such a model.

Once the productivity contribution to the profitability gain or shortfall has been determined, the price recovery portion can be found by differencing. That is:

$$\Delta PR_t = \Delta G_t^A - \Delta G_t^D \qquad (15.5)$$

Example 15.1

This approach to a "profitability = productivity + price recovery" analysis was performed on a large, Fortune 500 textile fiber firm, the Celanese Corporation, for the period from 1976 to 1981. The input data used in the analysis were taken from the company's annual report. The result is shown graphically in Figure 15.3. The margin and productivity level in effect in 1976 is used as the base for each year's cumulative growth comparison. The basic calculations involved for one particular year's analysis, 1978, are given in Table 15.1.

As can be seen in Figure 15.3, Celanese had a slightly favorable profitability growth from 1976 through 1978. As indicated in Table 15.1, *if* the profit margin that existed in 1976 (period 1) *had been* achieved in 1978 (period 2), a gross profit of \$506.3 million would have been earned. However, an actual profit of \$525.0 million was achieved in 1978, implying a favorable growth in profitability worth \$18.7 million. As indicated in the graph, the reason for this profit growth was an excellent growth in productivity. Specifically, Celanese's total factor productivity in 1976 was \$2,123 million/\$1,711 million = 1.24, and it improved by 9.7 percent to a level of 1.36 in 1978. As shown in Table 15.1, this 9.7 percent increase translates into a \$173.4 million growth in profits. Unfortunately, not all of this growth was realized due to poor price recovery performance. In fact, because the cost increases due to inflation during the

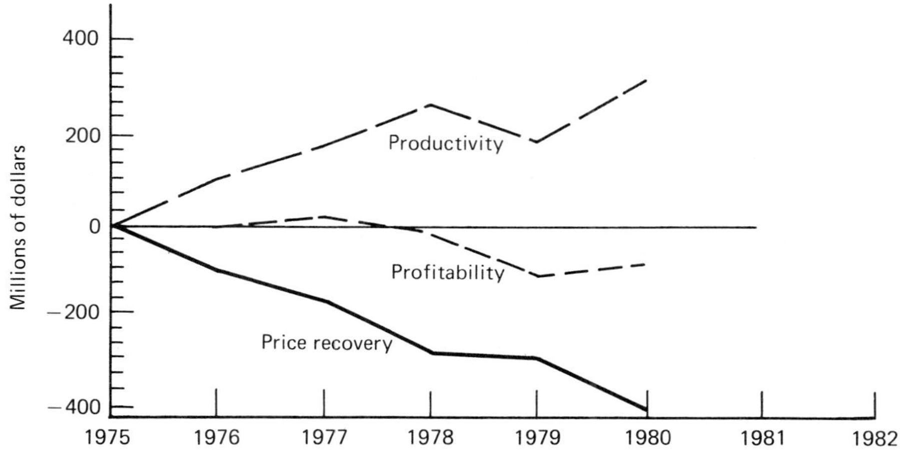

Figure 15.3 The "Profitability = Productivity + Price Recovery" Analysis for the Celanese Corporation

Table 15.1 Sample Calculations for a "Profitability = Productivity + Price Recovery" Analysis

	Period 1	Period 2
Reported net sales ($Million)	2123	2609
Deflation factor	1.000	1.045
Base period value of sales	2123	2497
Reported cost of sales ($Million)[a]	1711	2084
Deflation factor	1.000	1.133
Base period value of cost	1711	1839
Gross profit ($Million)	412	525
Gross profit margin (%)	19.4	20.1
Productivity level	2123/1711 = 1.24	2497/1839 = 1.36

Calculations

(1) Profitability growth:

$$\underbrace{(2609 - 2084)}_{\substack{525 \\ \text{Actual period 2 profit}}} - \underbrace{(2609)\,[(2.23 - 1711)/2123]}_{\substack{506.3 \\ \text{Anticipated period 2} \\ \text{profit at period 1} \\ \text{margin rate}}} = \underbrace{\$18.7 \text{ million}}_{\text{Profit gain}}$$

(2) Productivity contribution:

$$\underbrace{(2497 - 1839)}_{\substack{658 \\ \text{Deflated period 2 profit}}} - \underbrace{(2497)\,[(2123 - 1711)/2123]}_{\substack{484.6 \\ \text{Anticipated period 2} \\ \text{deflated profit at} \\ \text{period 1 productivity rate}}} = \underbrace{\$173.4 \text{ million}}_{\substack{\text{Profit contribution} \\ \text{due to gain in} \\ \text{productivity}}}$$

(3) Price recovery contribution: $18.7 − $173.4 = −$154.7 million

[a]Cost of Sales is made up of material and energy, wages and salaries, depreciation, and miscellaneous expenditures such as supplies; each category is deflated with a separate inflation factor.

period 1976 to 1978 outstripped increases in revenue as generated by increases in sales prices, net price recovery performance caused Celanese to lose the opportunity to gain $154.7 million in profit. The cumulative price increases underlying this poor price recovery performance are shown in Figure 15.4 (where GNP is the Gross National Product implicit Price Deflator). The net or combined effect of the 1978 favorable productivity profit contribution and the shortfall caused by negative price recovery was the $18.7 million growth in profitability. ■

15.3 PRODUCTIVITY IMPROVEMENT

The measurement and analysis of how productive an organization is serves an important management function. It provides insight as to ways in which "the bottom line"—profits—can be improved. And it provides motivation for and monitoring of this improvement.

But how exactly do we improve productivity? The answer is simple, yet complex. There are many activities and mechanisms available that have an impact on the productivity of a specific operation or set of operations. These range from better interpersonal skills-type training for foremen to the replacement of manual material-transfer operations with robots. In general, anything we do to reduce the time, materials, effort, and so forth required to accomplish a given task contributes to improving productivity. Improvements can also be had by increasing the output rate of a task while holding inputs constant—such as by replacing an existing raw material with one that leads to fewer scrapped finished pieces and therefore a higher output of *good* units.

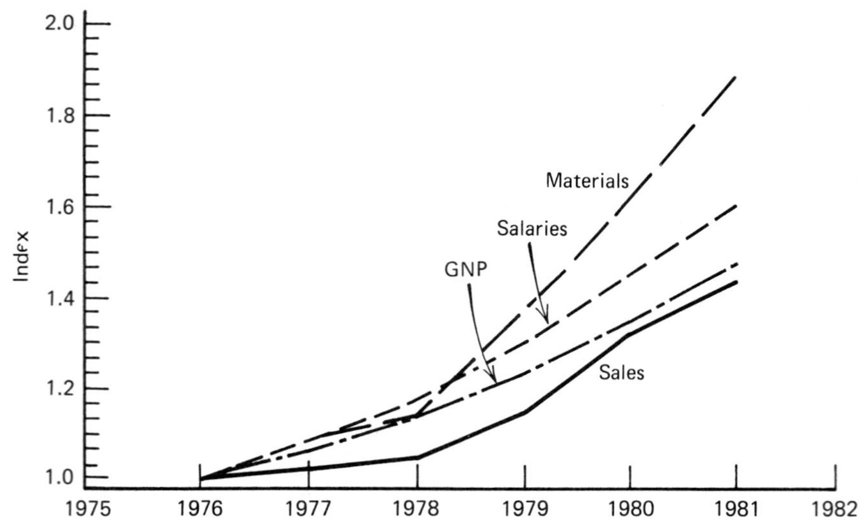

Figure 15.4 Inflation Factors Underlying Celanese's Price Recovery Performance

The discipline of industrial engineering and operations research is devoted to, and has as its primary function, productivity improvement. Fredrick Taylor's pioneering work in job simplification and task standardization is an early example of the application of IE methodology to improve the output-over-input ratio of factory operations. Such efforts were manifested in the first assembly lines of Henry Ford—resulting in such an increase in productivity that automobiles became affordable for people other than just the very wealthy. In the next sections, we discuss how several IE/OR tools are used to make specific impacts on productivity.

15.3.1 Traditional IE Productivity Improvement Tools

All IE tools and techniques could be listed as productivity improvers. For example, graphical models such as flow process charts and left-hand, right-hand charts lead to productivity gains. Specifically, they are used to help find areas in which time and effort are being wasted. Several other specific examples of the impact of traditional IE tools on productivity are cited below.

1. *Product and process planning* (*Chapter 3*). When a product is designed (or redesigned) with a concern for how it is to be produced, it is possible significantly to enhance the labor, capital, energy, and material productivity associated with this production. For instance, proper specification of low-cost material or, alternatively, highly reliable material or purchased subassemblies can reduce the dollar value of raw materials consumed in producing a given number of good output units. Downgrading an excessively high requirement for a smooth surface finish on a machine part can improve labor and capital productivity by eliminating an unnecessary polishing operation, or allowing a low-cost machine such as a wheel grinder to be substituted for a higher-cost polishing machine.

The specification of the size, shape, appearance, and reliability of a product clearly has a direct impact on how easily and how cheaply the product can be made. That is, the design has a direct bearing on the overall productivity with which that product is made. The specification of which *manufacturing processes* are to be used to produce that given design also has a direct bearing on its productivity. As was pointed out in Chapter 3, there are often alternative production operations or processes that can be employed to accomplish a given manufacturing task. For example, it is often possible to use a completely cast part instead of an assembly of several smaller machined parts. This improves labor and capital productivity by eliminating several machining operations. In joining subassemblies together, spot welding can often be used in lieu of arc welding thereby saving time and hence increasing labor productivity. Buffa [3, Chapter 9] provides a good coverage of illustrations of such improvements.

2. *Work measurement/work methods* (*Chapter 4*). Work measurement techniques are data-gathering techniques. In themselves, they are not productivity improvement techniques. However, they are the means by which information is obtained to identify activities that are not being performed as productively or efficiently as possible. For example, a stop-watch time study produces a standard for how long a particular job *should* take. If that job is currently consuming more time than the standard, it is a prime candidate for improvement.

As mentioned in Chapter 4, such improvement can be accomplished by redesigning the *method* by which the job is currently being performed. Through work methods improvements, more efficient means of assembly, machine loading and unloading, material handling, and the like, can be obtained. Such efficiencies lead directly to either (or both) reduced useage of labor hours, or to increased throughput or output—with a resulting improvement in productivity.

3. *Wage incentives* (*Chapter 5*). The major reason that a company installs a wage incentive plan is to *motivate* the workforce to increase the production it achieves in its normal workday. This leads directly to increases in labor productivity since output goes up while labor input (man-hours worked) remains constant. In fact, many wage incentive plans base the extra financial incentives earned by a worker on typical productivity measures such as Eq. (15.1) as it would be used for labor. That is, each period the actual productivity level is calculated for a job or perhaps for an entire department and is compared to the level in a base period (the "standard" productivity level). An increase in performance means an increase in pay to the associated worker or workers. Therefore, the workforce is usually motivated to improve their own productivity—by working smarter, working harder, minimizing their idle time, doing the job right the first time, and so forth. A good discussion of the connection between wage incentive systems and productivity can be found in Ref. 9.

15.3.2 Operations Research Productivity Improvement Tools

The Operations Research methods discussed in Section III of this book generally share a common purpose, which is to generate the best decision possible as to how to operate a given system. In other words, OR methods are employed as aids for determining the most effective, the most efficient way to use our available resources (labor, material, etc.) in accomplishing a required task or operation. In so doing, these methods lead to operating decisions that minimize the use of the resources required to perform the operation. That is, they lead to maximizing the productivity of the operation. The specific impact of several OR methods on productivity improvement is cited below.

1. *Classical optimization* (*Chapter 7*). Those mathematical methods categorized as classical or unconstrained optimization techniques are generally used to derive the minimum-cost or maximize-effectiveness answer to a mathematical model. This model represents how an operation should be designed or run, such as when and to what degree to restock an inventory of spare parts. The solution to the decision model in the inventory case would be a restocking policy such as "reorder 15 parts whenever the on-hand inventory gets down to 35 parts." This solution is "optimal" in the sense that no alternative restocking policy can be found that has a lower cost of inventoring this spare part.

The cost of inventoring normally consists of factors such as the time required to place resupply orders and the "idle" cost of the money tied up in inventoried parts. The optimal restocking policy is the one that minimizes the combined consumption of time and idle capital yet meets the required inventory throughput (or output). Therefore, such a policy is optimal in the productivity sense due to the fact that it minimizes the resource inputs required to obtain a given output. In general, optimization methods, such as search techniques and application of necessary and sufficient conditions, are constructed for the immediate purpose of maximizing the output of an operation or minimizing the use of one or more resources required by that operation.

2. *Mathematical programming* (*Chapter 8*). The main difference between math programming methods and classical optimization techniques is that the latter do not take into explicit account any constraints or restrictions on the decision. Both groups of methods consist of techniques to solve mathematical models representing decisions as to how best to operate a particular system. And therefore both groups produce operating strategies that maximize one or more productivity measures of the system.

As a specific illustration of the impact of math programming methods on productivity, consider the use of linear programming in the oil-refining industry. Linear programming is one particular type of math programming method, and, as applied to an oil refinery, produces a set of operating rules that tell management the most economical proportions of various input materials to use in order to produce a required number of gallons of finished gasoline. In this manner, the demands in a given time period for each type of gasoline are satisfied (i.e., a required output is achieved), and at the same time the consumption of raw materials is minimized. In other words, material productivity is maximized.

3. *Queueing theory* (*Chapter 9*). Queueing theory consists of mathematical techniques designed to infer information concerning the operations of systems of waiting lines. Part of queueing theory also addresses how to design waiting-line systems so as to minimize factors such as the number of servers used or the average amount of time spent waiting in line for service. The impact of queueing theory on productivity is similar to that of math programming. The most efficient means of operating the waiting-line system is found that, at the same time, meets preconditions or restrictions on factors such as ensuring a certain throughput of customers in a given time period. In this manner, minimum consumption takes place of resources such as the manpower used as servers and idle time lost by customers or workers waiting in line for service. The system is designed and operated so that a preset number of customers is served in each time period. Therefore, output is held constant and labor inputs are minimized. Hence, queueing theory leads to the maximization of labor productivity in the operation of the waiting line.

15.3.3 Quality Circles

There are many techniques for improving productivity that are not usually considered part of the IE/OR discipline. Automation of manual operations is an example of such a tool, as is better management training for plant foremen. Another such tool that has proved to be quite successful is the routine involvement of workers in the decision-making and problem-solving aspect of the management of an organization.

Employee involvement programs can be structured in a variety of ways. One approach is to create teams of 6 to 10 co-workers who meet together on a regularly scheduled basis to identify and solve problems in their work area. This approach to employee involvement is generally called ''quality circles.''

The problems addressed by a team or circle usually are issues that inhibit the work performance in the area or department in which the circle members work. These issues range from ''quality of work life'' factors, such as noise problems and area cleanliness, to problems affecting production output and product quality. Typically, the circle selects its own problems to work on, analyzes the problem itself, and makes a formal presentation of its recommendations to management. In analyzing a problem, a circle might employ techniques such as brainstorming and basic statistical tools like a historgram, an ABC or Pareto chart, and cause-and-effect relationship diagrams. The members of the circle receive training in the use of these techniques as well as in general group problem solving.

As an illustration of the type of problems addressed by a circle, consider the following case study. A circle of machinists determined that installation of travel-dial indicators on four machine tools would increase product quality. Management followed their recommendation, and the indicators were installed. Four months later it was found that the indicators did indeed improve quality, as well as improving machine output by 20 percent.

Most circles are formed at the hourly worker level and use the associated foreman as the circle leader. However, circles have also successfully operated in white-collar areas such as in engineering groups. Membership is usually voluntary.

As is the case with many successful production management tools, quality circles were developed by the Japanese using American ideas and concepts. Quality circles began in Japan in 1961 primarily as a tool to improve the quality of their exports. The growth of circles in Japan was very rapid, with approximately 700,000 circles in operation in 1975 [2]. The use of quality circles is now worldwide, although the concept did not become widely accepted in the United States until approximately 1980.

Quality circle programs have both a direct and an indirect impact on productivity. Direct improvement comes from the specific problem-solving recommendations of a circle, such as the machinist case study mentioned earlier. Indirect productivity improvements result because of the existence of the program itself. This is because participation in a circle is an excellent motivational tool. Workers feel more a part of the organization. They can influence their own job design and environment. Consequently, they enjoy their work more and tend to reduce their absenteeism, accidents, and turnover.

For instance, a plant that produces aluminum extrusions installed a quality circle program at a time when employee morale was at a very low level. Three years later, employee turnover had improved by 92 percent, safety improved by nearly 100 percent, and absenteeism improved by 52 percent. Obviously, such changes have a very favorable impact on productivity—fewer overtime hours and fewer ''extra'' workers are needed to produce the same volume of output. In fact, this plant experienced a 16 percent increase in labor productivity as measured by pounds of good extrusions shipped divided by total plant man-hours worked. A quality circle program can be an effective productivity improvement tool and should be added to the IE/OR tool kit.

15.4 SUMMARY

This chapter presents an overview of productivity measurement and productivity improvement. Because of the orientation of IE/OR toward the evaluation and improvement of *efficiency* of operations, *productivity* measurement and improvement is an integral part of IE/OR.

The most common approach to forming productivity measures is to use ratio models. Specifically, some indicator of the output of the operation or system is divided by some indication of one or more inputs used in generating that output. Inputs are usually resources like labor, energy, raw material, and capital. Output is usually production volume or sales volume or some measure of service provided. When only a single input resource is used in the ratio, the model is called a partial productivity measure. When several inputs are used, the ratio is called a multifactor measure.

The output process often involves multiple products, each defined in perhaps different units such as gallons or pounds. The different output elements must be added together using a common denominator such as dollars. Similarly, in multifactor measures, resource inputs must be added together using dollars since we cannot add man-hours to BTUs. In such cases, the dollars used must be deflated or constant-value dollars. Taking inflation effects out of productivity measures is a critical part of the measurement process. Other keys to remember in developing and using productivity measures are that (1) the measure should be defined with its end use in mind, (2) the reason for changes in a measure is not always what it at first seems to be, (3) partial measures should not be used on a stand-alone basis, and (4) productivity measures are descriptive mathematical models, subject to all of the pros and cons of any models.

The most common models used to measure productivity are partial factor models. The most common single resource tracked with these models is labor. However, multiple-factor models are more useful in shedding light on the overall productivity of an activity. Examples of such models are the ones suggested by Craig and Harris, Taylor and Davis, and the American Productivity Center. The advantage of the latter model is that it relates productivity measurement *directly* to the firm's profit picture.

Measurement of productivity is an important activity, particularly as it should lead to *improvement* of productivity. Once measurement and analysis techniques have indicated areas that are candidates for productivity enhancement, tools such as those scattered throughout the IE/OR discipline can be used to improve productivity. In fact, most every technique discussed in this book is a productivity improver. In addition, other tools outside the scope of the IE/OR discipline are effective in improving productivity. One such method, discussed in this chapter, is "quality circles," an employee involvement technique in which workers address and solve production problems.

REFERENCES AND SUGGESTED READING

1. R. T. Amsden and D. M. Amsden (Ed.), *QC Circles: Applications, Tools, and Theory* (a collection of Articles), American Society for Quality Control, Milwaukee, Wisc., 1976.
2. R. T. Amsden and D. M. Amsden, ''Statistics Applied to and by QC Circles,'' *ASQC Technical Conference Transactions,* 1978.
3. E. S. Buffa, *Modern Production Management,* 2nd ed., Wiley, New York, 1973.
4. C. E. Craig and R. C. Harris, ''Total Productivity Measurement at the Firm Level'', *Sloan Management Review,* 13–28 (Spring 1973).
5. J. W. Kendrick and D. Creamer, *Measuring Company Productivity,* The Conference Board, Studies in Business Economics, No. 89, New York (1965).
6. P. Mali, *Improving Total Productivity; MBO Strategies for Business Government, and Not-for Profit Organization,* Wiley, New York, 1978.
7. M. E. Mundel (Ed.), ''Productivity: A Series (of Articles) from *Industrial Engineering* Magazine,'' *Am. Inst. of Ind. Engs.,* Atlanta, Ga., Monograph Series (1978).
8. ''National Productivity Report,'' **10**, No. 11 (June 1981).
9. C. S. O'Dell, *Gainsharing: Involvement, Incentives, and Productivity,* AMACOM, American Management Association Briefing (1981).
10. I. H. Siegal, ''Productivity Measurement: An Evolving Art; Highlights of the Literature,'' Work in Am. Inst., Inc., Scarsdale, NY, Series III, No. 17 (1980).
11. D. J. Sumanth and N. G. Einspruch, ''Productivity Awareness In the U.S.: A Survey of Some Major Corporations,'' *Industrial Engineering* (October 1980).
12. B. W. Taylor and K. R. Davis, ''Corporate Productivity—Getting It All Together,'' *Industrial Engineering* (March 1977).
13. ''Your Key for Planning for Profits,'' *Productivity Brief* No. 6, American Productivity Center, Houston, Tex. (October 1981).
14. D. M. Miller, ''Profitability = Productivity + Price Recovery,'' *Harvard Business Review* (May–June 1984).

REVIEW QUESTIONS AND PROBLEMS

15.1 Briefly define or describe the following terms and concepts.

(a) Productivity
(b) Deflation
(c) Quality circles
(d) A productivity measure

15.2 What is the difference, if any, between ''productivity'' and ''efficiency''?

15.3 Discuss the connection between classical industrial engineering and productivity. Do the same for the connection between operations research and productivity.

15.4 Discuss the statement "Industrial engineering and operations research *is* productivity."

15.5 Compare and contrast (1) partial productivity measures and (2) multifactor productivity measures.

15.6 Discuss the difference between "overall" productivity ratios and "objective" productivity ratios.

15.7 Suggest several useful indicators for measuring the productivity of each of the following:

(a) The operations of the lathes in a job shop
(b) The performance of a policeman walking a "beat"
(c) The work of a janitor
(d) A service station
(e) The combat operations of an infantry company
(f) The operations of a commercial trucking firm
(g) A university
(h) A professional football team
(i) A multinational, diverse corporation
(j) A single product plant

15.8 One approach to measuring productivity is to use a "production function," such as the Cobb–Douglas model:

$$Q = aL^{d}K^{f}e^{u}$$

where Q is the output in a period, L is the labor input, K is the capital input, u is a random measurement error component, and a, d, and f are parameters to be estimated. Critique this productivity measure. Compare its utility to the Craig–Harris model in Eq. (15.2).

15.9 Compare and contrast the Craig–Harris model, Eq. (15.2), with the "profitability = productivity + price recovery" approach to productivity measurement.

15.10 Provide justification for the use of (and definition of) the "economic depreciation" term in the Craig–Harris model.

15.11 Consider the following actual data from a major industrial firm based on data reported in Ref. 12.

	1967	1968	1969	1970
Reported sale ($Million)	91.11	92.64	60.12	78.55
Annual sales inflation factor (%)	100.0	86.4	78.8	70.3
Inventory change[a]	6.96	.34	.57	4.39
Manufacturing plant[a]	5.39	4.78	4.53	5.76
Depreciation[a]	63.33	65.16	47.26	72.31
Wages and salaries ($Million)	34.17	35.63	31.49	36.85
Consumer Price Index (%)	100.0	101.8	103.5	105.2
Working and fixed capital[a]	67.27	79.16	81.62	99.06
Investor contributions[a]	11.86	13.71	13.89	16.43
Materials and other cost of sales[a]	63.33	65.16	47.26	72.31

[a]These figures are in constant, 1967 millions of dollars.

For the firm represented by these data, answer the following questions.

(a) Provide a reasonable explanation for the *decline* in average sales price—as represented by the declining inflation factor—as opposed to a more typical increase in prices.
(b) Is the Consumer Price Index a good means of deflating wage and salary costs?
(c) Develop a partial labor productivity measure for each year (1967–1970). Interpret the results.
(d) What other useful partial productivity measures can you suggest using? Develop these for the data shown and interpret the results.
(e) Develop a multiple-factor productivity measure for each year using the Craig-Harris model.
(f) Apply the "profitability = productivity + price recovery" analysis to this firm; develop a graph like Figure 15.2.
(g) Compare the results of productivity measurement via e and f.

15.12 How is each of the following useful in analyzing or (and) improving productivity?

(a) Work methods
(b) Wage incentives
(c) Mathematical programming
(d) Simulation
(e) Queueing theory
(f) Quality circles

15.13 Discuss quality circles from the following standpoints.

(a) Structure (e.g., makeup of each circle, use of a circle leader, etc.)
(b) Likely effectiveness as a productivity improvement tool
(c) Applicable situations (e.g., hourly workers in a job shop, policemen, research engineers)

15.14 Write a computer program to carry out the calculations involved in the "profitability = productivity + price recovery" analysis. Check your code by reproducing the results of Example 16.1. (Note that your program should be general enough to handle the deflation of several elements of cost of sales on an individual basis; for this exercise, let all but one element have an inflation factor of 1.0 for each year.)

Index

Index